ADVANCED SOLID STATE PHYSICS

Philip Phillips

University of Illinois, Urbana-Champaign

Routledge
Taylor & Francis Group

LONDON AND NEW YORK

First published 2003 by Westview Press

Published 2018 by Routledge
52 Vanderbilt Avenue, New York, NY 10017
2 Park Square, Milton Park, Abingdon, Oxon OX14 4RN

Routledge is an imprint of the Taylor & Francis Group, an informa business

A cataloging-in-publication data record for this book is available from the Library of Congress.

ISBN 13: 978-0-367-00738-6 (hbk)
ISBN 13: 978-0-367-15725-8 (pbk)

For Angeliki

Contents

Foreword

The problem of communicating, in a coherent fashion, recent developments in the most exciting and active fields of physics continues to be with us. The enormous growth in the number of physicists has tended to make the familiar channels of communication considerably less effective. It has become increasingly difficult for experts in a given field to keep up with current literature; the novice can only be confused. What is needed is both a consistent account of a field and the presentation of a definite point of view concerning it. Formal monographs cannot meet such a need in a rapidly developing field, while review articles seem to have fallen into disfavor. Indeed, it would seem that the people who are most actively engaged in developing a given field are the people least likely to write at length about it.

The Frontiers in Physics series was conceived in 1961 in an effort to improve the situation in several ways. Leading physicists frequently give lectures, graduate seminars, or graduate courses in their special fields of interest. Such lectures serve to summarize the present status of a rapidly developing field and may well constitute the only coherent account available at the time. One of the principal purposes of the Frontiers in Physics series is to make notes on such lectures available to the wider physics community.

As Frontiers in Physics has evolved, a second category of book, the informal text or monograph—an intermediate step between lecture notes and formal text or monographs—has played an increasingly important role in the series. In an informal text or monograph, an author has reworked his or her lecture notes into a coherent summation of a newly developed field, complete with references and problems, suitable for either classroom teaching or individual study.

Philip Phillips' *Advanced Solid State Physics* is just such a book. The author provides a much-needed lucid introduction to the significant topics at the frontiers of condensed matter physics, while at the same time providing, in clear pedagogical fashion, background material on the well-established fundamentals of the subject. His success in integrating introductory and advanced topics makes this volume particularly appealing to graduate students and advanced researchers alike and should make it of considerable interest, as well, to mature scholars in other subfields of science who wish to obtain an overview of the very considerable intellectual challenge of contemporary solid state physics. It gives me great pleasure to welcome Philip Phillips to the Frontiers in Physics series.

DAVID PINES

Preface

Solid state physics continues to be the most rapidly growing subdiscipline in physics. As a result, entering graduate students wishing to pursue research in this field face the daunting task of not only mastering the old topics but also gaining competence in the problems of current interest, such as the fractional quantum Hall effect, strongly correlated electron systems, and quantum phase transitions. This book is written to serve the needs of such students. I have attempted in this book to present some of the standard topics in a way that makes it possible to move smoothly to current material. Hence, all the interesting topics are not presented at the end of the book. For example, immediately after the first 50 pages, Anderson's analysis of local magnetic moments is presented as an application of Hartree-Fock theory; this affords a discussion of the relationship with the Kondo model and how scaling ideas can be used to uncloak low-energy physics. As the key problems of current interest in solid state involve some aspects of electron-electron interactions or disorder or both, I have focused on the archetypal problems in which such physics is central. However, only those problems in which there is a consensus view are discussed extensively. In addition, I have placed the emphasis on physics rather than on techniques. Consequently, I focus on a clear presentation of the phenomenology along with a pedagogical derivation of the relevant equations. A key goal of the detailed derivations is to make it possible for the students who have read this book to immediately comprehend research papers on related topics. A key omission in this book is magnetism beyond the Stoner criterion and local magnetic moments. This omission has arisen primarily because the topic is adequately treated in the book by Assa Auerbach.

Most of this book grew out of lectures I have given in the one-semester advanced solid state physics class (Physics 490) taught here at the University of Illinois. In teaching this course, I have relied on the advice of those who taught the course before me, in particular my colleague Gordon Baym. Gordon had a significant impact in shaping this book: we have had numerous stimulating conversations, he collaborated in extensively revising the first seven chapters of this book, and he shared with me his lecture notes, which influenced the presentation of the material for Chapters 1, 2, 4, 5, and parts of 8 and 11, in particular the derivative of the polarization function (Chapter 8), the pair amplitude formulation of T_c (Chapter 11), and sound propagation in Chapter 10.

Numerous others have been instrumental in the writing of this book. I start by acknowledging all the students in Physics 490 during the semesters Spring '94, '95, '97, and '98, who labored through various unfinished versions of the chapters that constitute this book. It is for these students that this book is written. While all of these students have pointed out typos or logical inconsistencies over the years, four stand out in particular for their copious and careful suggestions and revisions: Johannes Walcher, Julian Velev, Adrian Gozar, and Ke Dong. I also thank Harry Westfahl, who has been a constant sounding board and has been most willing to read and correct vast sections of this book, in particular Chapters 8, 9, 11, and 13. Harry also assisted in the drawing of several figures. In a similar vein, I thank Revaz Ramazashvili, who carefully read Chapters 6 and 12 and convinced me that a much-expanded version of the latter was necessary. I would also like to thank David Pines for providing the final encouragement I needed to write the chapter on quantum phase transitions. This chapter was written last and would truly not be included here had it not been for a conversation I had with David on what might be lacking from the book. In this context, I thank my student Denis Dalidovich for his numerous insights, which have shaped my approach to this field. Ali Yazdani and his students also offered useful comments on the Kondo and localization chapters. In addition, George Paroanou, Michael Stone, and Bob Laughlin provided a sound critique of the quantum Hall chapter. I wish to thank Greg Whitlock for his masterful copyediting of the book as well as the staff at Publication Services for their willingness to accommodate my numerous revisions, especially Jason Brown, Ben Coblentz, and Susie Yates. I would also like to thank Michael Baym for the caring effort he took in drawing the figures in Chapters 4 through 9.4, 10 through 11, Figures 12.3, 12.6, and those in Chapter 14. Regarding the typesetting and layout of the manuscript, I have depended heavily on Craig Copi (Physics Department, Case Western Reserve University), who wrote all of the LaTeX macros that are responsible for the design of each chapter. Craig also prepared the Table of Contents and the Index. I also thank Damian Menscher for typesetting assistance and Sandy Chancy and April Orwick for learning LaTeX and typing the initial versions of Chapters 1–7 and 9–11. Finally, I thank Nathan Hearne for scanning in and tweaking the experimental figures into the final format they now have.

Above all, my enduring thanks go to Angeliki Tzanetou, who has encouraged me to stay focused on this manuscript, when its completion and ultimate utility seemed doubtful, at best. I have inflicted many sentences of this book on her and she has always told me when they were unclear. It is to her that this book is dedicated.

– 1 –

Noninteracting Electron Gas

To a surprising extent, one can understand the elementary properties of metals in terms of noninteracting electrons and phonons-quantized lattice vibrations. For example, the low-temperature specific heat of a metal is the sum of a term linear in the temperature, T, from the electrons and a term proportional to T^3 from the phonons. This result follows from a noninteracting particle picture. The electrical conductivity limited by nonmagnetic impurity scattering is also well described by a noninteracting electron gas. In addition, from a knowledge of single-electron band theory, one can qualitatively discern the differences between metals, insulators, and semiconductors. The remarkable success of the noninteracting model is paradoxical because electrons and ions strongly interact both with themselves and with one another. Along with its successes, the noninteracting picture has colossal shortcomings, most notably its inability to describe old problems, such as cohesive energies, superconductivity, magnetism, and newer phenomena, such as the Kondo and fractional quantum Hall effects. We first review the physics of the noninteracting electron gas. It is only after we develop methodology for dealing with electron interactions that we can lay plain the reasons why the noninteracting model works so well.

Electrons in metals are quantum mechanical particles with spin $\hbar/2$, obeying Fermi-Dirac statistics. The Hamiltonian of a single electron is $\hat{p}^2/2m$ where $\hat{\mathbf{p}}$ is the electron momentum (operator) and m the electron mass. Its eigenstates are plane waves of the form $e^{i\mathbf{p}\cdot\mathbf{r}/\hbar}/\sqrt{V}$ times a spinor that specifies the electron spin projection on a convenient axis (usually $\hat{z}$), $\hbar\sigma/2$ where $\sigma = \pm1$; here V is the system volume. The Hamiltonian (operator) for N such noninteracting

1

electrons,

$$\widehat{H} = \sum_{i=1}^{N} \frac{\widehat{p}_i^2}{2m}, \tag{1.1}$$

is simply the sum of the kinetic energies of the individual particles. In this case, the eigenstates are products of the occupied single-particle plane-wave states. Each plane-wave state can be occupied at most by one electron of a given spin. We label these eigenstates by the distribution function $f_{\mathbf{p}\sigma}$, which is 1 if the single-particle momentum-spin state is occupied and 0 otherwise. In the ground state, the lowest $N/2$ single-particle states are doubly occupied with electrons of opposite spin. Consequently, in the ground state (temperature $T = 0$), the distribution function is

$$f_{\mathbf{p}\sigma} = \Theta(\mu_0 - p^2/2m), \tag{1.2}$$

where $\Theta(x)$ is the Heaviside function, $\Theta(x > 0) = 1$, and 0 otherwise. Here, μ_0 is the zero-temperature electron chemical potential, which in this case is simply the Fermi energy, the energy of the highest occupied state, $p_F^2/2m$, where p_F is the electron Fermi momentum. The Fermi temperature T_F equals μ_0/k_B.

In terms of $f_{\mathbf{p}\sigma}$, the total number of electrons is given by

$$N = \sum_{\mathbf{p},\sigma} f_{\mathbf{p}\sigma}. \tag{1.3}$$

In the ground state, we can replace the sum by an integral and find the electron density at $T = 0$:

$$n_e = \frac{N(T = 0)}{V} = 2 \sum_{p < p_F} = 2 \int_0^{p_F} \frac{d\mathbf{p}}{(2\pi\hbar)^3} = \frac{p_F^3}{3\pi^2\hbar^3}. \tag{1.4}$$

The average interparticle spacing is essentially the radius, r_e, of a sphere containing a single electron.

$$\frac{4\pi r_e^3}{3} n_e = 1. \tag{1.5}$$

Thus, from Eq. (1.4), the scale for the interparticle separation is

$$r_e = \left(\frac{9\pi}{4}\right)^{1/3} \frac{\hbar}{p_F} = 1.92 \frac{\hbar}{p_F}, \tag{1.6}$$

which is on the order of the lattice spacing. For $r_e \approx 1\text{Å}$, we find that the Fermi velocity $v_F = p_F/m \approx \hbar/mr_e \approx 10^8 \text{ cm/s} \approx c/300$, where c is the speed of light. (Relativistic effects are generally not important for the motion of electrons in the ground state.) It is conventional to work with the dimensionless ratio $r_s = r_e/a_0$, where $a_0 = \hbar^2/me^2$ is the Bohr radius; this quantity provides a measure of the electron density. The dense limit corresponds to $r_s \ll 1$ and the dilute regime to $r_s \gg 1$. In metals, r_s varies between 2 and 6. Listed below are values of r_s for the alkali metals.

	Li	Na	K	Rb	Cs
r_s	3.25	3.93	4.86	5.2	5.62

Cesium is, in fact, the most dilute of all metals. It is this large value of r_s that is responsible for the inhomogeneities in the density of Cs. In Chapter 5, we will discuss further physics associated with large r_s, such as eventual formation of a Wigner crystal. We can also use Eq. (1.4) to solve for the zero-temperature chemical potential

$$\mu_0 = (3\pi^2)^{2/3}\frac{\hbar^2}{2m}n_e^{2/3}. \tag{1.7}$$

For Na, $\mu_0 = 3.1$ eV; typically in a metal, μ_0 ranges between 1 and 5 eV.

The total energy of the system is the sum over the occupied states weighted by the single-particle energies, $\epsilon_p = p^2/2m$:

$$E = \sum_{\mathbf{p},\sigma} \epsilon_{\mathbf{p}} f_{\mathbf{p}\sigma}. \tag{1.8}$$

At $T = 0$, the energy is given by

$$E_0 = \frac{p_F^5 V}{10m\pi^2\hbar^3} = \frac{3}{5}N\mu_0. \tag{1.9}$$

From the thermodynamic relation $P = -(\partial E/\partial V)_{T,N}$, we find the pressure in the ground state, $P_0 = 2\mu_0 n_e/5$. This quantity is of order 10^6 atm and arises entirely from the exclusion principle between the particles.

At finite temperature, we define the distribution function that ranges between 0 and 1 and measures the average occupation of the single-particle states. For a system in equilibrium at chemical potential μ,

$$f_{\mathbf{p}\sigma} = \frac{1}{e^{\beta(\epsilon_{\mathbf{p}} - \mu)} + 1} \tag{1.10}$$

is the Fermi-Dirac distribution, where $\beta = 1/k_B T$. The Fermi-Dirac distribution function maximizes the entropy at a given energy and electron number. In general, the entropy, S, is given by the log of the number of microscopic states W,

$$S = k_B \ln W, \tag{1.11}$$

consistent with the macroscopic thermodynamic state of the system. For the electron problem, the microscopic states are indexed by a momentum $\mathbf{p}$ with occupancy 0 or 1. The distribution function $f_{\mathbf{p}\sigma}$, however, is a smooth function over all the momentum states. To construct this function, we group the momentum states into cells, each cell containing g_i momentum states and n_i particles. Because each cell contains g_i states,

$$\sum_i g_i \cdots = \sum_{\mathbf{p},\sigma} \cdots. \tag{1.12}$$

For the ith cell, the number of distinct ways of distributing n_i particles in g_i states is given by the combinatoric factor $W_i = g_i!/n_i!(g_i - n_i)!$. Applying Stirling's approximation

$$\ln N! \approx N(\ln N - 1) \tag{1.13}$$

to W_i, we obtain

$$\ln W_i \approx -n_i \ln \frac{n_i}{g_i} - (g_i - n_i) \ln \frac{(g_i - n_i)}{g_i} \tag{1.14}$$

$$= -g_i \left[\frac{n_i}{g_i} \ln \frac{n_i}{g_i} + \left(1 - \frac{n_i}{g_i}\right) \ln \left(1 - \frac{n_i}{g_i}\right) \right], \tag{1.15}$$

where n_i/g_i is the fraction of states occupied in cell number i. In fact, $n_i/g_i = f_i$ is the smooth distribution function we seek. If we substitute this expression into the equation for $\ln W_i$, we recover the familiar result for the entropy,

$$S = -k_B \sum_{\mathbf{p},\sigma} [f_{\mathbf{p}\sigma} \ln f_{\mathbf{p}\sigma} + (1 - f_{\mathbf{p}\sigma}) \ln (1 - f_{\mathbf{p}\sigma})], \tag{1.16}$$

where we have converted the sum over cells to a sum over spin and momentum states by using Eq. (1.12). To obtain the distribution function, we maximize the entropy subject to the constraint that the particle number and energy are fixed. Extremizing $S - E/T + \mu N/T$ with respect to $f_{\mathbf{p}\sigma}$, we find

$$0 = \epsilon_{\mathbf{p}\sigma} - \mu - k_B T \ln (f_{\mathbf{p}\sigma}^{-1} - 1), \qquad (1.17)$$

which implies that $f_{\mathbf{p}\sigma}$ is the Fermi distribution function in Eq. (1.10).

Let us now calculate the heat capacity from the thermodynamic relationship

$$C_V = T \left(\frac{\partial S}{\partial T} \right)_V. \qquad (1.18)$$

To compute the temperature derivative of the entropy, we consider the general variation of the entropy

$$\delta S = -2k_B \sum_{\mathbf{p}} \delta f_{\mathbf{p}} \ln \frac{f_{\mathbf{p}}}{1 - f_{\mathbf{p}}} = 2k_B \sum_{\mathbf{p}} \delta f_{\mathbf{p}} \frac{\epsilon_{\mathbf{p}} - \mu}{k_B T} \qquad (1.19)$$

$$= 2Vk_B \int \frac{d\mathbf{p}}{(2\pi\hbar)^3} \delta f_{\mathbf{p}} \frac{\epsilon_{\mathbf{p}} - \mu}{k_B T} \qquad (1.20)$$

with respect to the distribution function $f_{\mathbf{p}\sigma}$. We can simplify this expression further by introducing the single-particle density of states per unit volume

$$N(\epsilon) = 2 \int \frac{d\mathbf{p}}{(2\pi\hbar)^3} \delta (\epsilon - \epsilon_{\mathbf{p}}) \qquad (1.21)$$

$$= \frac{1}{\pi^2 \hbar^3} \int p^2 \frac{dp}{d\epsilon_{\mathbf{p}}} \delta(\epsilon - \epsilon_{\mathbf{p}}) d\epsilon_{\mathbf{p}} \qquad (1.22)$$

$$N(\epsilon_{\mathbf{p}}) = \frac{mp}{\pi^2 \hbar^3}. \qquad (1.23)$$

The key point here is that the single-particle density of states is a linear function of momentum. An equivalent way of writing this quantity is $N(\epsilon_{\mathbf{p}}) = (dp/d\epsilon_{\mathbf{p}})(p^2/2\pi\hbar^3)$. If we rewrite δS in terms of the density of states

$$\delta S \; = \; \frac{2V}{(\pi^2\hbar)^3 T} \int d\epsilon_{\mathbf{p}} d\,\Omega \frac{p^2 dp}{d\epsilon_{\mathbf{p}}} \; \delta f_{\mathbf{p}}(\epsilon_{\mathbf{p}} - \mu) \tag{1.24}$$

$$= \; \frac{V}{T} \int d\epsilon_{\mathbf{p}} \delta f_{\mathbf{p}} N(\epsilon_{\mathbf{p}})(\epsilon_{\mathbf{p}} - \mu), \tag{1.25}$$

we obtain a single integral that can be evaluated using the Sommerfeld expansion. Although this expansion is standard, we will review it.

Sommerfeld expansion: Consider an integral of the form

$$I \; = \; \int_0^\infty d\epsilon f(\epsilon) h(\epsilon), \tag{1.26}$$

where $h(\epsilon)$ is any smooth function and $f(\epsilon) = 1/(e^{\beta(\epsilon-\mu)} + 1)$. We integrate I by parts:

$$I \; = \; \int_0^\infty f'(\epsilon) H(\epsilon) \, d\epsilon, \tag{1.27}$$

where

$$H \; = \; -\int_0^\epsilon h(x) \, dx. \tag{1.28}$$

Because $f'(\epsilon)$ is strongly peaked at the chemical potential, we can expand H in a Taylor series around $\epsilon = \mu$,

$$H(\mu) + (\epsilon - \mu)\left(\frac{\partial H}{\partial \epsilon}\right)_{\epsilon=\mu} + \frac{1}{2}(\epsilon - \mu)^2 \left(\frac{\partial^2 H}{\partial \epsilon^2}\right)_{\epsilon=\mu} + \cdots, \tag{1.29}$$

which gives us a series of integrals of the form

$$L_j \; = \; -\int_0^\infty (\epsilon - \mu)^j f'(\epsilon) \, d\epsilon \tag{1.30}$$

to calculate. For $j = 0$, $L_0 = -f(\infty) + f(0) = 1$. In the remaining integrals, we can replace the lower limit with $-\infty$. Letting $x = \beta(\epsilon - \mu)$, we have

$$L_j \; = \; \frac{1}{\beta^j} \int_{-\infty}^\infty x^j \frac{e^x}{(e^x + 1)^2} \, dx. \tag{1.31}$$

Since the integrand is odd in x for j odd, only the even j's survive. The first several values of L_j are

$$L_0 = 1, \tag{1.32}$$

$$L_2 = \frac{\pi^2}{3}(k_B T)^2, \tag{1.33}$$

$$L_4 = \frac{7\pi^4}{30}(k_B T)^4. \tag{1.34}$$

Consequently, we systematically obtain the series expansion for I:

$$I = \int_0^\mu h(\epsilon)d\epsilon + \frac{\pi^2}{6}(k_B T)^2 h'(\mu) + \frac{7\pi^4}{720}(k_B T)^4 h'''(\mu) + \cdots. \tag{1.35}$$

For functions h independent of temperature, the first term in this expansion is independent of temperature and hence, at fixed μ,

$$\int_0^\infty d\epsilon\, \delta f(\epsilon)h(\epsilon) = \frac{\pi^2}{3}h'(\epsilon = \mu)k_B^2 T \delta T \tag{1.36}$$

is the leading term, when f_p is varied as a function of temperature.

Now let us return to the calculation of the low-temperature entropy. We want to calculate the temperature variation of the entropy at fixed-particle number. We first calculate at fixed μ and then show that the change in chemical potential with temperature for fixed-particle number can be neglected here. If we now substitute Eq. (1.36) into Eq. (1.25), we find that the variation of the entropy per unit volume

$$\delta s = \frac{\pi^2}{3}k_B^2 N(\epsilon_F)\delta T \tag{1.37}$$

is a constant independent of temperature. The heat capacity per unit volume

$$c_V = T\left(\frac{\delta s}{\delta T}\right)_V = \frac{\pi^2}{3}k_B^2 N(\epsilon_F)T \tag{1.38}$$

scales as a linear function of temperature. This contribution arises entirely from the conduction electrons. The contribution per electron is

$$\frac{c_V}{n_e} = \frac{\pi^2 k_B^2 m T}{p_F^2} = \frac{\pi^2}{2} k_B \frac{T}{T_F}. \qquad (1.39)$$

Comparing this result to the classical heat capacity, $3k_B T/2$, we find that the quantum mechanical value is smaller by a factor of $\pi^2/3T_F$. In a metal, only a fraction of the electrons are at the Fermi level. The ratio T/T_F defines this fraction of electrons within $k_B T$ of the Fermi energy. The further an electron is below the Fermi level, the smaller is its contribution to the heat capacity.

We may use the Sommerfeld expansion to show that the chemical potential as a function of density and temperature is given, at low T, by

$$\mu(n_e, T) = \mu(n_e, 0)\left(1 - \frac{\pi^2}{12}\left(\frac{T}{T_F}\right)^2\right), \qquad (1.40)$$

where $\mu(n_e, 0) = \epsilon_F$. The proof of this result is similar to that used to derive the entropy, and we leave the derivation as an exercise (see Problem 1.2). From Eq. (1.40), we see that the first correction to the chemical potential at fixed n_e is of order T^2, and hence this correction can be ignored in calculating the low-temperature entropy, Eq. (1.37).

PROBLEMS

1. Evaluate the integral

$$L_j = \int_{-\infty}^{\infty} x^j \frac{e^x}{(e^x + 1)^2} dx.$$

Show in particular that only the integrals for even j's survive.

2. Use the Sommerfeld expansion to compute the first temperature correction, Eq. (1.40), to the chemical potential of a Fermi gas as a function of the density.

– 2 –

Born-Oppenheimer Approximation

In this chapter, we develop the basic framework to see how electron-electron (e-e), electron-ion (e-i), and ion-ion (i-i) interactions affect the properties of solids. We first show that the electron and ion degrees of freedom can be decoupled. Such a separation arises because ions and electrons have vastly different velocities in a solid; roughly, electron velocity is 1000 times larger than the ion velocity. As a consequence, the electrons view the ions as providing a static background in which they move. This physical picture is at the heart of the Born-Oppenheimer approximation [BO1927].

2.1 BASIC HAMILTONIAN

It is useful at the outset to consider the total Hamiltonian to understand how to separate the electronic from the ionic motion. We denote the position and momentum of the ith ion by $\mathbf{R}_i$ and $\mathbf{P}_i$, respectively, and that of the jth electron by $\mathbf{r}_j$ and $\mathbf{p}_j$. We assume that all the ions have mass M and nuclear charge Ze. The total Hamiltonian is, then,

$$
H = \sum_i \frac{\mathbf{P}_i^2}{2M} + \sum_j \frac{\mathbf{p}_j^2}{2m} + \frac{(Ze)^2}{2} \sum_{i,i'} \frac{1}{|\mathbf{R}_i - \mathbf{R}_{i'}|}
$$

$$
+ \frac{e^2}{2} \sum_{j,j'} \frac{1}{|\mathbf{r}_j - \mathbf{r}_{j'}|} - Ze^2 \sum_{i,j} \frac{1}{|\mathbf{r}_j - \mathbf{R}_i|}. \tag{2.1}
$$

This Hamiltonian does not include external electric or magnetic fields or magnetic interactions among the constituents. The last three terms are the i-i, e-e, and e-i interactions, respectively. To progress with this Hamiltonian, we divide the electrons into two groups—the core and the valence or conduction electrons—depending on the degree to which they are bound. This separation is useful because core electrons move with the nuclei. Conduction or valence electrons transport throughout the solid. With this separation in mind, we obtain

$$
H = \sum_i \frac{\mathbf{P}_i^2}{2M} + \sum_{j=\text{cond.elec.}} \frac{\mathbf{p}_j^2}{2m} + \sum_{i,i'} V_{i,i'}(|\mathbf{R}_i - \mathbf{R}_{i'}|)
$$

$$
+ \frac{e^2}{2} \sum_{j,j'=\text{cond.elec.}} \frac{1}{|\mathbf{r}_j - \mathbf{r}_{j'}|} + \sum_{i,j} V_{ei}(|\mathbf{r}_j - \mathbf{R}_i|) + E_{\text{core}} \qquad (2.2)
$$

as the partitioned Hamiltonian. In Eq. (2.2), V_{ii} and V_{ei} represent the effective potential between the ions and between the valence electrons with the ions, respectively. The energy of the core electrons is E_{core}. In the example of Na, the total Z is 11, with the orbital filling $1s^2\,2s^2\,2p^6\,3s$. There is only one unpaired valence electron. The effective charge of the ion in this representation is $Z = 1$.

2.2 ADIABATIC APPROXIMATION

To simplify Eq. (2.2), we separate the nuclear motion from that of the electrons, a separation that makes sense because the ions are much more massive than the electrons. Typically, $m/M \sim 1/2000$ to 1/500,000. The small parameter characterizing the expansion is $(m/M)^{1/4}$. We now show that the ion velocity is related to the Fermi velocity by the ratio $(m/M)^{3/4}$. As a result, the ions can be treated as essentially static relative to the electrons. This allows us to solve for the electron motion, assuming first that the ions are fixed at their equilibrium positions. The effects of the ion motion can be treated as a perturbation; that is, the electrons adjust adiabatically to the ion motion. From the perspective of the ions, the rapid motion of the electrons creates an overall average electron potential that they feel. This separation is the essence of the Born-Oppenheimer approximation.

To understand the relative orders of magnitude of the ionic and electron velocities and energies, we assume, for the sake of the argument, that ions individually move in harmonic wells (see Fig. 2.1) of the form $V_{\text{osc}} = M\omega^2 R^2/2$, where R measures the deviation of an ion

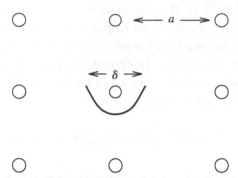

Figure 2.1 Ion lattice with spacing a. Each ion oscillates in a harmonic well with a deviation from its home position that is small relative to the ion spacing. The deviation is roughly $\delta \sim (m/M)^{1/4}a \sim 10^{-4}a$, where m is the electron mass and M the mass of the ions. It is for this reason that the ions can be treated essentially as fixed relative to the electronic degrees of freedom. That the ions constitute a fixed, almost rigid background for the electron motion is the essence of the Born-Oppenheimer approximation.

from its home (or equilibrium) position. Consider displacing an ion by a lattice spacing a. The energy required to do so, $\sim M\omega^2 a^2/2$, is essentially that required to distort the electron wave function, and thus the energy is of order $\hbar^2/2ma^2$, which in turn is of the order of the electron kinetic energy, $p_F^2/2m$. Thus, $M\omega^2 a^2/2 \sim p_F^2/2m$. Solving for ω, we find that $\omega \sim (m/M)^{1/2}\hbar/ma^2$. However, for an ion in a harmonic well, $P^2/2M = \hbar\omega/2$, or equivalently, the square of the ion velocity is $\hbar\omega/M$. Combining this result with those for ω, we see that $v_{\text{ion}} \sim (m/M)^{3/4}v_F \sim (10^{-2} \text{ to } 10^{-3})v_F$.

Let us estimate how far the ions move from their equilibrium positions. For a displacement δ, $M\omega^2\delta^2/2 \sim \hbar\omega/2$. Substituting $\omega \sim (m/M)^{1/2}\hbar/ma^2$, we find an ion displacement $\delta \sim a(m/M)^{1/4} \sim 10^{-4}a$, which is negligible, as illustrated in Fig. (2.1). As far as the electrons are concerned, the ions are static. We can calculate the role of the electronic degrees of freedom by developing a perturbation series in the small quantity δ/a or $(m/M)^{1/4}$. Note also that because $P^2/2M \sim \hbar\omega/2 \sim \epsilon_F(m/M)^{1/2} \ll \epsilon_F$, the ion kinetic energy is relatively small.

The formal development of the Born-Oppenheimer approximation begins with the assumption that the full wave function is a function of the many-electron positions $\mathbf{r} \equiv \{\mathbf{r}_j\}$, and ionic positions $\mathbf{R} \equiv \{\mathbf{R}_i\}$ can be expanded as

$$\Psi(\mathbf{r}, \mathbf{R}) = \sum_n \Phi_n(\mathbf{R})\Psi_{e,n}(\mathbf{r}, \mathbf{R}), \qquad (2.3)$$

where the $\Psi_{e,n}(\mathbf{r}, \mathbf{R})$ (indexed by n) are the solutions to the electron-ion problem for a fixed set of ion positions $\mathbf{R}$. The ionic wave functions, $\Phi_n(\mathbf{R})$, on the one hand, describe the amplitude for the ions to be found at positions $\mathbf{R}$; on the other hand, they can be regarded as expansion coefficients of the electronic wave functions. The $\Psi_{e,n}$'s form a complete orthonormal set. Consequently,

$$\int d\mathbf{r} \Psi_{e,n}^*(\mathbf{r}, \mathbf{R}) \Psi_{e,m}(\mathbf{r}, \mathbf{R}) = \langle en|em \rangle = \delta_{nm}. \tag{2.4}$$

Coupled with the orthogonality condition on the nuclear wave functions

$$\int d\mathbf{R} \Phi_n^*(\mathbf{R}) \Phi_m(\mathbf{R}) = \delta_{nm}, \tag{2.5}$$

the complete electron-ion wave function is normalized,

$$\langle \Psi(\mathbf{r}, \mathbf{R}) | \Psi(\mathbf{r}, \mathbf{R}) \rangle = 1.$$

We determine each of the expansion coefficients from the equations of motion obeyed by Φ_n and $\Psi_{e,n}$. To proceed, we rewrite Eq. (2.2) as

$$H = T_i + T_e + V_{ii} + V_{ee} + V_{ei} + E_{\text{core}}, \tag{2.6}$$

where there is a one-to-one correspondence between the terms in (3.9) and those in (2.2). The eigenvalue equation for (3.9) is

$$(T_i + T_e + V_{ii} + V_{ee} + V_{ei} + E_{\text{core}}) \Psi = E \Psi \tag{2.7}$$

$$(T_i + V_{ii} + E_{\text{core}}) \Psi + \sum_n \Phi_n (T_e + V_{ee} + V_{ei}) \Psi_{e,n}(\mathbf{r}, \mathbf{R}) = E \Psi. \tag{2.8}$$

We simplify this equation by noting that $T_e + V_{ee} + V_{ei}$ only operates on the electron part of the product wave function. Let $E_{e,n}(\mathbf{R})$ be the energy of the electron system for a fixed set of nuclear coordinates. As a consequence, the nuclear and the electronic eigenvalue equations

$$\sum_n (T_i + V_{ii} + E_{\text{core}} + E_{e,n} - E) \Phi_n \Psi_{e,n} = 0 \tag{2.9}$$

and

$$(T_e + V_{ee} + V_{ei}) \Psi_{e,n}(\mathbf{r}, \mathbf{R}) = E_{e,n}(\mathbf{R}) \Psi_{e,n} \tag{2.10}$$

can be separated.

Let us multiply the nuclear eigenvalue equation by $\Psi^*_{em} = \langle em | \mathbf{r}, \mathbf{R} \rangle$ and integrate:

$$\sum_n \int d\mathbf{r} \Psi^*_{e,m} (\mathbf{r}, \mathbf{R}) \, T_i \Phi_n(\mathbf{R}) \Psi_{e,n}(\mathbf{r}, \mathbf{R})$$

$$+ (V_{ii} + E_{\text{core}} + E_{e,m}(\mathbf{R}) - E) \Phi_m(\mathbf{R}) = 0. \qquad (2.11)$$

Because the matrix element $\langle en | V_{ii}(\mathbf{R}) | em \rangle$ involves purely algebraic operators, it has only diagonal elements. The kinetic energy term, however, has off-diagonal elements; explicitly,

$$\sum_i \langle em \, | \, \frac{\mathbf{P}_i^2}{2M} \Phi_n(\mathbf{R}_i) \, | \, en \rangle = -\frac{\hbar^2}{2M} \sum_i \int d\mathbf{r} \Psi^*_{e,m}(\mathbf{r}, \mathbf{R}) [(\nabla^2_{\mathbf{R}_i} \Phi_n(\mathbf{R}))$$

$$+ 2(\nabla_{\mathbf{R}_i} \Phi_n(\mathbf{R})) \cdot \nabla_{\mathbf{R}_i} + \Phi_n(\mathbf{R}) \nabla^2_{\mathbf{R}_i}] \Psi_{e,n}(\mathbf{r}, \mathbf{R}). \qquad (2.12)$$

Because $\nabla^2_{\mathbf{R}}$ acts exclusively on $\Phi_n(\mathbf{R})$ in the first term in the integral, the resultant matrix element is purely diagonal. For the moment, we ignore the last two terms in the kinetic energy matrix element and obtain

$$\sum_n (T_i + V_{ii} + E_{\text{core}} + E_{e,n}(\mathbf{R})) \Phi_n(\mathbf{R}) = \sum_n E_n \Phi_n(\mathbf{R}), \qquad (2.13)$$

or equivalently,

$$[T_i + V_{ii} + E_{\text{core}} + E_{e,n}(\mathbf{R})] \Phi_n(\mathbf{R}) = E_n \Phi_n(\mathbf{R}), \qquad (2.14)$$

as the eigenvalue equation for the nuclear degrees of freedom. Equations (2.10) and (2.14) are the principal results in the Born-Oppenheimer method. In the nuclear eigenvalue equation, $E_{e,n}(\mathbf{R})$ serves as the effective nuclear potential that results when the electronic degrees of freedom are integrated out. The solutions to (2.14) will describe the phonon modes of the ions.

To justify this treatment of the ion kinetic energy term, we analyze the relative magnitude of the three contributions in Eq. (2.12). We first need a reasonably accurate form for the nuclear wave functions. For our purposes, the harmonic approximation we made in conjunction with the derivation of the ion velocity is adequate. In this case,

$$\Phi_n \sim e^{-M\omega(\mathbf{R}-\mathbf{R}^0)^2/2\hbar}, \qquad (2.15)$$

where $\mathbf{R}^0$ is the equilibrium position of the ion. Consequently,

$$\frac{\hbar^2}{2M} \nabla^2_{\mathbf{R}_i} \Phi_n \cdot \Psi_{e,n} \sim \frac{\hbar^2}{2M} \left(\frac{M\omega}{\hbar} \delta\right)^2 \Phi_n \Psi_{e,n}$$

$$\sim \left(\frac{m}{M}\right)^{1/2} \epsilon_F \Phi_n \Psi_{e,n}. \tag{2.16}$$

Consider now the second term in Eq. (2.12). The inverse length-scale on which the electron wave functions change is $\nabla_R \sim 1/a$. As a consequence,

$$\frac{\hbar^2}{2M} \nabla_{\mathbf{R}_i} \Phi_n \cdot \nabla_{\mathbf{R}_i} \Psi_{e,n} \sim \frac{\hbar^2}{2M} \frac{M\omega}{\hbar} \frac{\delta}{a} \Psi_{e,n} \Phi_n$$

$$\sim \left(\frac{m}{M}\right)^{3/4} \epsilon_F \Phi_n \Psi_{e,n}, \tag{2.17}$$

and

$$\frac{\hbar^2}{2M} \Phi_n \nabla^2_{\mathbf{R}_i} \Psi_{e,n} \sim \frac{\hbar^2}{2M} \frac{1}{a^2} \Psi_{e,n} \Phi_n \sim \frac{m}{M} \epsilon_F \Phi_n \Psi_{e,n}. \tag{2.18}$$

As is evident, the largest contribution to the nuclear kinetic energy matrix element arises from $\nabla^2_{\mathbf{R}_i} \Phi_n$. The primary reason for this is that gradients of Φ_n exceed those of $\Psi_{e,n}$ by a factor of $(M/m)^{1/4}$. Consequently, dropping the last two terms in Eq. (2.12) incurs a negligible error on the order of $(m/M)^{1/4}$.

To lowest order in m/M, we then neglect the ion kinetic energy and assume $\mathbf{R} = \mathbf{R}^0$. The ground electronic wave function is $\Psi_{e,n=0}(\mathbf{r}, \mathbf{R}^0)$; its corresponding energy, $E_{e,n=0}(\mathbf{R}^0)$, obeys the Schrödinger equation

$$[T_e + V_{ee} + V_{ei}(\mathbf{r} - \mathbf{R}^0)]\Psi_{e,n=0}(\mathbf{r}, \mathbf{R}^0) = E_{e,n=0}(\mathbf{R}^0)\Psi_{e,n=0}(\mathbf{r}, \mathbf{R}^0). \tag{2.19}$$

The subsequent ground-state nuclear wave functions and energies can be found with the effective potential $E_{e,n=0}(\mathbf{R}^0)$. If deviations about the equilibrium ion positions are considered within a simple harmonic oscillator model, the total energy of each low-energy state is given by

$$E_n = E_{e,n}(\mathbf{R}^0) + E_{core} + V_{ii}(\mathbf{R}^0)$$

$$+ \sum_q \hbar\omega_q \left(n_q + \frac{1}{2}\right) + \text{anharmonic terms.} \tag{2.20}$$

2.3   REDUCED-ELECTRON PROBLEM

From our preceding discussion, the effective-electron problem is

$$H_e = T_e + V_{ee} + V_{\text{ion}}(\mathbf{r}), \qquad (2.21)$$

where

$$V_{\text{ion}}(\mathbf{r}) = \sum_{i,j} V_{ei}(\mathbf{r}_j - \mathbf{R}_i^0). \qquad (2.22)$$

Here, $V_{\text{ion}}(\mathbf{r})$ is the potential felt by the electrons produced by the ions in their equilibrium positions. To a good approximation, this potential is periodic. Let us group all the one-body terms together as $h_e(\mathbf{r}) = T_e + V_{\text{ion}}(\mathbf{r})$. The reduced electronic Hamiltonian is, then,

$$H_e = h_e + V_{ee}. \qquad (2.23)$$

It is this Hamiltonian that we will primarily discuss in the remainder of this book.

REFERENCES

[BO1927] M. Born, J. R. Oppenheimer, *Ann. d. Physik* **84**, 457 (1927).

– 3 –

Second Quantization

To solve many-particle problems, it is expedient to introduce the language of second quantization. This approach does not add new conceptual baggage to quantum mechanics. Rather, it provides a convenient bookkeeping method for dealing with many-particle states. As a result, we will be brief in the formal development of this technique. For a more complete development, see, for example, Baym [B1969]. We develop this approach for fermions, that is, particles that are antisymmetric with respect to interchange of any two of their coordinates, and for bosons, particles for which such interchanges do not incur any sign change in the many-particle wave function.

3.1 BOSONS

Consider the many-body state of a boson system containing n identical particles:

$$\varphi_0(\mathbf{r}) = \langle \mathbf{r}_1, \mathbf{r}_2 \cdots \mathbf{r}_n | n \rangle. \tag{3.1}$$

For states containing different numbers of particles, the orthogonality condition

$$\langle n | m \rangle = \delta_{nm} \tag{3.2}$$

holds. As in the case of the harmonic oscillator, we define the operator a_0 such that

$$a_0 | n \rangle = \sqrt{n} | n - 1 \rangle. \tag{3.3}$$

That is, the operator a_0 annihilates a single particle from the state $| n \rangle$ and produces the corresponding $n - 1$ particle state. The adjoint operation in Eq. (3.2) suggests that

$$\langle n|a_0^\dagger = \langle n - 1| \sqrt{n}. \tag{3.4}$$

As a consequence, the matrix element

$$\langle n|a_0^\dagger|n - 1\rangle = \sqrt{n}\langle n - 1|n - 1\rangle = \sqrt{n}. \tag{3.5}$$

An equivalent statement of the result in Eq. (3.5) is that

$$a_0^\dagger|n - 1\rangle = \sqrt{n}|n\rangle. \tag{3.6}$$

The operator $a_0^\dagger$ acts on the $|n - 1\rangle$ particle state producing the original state $|n\rangle$.

The choice of the factor of $\sqrt{n}$ will now become clear when we consider the operator $\hat{N}_0 = a_0^\dagger a_0$. $\hat{N}_0$ is a Hermitian operator that, when acting on $|n\rangle$

$$\hat{N}_0|n\rangle = a_0^\dagger a_0|n\rangle = a_0^\dagger \sqrt{n}|n - 1\rangle = n|n\rangle, \tag{3.7}$$

counts the number of particles in state $|n\rangle$. Consequently, $\langle n|\hat{N}_0|n\rangle = n$, and $\hat{N}_0$ is called the *number operator*. Consider now the creation of the state $|n\rangle$ from the vacuum state $|0\rangle$. It follows from Eqs. (3.1) through (3.7) that

$$|n\rangle = \frac{(a_0^\dagger)^n}{\sqrt{n!}}|0\rangle. \tag{3.8}$$

We see, then, that the construction of an n-particle state of bosons is directly analogous to the rules for creating harmonic oscillator states. As such, the operators a_0 and $a_0^\dagger$ must obey the commutation relation

$$[a_0, a_0^\dagger] = 1. \tag{3.9}$$

To see how this comes about, consider the operation $[a_0 a_0^\dagger - a_0^\dagger a_0]|n\rangle$. From Eqs. (3.1)–(3.7), we have found that

$$[a_0 a_0^\dagger - a_0^\dagger a_0]|n\rangle = (n + 1 - n)|n\rangle = |n\rangle, \tag{3.10}$$

which implies that $[a_0, a_0^\dagger] = 1$.

Let us now generalize to a boson state of the form $|n_0, n_1, n_2 \cdots\rangle$, where n_i represents the occupation number for the boson state i.

Application of the boson rules outlined previously yields

$$a_\ell |\cdots n_\ell, \cdots\rangle = \sqrt{n_\ell} |\cdots n_\ell - 1 \cdots\rangle$$
$$a_\ell a_j |\cdots n_\ell \cdots n_j \cdots\rangle = \sqrt{n_\ell}\sqrt{n_j} |\cdots n_\ell - 1 \cdots n_j - 1 \cdots\rangle. \quad (3.11)$$

Analogous relationships hold for the creation operators. The generalized commutation relations are

$$[a_j, a_\ell^\dagger] = \delta_{j\ell}$$
$$[a_j, a_\ell] = [a_j^\dagger, a_\ell^\dagger] = 0 \quad (3.12)$$

A final note on the application of products of boson operators to the state $|n_0 \cdots n_j \cdots\rangle$ is in order. Because bosons commute, operations of the type $a_i^\dagger a_j^\dagger |\cdots n_i \cdots n_j \cdots\rangle$ yield the same result, irrespective of the order of the operators $a_i^\dagger$ and $a_j^\dagger$. As we will see, this is not the case with fermions.

3.2 FERMIONS

Let us assume now that the particles occupying the state $|n_0\rangle$ are fermions. We are restricted by the Pauli exclusion principle that $n_0 = 0, 1$. Analogously,

$$a_0 |1\rangle = |0\rangle$$
$$a_0^\dagger |0\rangle = |1\rangle. \quad (3.13)$$

Note also that $a_0 |0\rangle = a_1^\dagger |1\rangle = 0$, because no particle can be annihilated from the vacuum and each state can have at most one particle. To illustrate the inherent antisymmetry of fermions, consider the state $|n_0, n_1 \cdots\rangle$. The sign convention for the application of creation and annihilation operators to this reference state is

$$a_j^\dagger |\cdots n_{j-1}, n_j, n_{j+1} \cdots\rangle = \begin{cases} (-1)^{\eta_j} |\cdots n_{j-1}, 1, n_{j+1} \cdots & n_j = 0 \\ 0 & n_j = 1 \end{cases}$$

$$(3.14)$$

$$a_j |\cdots n_{j-1}, n_j, n_{j+1} \cdots\rangle = \begin{cases} (-1)^{\eta_j} |\cdots n_{j-1}, 0, n_{j+1} \cdots & n_j = 1 \\ 0 & n_j = 0, \end{cases}$$

$$(3.15)$$

where η_j is the number of occupied states to the left of state j.

Let Ψ represent an occupation number wave function that contains the state j but not the state ℓ. Let's also assume $j > \ell$. Also, let η_j and

η_ℓ represent the number of occupied states immediately to the left of j and ℓ, respectively. By the rules in Eqs. (3.14) and (3.15), we have that

$$a_\ell^\dagger a_j \Psi = (-1)^{\eta_j + \eta_\ell} | \cdots n_{\ell-1}, 1, \cdots n_{j-1}, 0, \cdots \rangle \tag{3.16}$$

$$a_j a_\ell^\dagger \Psi = (-1)^{\eta_\ell + \eta_j + 1} | \cdots n_{\ell-1}, 1, \cdots n_{j-1}, 0, \cdots \rangle. \tag{3.17}$$

If we now add (3.16) and (3.17), we obtain

$$[a_\ell^\dagger a_j + a_j a_\ell^\dagger] \Psi = (-1)^{\eta_j + \eta_\ell} [1 - 1] | \cdots n_{\ell-1}, 1, \cdots n_{j-1}, 0, \cdots \rangle, \tag{3.18}$$

or, equivalently,

$$[a_\ell^\dagger, a_j]_+ = a_\ell^\dagger a_j + a_j a_\ell^\dagger = 0 \text{ for } j \neq \ell. \tag{3.19}$$

It is straightforward to show that, if $j = \ell$ in Eq. (3.16), $a_j^\dagger a_j + a_j a_j^\dagger = 1$. The general anticommutation relation is

$$[a_j^\dagger, a_\ell]_+ = \delta_{j\ell}. \tag{3.20}$$

The remainder of the anticommutation relations, namely,

$$[a_j, a_\ell]_+ = [a_j^\dagger, a_\ell^\dagger]_+ = 0, \tag{3.21}$$

can be derived in an analogous fashion. The final quantity we should define is the number operator,

$$\hat{N} = \sum_j a_j^\dagger a_j. \tag{3.22}$$

As in the boson case, $a_j^\dagger a_j$ counts the number of particles in state j.

3.3   FERMION OPERATORS

We now turn to the task of writing one- and two-body fermion operators in their second quantized form. A straightforward way of accomplishing this is to consider their action on a fully antisymmetrized many-particle state. Consider two single-particle states ϕ_1 and ϕ_2. The normalized wave function that is antisymmetric with respect to particle interchange is

$$\langle \mathbf{r}_1, \mathbf{r}_2 | n_1, n_2 \rangle = \frac{1}{\sqrt{2}} [\phi_1(1)\phi_2(2) - \phi_1(2)\phi_2(1)]. \tag{3.23}$$

This state can be constructed from the determinant of ϕ_1 and ϕ_2:

$$D_2 = \frac{1}{\sqrt{2!}} \begin{vmatrix} \phi_1(1) & \phi_2(1) \\ \phi_1(2) & \phi_2(2) \end{vmatrix}. \tag{3.24}$$

Dirac [D1929, D1958] and Slater [S1929] showed that the general rule for constructing an antisymmetric wave function out of n single-particle states is

$$D_n = \langle \mathbf{r}_1, \mathbf{r}_2, \cdots | n_\alpha, n_\beta, \cdots \rangle = \frac{1}{\sqrt{n!}} \| \phi_1 \cdots \phi_n \|, \tag{3.25}$$

where $\| \; \|$ represents the determinant. D_n contains all antisymmetrized permutations of the orbital set $\phi_1 \cdots \phi_n$ and hence may be written as

$$D_n = \langle \mathbf{r}_1, \mathbf{r}_2, \cdots | n_1, n_2, \cdots \rangle = \frac{1}{\sqrt{n!}} \sum_P (-1)^P P[\phi_1 \cdots \phi_n], \tag{3.26}$$

with P as the permutation operator. The inherent advantage in using second-quantized notation is that a general many-particle fermionic state can be written compactly as

$$|n_1, n_2, \cdots \rangle = a_1^\dagger a_2^\dagger \cdots |0\rangle. \tag{3.27}$$

Complete antisymmetry under particle interchange is built into this many-body state as a result of the anticommuting property of the fermion operators. Note there is no $\sqrt{n!}$ normalization factor. In first quantization, however, an explicit $\sqrt{n!}$ factor appears, because particles are placed in particular single-particle states and all possible permutations are summed over. In second quantization, no labels are attached to the particles.

Let us now return to our initial goal. Consider the one-body operator $\widehat{H}_1$. The matrix element of a one-body operator between two many-particle states is nonzero only if the two many-particle states do not differ by more than two single-particle states (see Problem 3.1). Consequently, the most $\widehat{H}_1$ can do when it acts on a general many-body state is annihilate a particle from a particular single-particle state and fill the same or a previously vacant state. Hence, in second-quantized form, a general one-body operator is restricted to have a single creation-annihilation operator pair. Quite generally, then, we can write a one-body operator as

$$\widehat{H}_1 = \sum_{\nu,\lambda} c_{\lambda\nu} a_\lambda^\dagger a_\nu. \tag{3.28}$$

To determine the coefficient $c_{\lambda\nu}$, we simply evaluate the matrix element $\langle\mu|\widehat{H}_1|\gamma\rangle$. Orthogonality of the single-particle states implies immediately that $\langle\mu|\widehat{H}_1|\gamma\rangle = c_{\mu\gamma}$. Consequently, the most general way of writing a one-body operator in second quantization is

$$\widehat{H}_1 = \sum_{\nu,\lambda} \langle\lambda|\widehat{H}_1|\nu\rangle a_\lambda^\dagger a_\nu. \tag{3.29}$$

In the event that the single-particle states are eigenfunctions of $\widehat{H}_1$, then $\widehat{H}_1|\nu\rangle = \epsilon_\nu|\nu\rangle$, where ϵ_ν is a c-number. Equation (3.29) then reduces to

$$\widehat{H}_1 = \sum_\lambda \epsilon_\lambda a_\lambda^\dagger a_\lambda = \sum_\lambda n_\lambda \epsilon_\lambda. \tag{3.30}$$

Consequently, in the case that $\widehat{H}_1$ is a one-body energy operator, the average of $\widehat{H}(1)$ determines the average energy of the system. The resultant expression is analogous to Eq. (1.8).

Consider now a general 2-body operator

$$\widehat{H}_2 = \frac{1}{2} \sum_{i,j} \widehat{V}(i,j). \tag{3.31}$$

In the electron gas, $\widehat{V}(i,j) = e^2/|\hat{\mathbf{r}}_i - \hat{\mathbf{r}}_j|$, the Coulomb energy. A two-body operator can at most create two particle-hole excitations in a general many-body state. The general form of the operator that creates such excitations is $a_k^\dagger a_\ell^\dagger a_j a_i$. As a consequence, a general 2-body operator in second-quantized form can be written as

$$\widehat{H}_2 = \frac{1}{2} \sum_{i,j,k,\ell} V_{ij,k,\ell} a_k^\dagger a_\ell^\dagger a_j a_i. \tag{3.32}$$

The interacting electron Hamiltonian containing both one- and two-body terms can be recast as

$$\widehat{H}_e = \sum_{\nu,\lambda} \langle\nu|\widehat{H}_1|\lambda\rangle a_\nu^\dagger a_\lambda + \frac{1}{2} \sum_{i,j,k,\ell} \langle k\ell| \frac{e^2}{|\mathbf{r}_1 - \mathbf{r}_2|} |ij\rangle a_k^\dagger a_\ell^\dagger a_j a_i. \tag{3.33}$$

To make contact with the electron gas, it is customary to transform to momentum space, in which the single-particle plane-wave states,

$$\phi_p(\mathbf{r}) = \frac{e^{i\mathbf{p}\cdot\mathbf{r}/\hbar}}{\sqrt{V}}, \qquad (3.34)$$

diagonalize exactly the electron kinetic energy. These states are defined in a box of volume V with periodic boundary conditions imposed. Particles with spin σ are added or removed from these states by the operators $a_{\mathbf{p}\sigma}^{\dagger}$ or $a_{\mathbf{p}\sigma}$, respectively. We introduce the *field operator*

$$\Psi_{\sigma}^{\dagger}(\mathbf{r}) = \sum_{\mathbf{p}} \frac{e^{-i\mathbf{p}\cdot\mathbf{r}/\hbar}}{\sqrt{V}} a_{\mathbf{p}\sigma}^{\dagger}, \qquad (3.35)$$

which creates an electron at $\mathbf{r}$ with spin σ. The Hermitian conjugate field, $\Psi_{\sigma}(\mathbf{r})$, annihilates a particle with spin σ at $\mathbf{r}$. Field operators create and annihilate particles at particular positions. In so doing, they do not add or remove particles from a particular momentum state. Rather, they add or subtract particles from a superposition of momentum states with amplitude $e^{\pm i\mathbf{p}\cdot\mathbf{r}/\hbar}/\sqrt{V}$. The product of the creation and annihilation field operators

$$\Psi_{\sigma}^{\dagger}(\mathbf{r})\Psi_{\sigma}(\mathbf{r}) = \frac{1}{V}\sum_{\mathbf{p},\mathbf{p}'} e^{-i\mathbf{r}\cdot(\mathbf{p}-\mathbf{p}')/\hbar} a_{\mathbf{p}\sigma}^{\dagger} a_{\mathbf{p}'\sigma} \qquad (3.36)$$

defines the particle density operator. Consequently, if we integrate Eq. (3.36) over $\mathbf{r}$,

$$\begin{aligned}
\hat{n}_{\sigma} &= \int d\mathbf{r}\,\Psi_{\sigma}^{\dagger}(\mathbf{r})\Psi_{\sigma}(\mathbf{r}) \\
&= \frac{1}{V}\int d\mathbf{r}\sum_{\mathbf{p},\mathbf{p}'} e^{-i(\mathbf{p}-\mathbf{p}')\cdot\mathbf{r}/\hbar} a_{\mathbf{p}\sigma}^{\dagger} a_{\mathbf{p}'\sigma} \\
&= \sum_{\mathbf{p}} a_{\mathbf{p}\sigma}^{\dagger} a_{\mathbf{p}\sigma} = \sum_{\mathbf{p}} \hat{n}_{\mathbf{p}\sigma},
\end{aligned} \qquad (3.37)$$

we obtain the total particle density for electrons with spin σ.

In analogy with Eq. (6.47), we can construct a general many-body state $|\mathbf{r}_{1\sigma_1}\cdots\mathbf{r}_{n\sigma_n}\rangle$,

$$|\mathbf{r}_{1\sigma_1}\cdots\mathbf{r}_{n\sigma_n}\rangle = \frac{1}{\sqrt{n!}}\Psi_{\sigma_1}^{\dagger}(\mathbf{r}_1)\cdots\Psi_{\sigma_n}^{\dagger}(\mathbf{r}_n)|0\rangle, \qquad (3.38)$$

from the vacuum state, using the field operator $\Psi_{\sigma_i}^\dagger(\mathbf{r}_i)$. The rules for applying $\Psi_{\sigma_i}^\dagger$ and Ψ_{σ_i} to $|\mathbf{r}_{1\sigma_1} \cdots \mathbf{r}_{n\sigma_n}\rangle$ are

$$\Psi_{\sigma_{n+1}}^\dagger(\mathbf{r}_{n+1})|\mathbf{r}_{1\sigma_1} \cdots \mathbf{r}_{n\sigma_n}\rangle = \sqrt{n+1}(-1)^{\eta_{n+1}}|\mathbf{r}_{1\sigma_1} \cdots \mathbf{r}_{n+1\sigma_{n+1}}\rangle$$

$$(3.39)$$

and

$$\Psi_\sigma(\mathbf{r})|\mathbf{r}_{1\sigma_1} \cdots \mathbf{r}_{n\sigma_n}\rangle = \frac{1}{\sqrt{n}} \sum_\alpha \delta(\mathbf{r} - \mathbf{r}_\alpha)(-1)^{\eta_\alpha}|\mathbf{r}_1 \cdots \mathbf{r}_{\alpha-1}, \mathbf{r}_{\alpha+1} \cdots \mathbf{r}_n\rangle.$$

$$(3.40)$$

Here again, η_α is the number of occupied states to the left of $\mathbf{r}_\alpha$.

PROBLEMS

1. Show that the matrix element of a one-body operator is nonzero between two many-particle states, provided the two many-body states do not differ by more than two single-particle states.

2. Prove explicitly that a general two-body operator can be written in the form

$$\hat{V}_2 = \sum_{\alpha\beta\gamma\delta} \langle\gamma\delta|V_2|\alpha\beta\rangle a_\gamma^\dagger a_\delta^\dagger a_\beta a_\alpha,$$

where $\langle\gamma\delta|V_2|\alpha\beta\rangle$ is the matrix element of the two-body interaction between initial states ϕ_α and ϕ_β and final states ϕ_γ and ϕ_δ. Also, show that for fermions in a state of the form (6.47), one can factor the expectation value $\langle a_\gamma^\dagger a_\delta^\dagger a_\beta a_\alpha\rangle$ as

$$\langle a_\gamma^\dagger a_\delta^\dagger a_\beta a_\alpha\rangle = \langle a_\gamma^\dagger a_\alpha\rangle\langle a_\delta^\dagger a_\beta\rangle - \langle a_\gamma^\dagger a_\beta\rangle\langle a_\delta^\dagger a_\alpha\rangle.$$

3. Show that the field operators obey the commutation relations

$$[\Psi_\sigma^\dagger(\mathbf{r}), \Psi_\sigma(\mathbf{r}')]_\pm = \delta(\mathbf{r} - \mathbf{r}'),$$
$$[\Psi_\sigma(\mathbf{r}), \Psi_\sigma(\mathbf{r}')]_\pm = [\Psi_\sigma^\dagger(\mathbf{r}), \Psi_\sigma^\dagger(\mathbf{r}')]_\pm = 0, \qquad (3.41)$$

where $\pm$ denotes the commutator for bosons and the anticommutator for fermions.

REFERENCES

[B1969] G. Baym, *Lectures on Quantum Mechanics* (W. A. Benjamin, Inc., NY, 1969; Perseus Books, Cambridge, MA, 1969).

[D1929] P. A. M. Dirac, *Proc. Roy. Soc. Lond.* **A123**, 714 (1929).

[D1958] P. A. M. Dirac, *The Principles of Quantum Mechanics*, 4th ed. (Oxford Univ. Press, Oxford, England, 1958), p. 248.

[S1929] J. C. Slater, *Phys. Rev.* **34**, 1293 (1929).

– 4 –

Hartree-Fock Approximation

In the last chapter, we showed that the Hamiltonian for interacting electrons in a solid can be written in second-quantized form as

$$\widehat{H}_e = \sum_{\nu\lambda}\langle\nu|\widehat{h}(1)|\lambda\rangle a_\nu^\dagger a_\lambda + \frac{1}{2}\sum_{\nu\lambda\alpha\beta}\langle\nu\lambda|\frac{e^2}{|\mathbf{r}_1 - \mathbf{r}_2|}|\alpha\beta\rangle a_\nu^\dagger a_\lambda^\dagger a_\beta a_\alpha, \quad (4.1)$$

where $\widehat{h}(1) = \widehat{\mathbf{p}}_1^2/2m + \widehat{V}_{\text{ion}}(\mathbf{r}_1)$ and the Greek letters denote single-particle orbitals. In this chapter, we introduce the Hartree-Fock approximation to the correlated-electron problem. The basic assumption of this approximation is that the ground state is the same as that of the noninteracting system. The energy is taken to be the expectation value in this state.

4.1 NONINTERACTING LIMIT

Let us first reformulate the noninteracting problem, using second quantization. In the noninteracting electron problem at $T = 0$, all momentum states up to the Fermi level are doubly occupied. As a consequence, we represent the ground-state wave function for the filled Fermi sea as

$$|\psi_0\rangle = |\mathbf{p}_0\uparrow, \mathbf{p}_0\downarrow, \ldots \mathbf{p}_F\uparrow, \mathbf{p}_F\downarrow\rangle = a_{\mathbf{p}_0\uparrow}^\dagger a_{\mathbf{p}_0\downarrow}^\dagger \ldots a_{\mathbf{p}_F\uparrow}^\dagger a_{\mathbf{p}_F\downarrow}^\dagger|0\rangle. \quad (4.2)$$

We compute the occupancy in the pth level by acting on the ground-state wave function with the number operator $\widehat{n}_{\mathbf{p}\sigma}$ for a momentum state $\mathbf{p}$:

$$\widehat{n}_{\mathbf{p}\sigma}|\psi_0\rangle = n_{\mathbf{p}\sigma}|\psi_0\rangle. \quad (4.3)$$

With this equality in hand, we simplify the expectation value of the kinetic energy,

$$\langle \hat{T} \rangle = \langle \psi_0 | \sum_{\mathbf{p},\sigma} \frac{p^2}{2m} \hat{n}_{\mathbf{p}\sigma} | \psi_0 \rangle$$

$$= 2 \sum_{p=0}^{p_F} \frac{p^2}{2m} \tag{4.4}$$

to obtain the standard free-particle result,

$$\langle \hat{T} \rangle = \frac{3}{5} \frac{p_F^2}{2m} N. \tag{4.5}$$

In computing the expectation value of the ion term, we must evaluate an expression of the form $\langle \psi_0 | a_{\mathbf{p}\sigma}^\dagger a_{\mathbf{p}'\sigma} | \psi_0 \rangle$. Because all states with $p < p_F$ are full, $a_{\mathbf{p}\sigma}^\dagger | \psi_0 \rangle = 0$ for $p < p_F$. Similarly, $a_{\mathbf{p}'\sigma} | \psi_0 \rangle = 0$ if $p' > p_F$. When $a_{\mathbf{p}\sigma}^\dagger a_{\mathbf{p}'\sigma}$ acts on $|\psi_0 \rangle$, a new state is created that differs from ψ_0 by at most two states. The overlap of this state with $|\psi_0 \rangle$ will be zero because of the orthogonality of the momentum eigenstates unless, of course, $\mathbf{p} = \mathbf{p}', \sigma = \sigma'$:

$$\langle \psi_0 | a_{\mathbf{p}\sigma}^\dagger a_{\mathbf{p}'\sigma} | \psi_0 \rangle = \delta_{\mathbf{p}\mathbf{p}'} n_{\mathbf{p}\sigma}. \tag{4.6}$$

As a consequence,

$$\langle \hat{V}_{\text{ion}} \rangle = \sum_{\mathbf{p},\sigma} n_{\mathbf{p}\sigma} V_{\text{ion}}(0), \tag{4.7}$$

where

$$V_{\text{ion}}(0) = \frac{1}{V} \int d\mathbf{r} \, V_{\text{ion}}(\mathbf{r}). \tag{4.8}$$

The final result is that

$$\langle \hat{H}_1 \rangle = 2 \sum_{p < p_F} \left(\frac{p^2}{2m} + V_{\text{ion}}(0) \right). \tag{4.9}$$

As a preliminary to the Hartree-Fock approximation, it is useful to write this result in position space. To reformulate our problem in position space, we undo the Fourier transform and write the creation operators as

$$a_{\mathbf{p}\sigma}^\dagger = \int d\mathbf{r} \frac{e^{i\mathbf{p}\cdot\mathbf{r}/\hbar}}{\sqrt{V}} \psi_\sigma^\dagger(\mathbf{r}). \tag{4.10}$$

Using Eq. (4.10) for the second-quantized operators in the Hamiltonian and doing the sum over $\mathbf{p}$, we find that

$$\widehat{H}_1 = \sum_\sigma \int d\mathbf{r} \left[-\frac{\hbar^2}{2m} \psi_\sigma(\mathbf{r}) \nabla^2 \psi_\sigma^\dagger(\mathbf{r}) + \psi_\sigma^\dagger(\mathbf{r}) V_{\text{ion}}(\mathbf{r}) \psi_\sigma(\mathbf{r}) \right]$$

$$= \sum_\sigma \int d\mathbf{r} \left[-\frac{\hbar^2}{2m} |\nabla \psi_\sigma|^2 + \widehat{n}_\sigma(\mathbf{r}) V_{\text{ion}}(\mathbf{r}) \right], \tag{4.11}$$

the kinetic energy operator for a many-body system is proportional to the operator $|\nabla \psi_\sigma|^2$. Likewise, the one-body potential term is weighted with the density operator $\psi_\sigma^\dagger(\mathbf{r})\psi_\sigma(\mathbf{r})$.

It is instructive to connect the second-quantized expression (4.11) and the expectation value of $\widehat{H}_1$ in the orbital basis defined in Eq. (4.1):

$$\langle \widehat{H}_1 \rangle = \sum_\nu \langle \nu | \widehat{h}(1) | \nu \rangle n_\nu$$

$$= \sum_{\nu \text{ (occ.)}} \int \left[-\frac{\hbar^2}{2m} |\nabla \phi_\nu|^2 + \phi_\nu^*(\mathbf{r}) V_{\text{ion}}(\mathbf{r}) \phi_\nu(\mathbf{r}) \right] d\mathbf{r}. \tag{4.12}$$

The gradient and orbital density terms directly correspond to those in Eq. (4.11). The only difference is that in the second-quantized expression (4.11), field operators $\phi_\nu(\mathbf{r})$, which create or annihilate particles, replace the single-particle orbitals.

4.2   HARTREE-FOCK APPROXIMATION

We now turn to the two-body part of H. The additional term we must evaluate is

$$\langle \widehat{H}_2 \rangle = \langle \psi_0 | \widehat{V}_{ee} | \psi_0 \rangle \tag{4.13}$$

$$= \frac{1}{2} \sum_{\alpha,\beta,\nu,\lambda} \langle \nu\lambda | \frac{e^2}{|\mathbf{r}_1 - \mathbf{r}_2|} | \alpha\beta \rangle \langle a_\nu^\dagger a_\lambda^\dagger a_\beta a_\alpha \rangle, \tag{4.14}$$

where $\langle ... \rangle = \langle \psi_0 | ... | \psi_0 \rangle$ and $\langle r | \psi_0 \rangle$ is the ground-state wave function in the orbital basis. We treat the spin indices as implicit in the $\alpha, \beta, ...$; bear in mind, though, that the spin associated with the index ν is the same as that associated with α, whereas the spin associated with the index λ is the same as that associated with β.

To simplify Eq. (4.13), we must learn how to evaluate expectation values of strings of second-quantized operators in the ground state $|\psi_0\rangle$. Consider first the expectation value in the state $|n_\alpha n_\beta\rangle$. Using the relation $a_\alpha |n_\alpha\rangle = \sqrt{n_\alpha - 1}|n_\alpha - 1\rangle$, we have

$$\langle n_\alpha n_\beta | a_\nu^\dagger a_\lambda^\dagger a_\beta a_\alpha | n_\alpha n_\beta \rangle = \sqrt{n_\alpha n_\beta} \langle n_\alpha n_\beta | a_\nu^\dagger a_\lambda^\dagger | n_\alpha - 1, n_\beta - 1 \rangle.$$
(4.15)

Now this quantity is nonvanishing only if $\nu = \beta$ and $\lambda = \alpha$, or $\nu = \alpha$ and $\lambda = \beta$. Because of the anticommutation relations of the α's, these two cases will give contributions differing by a minus sign. We find, then,

$$\langle n_\alpha n_\beta | a_\nu^\dagger a_\lambda^\dagger a_\beta a_\alpha | n_\alpha n_\beta \rangle = (\delta_{\nu\alpha}\delta_{\lambda\beta} - \delta_{\nu\beta}\delta_{\lambda\alpha})n_\alpha n_\beta,$$
(4.16)

or, for the ground state given in Eq. (4.2),

$$\langle a_\nu^\dagger a_\lambda^\dagger a_\beta a_\alpha \rangle = (\delta_{\nu\alpha}\delta_{\lambda\beta} - \delta_{\nu\beta}\delta_{\lambda\alpha})\langle \hat{n}_\alpha \rangle \langle \hat{n}_\beta \rangle.$$
(4.17)

The general rules for factoring such expectation values were laid down by Wick [W1950], who showed that $\langle a_\nu^\dagger a_\lambda^\dagger ... a_l^\dagger a_\rho a_\alpha ... a_\gamma \rangle$ is given by the sum of all contractions, with the sign of each term determined by the number of particle interchanges necessary to line up side-by-side the contracted operators. Note, when two operators are contracted, their indices become equal.

With this result in hand, we rewrite $\langle \widehat{H}_2 \rangle$ as a sum of two integrals,

$$\langle \widehat{H}_2 \rangle = \frac{1}{2} \sum_{\lambda,\nu} \langle \nu\lambda | \frac{e^2}{|\mathbf{r}_1 - \mathbf{r}_2|} | \nu\lambda \rangle n_\lambda n_\nu - \frac{1}{2} \sum_{\lambda,\nu} \langle \nu\lambda | \frac{e^2}{|\mathbf{r}_1 - \mathbf{r}_2|} | \lambda\nu \rangle n_\lambda n_\nu$$

$$= \frac{1}{2} \sum_{\lambda,\nu \, (\text{occ.})} (U_{\nu\lambda} - \delta_{\sigma_\nu \sigma_\lambda} J_{\nu\lambda}).$$
(4.18)

The direct Coulomb interaction,

$$U_{\nu\lambda} = \int d\mathbf{r}_1 d\mathbf{r}_2 |\phi_\nu(\mathbf{r}_1)|^2 \frac{e^2}{|\mathbf{r}_1 - \mathbf{r}_2|} |\phi_\lambda(\mathbf{r}_2)|^2,$$
(4.19)

is a measure of the repulsion between two electrons at positions $\mathbf{r}_1$ and $\mathbf{r}_2$; the exchange interaction,

$$J_{\nu\lambda} = \int d\mathbf{r}_1 d\mathbf{r}_2 \; \phi_\nu^*(\mathbf{r}_1)\phi_\lambda^*(\mathbf{r}_2) \frac{e^2}{|\mathbf{r}_1 - \mathbf{r}_2|} \phi_\nu(\mathbf{r}_2)\phi_\lambda(\mathbf{r}_1), \quad (4.20)$$

arises solely when electrons of like spin exchange spatial coordinates. The negative sign accompanying the exchange term leads to an effective attraction between electrons of like spin. This attraction is the basis of Heisenberg's explanation of ferromagnetism. The Pauli exclusion principle, which results from the Fermi-Dirac statistics obeyed by electrons, leads to a depletion of electrons of the same spin in the neighborhood of an electron of given spin. The depletion is commonly called the *exchange hole* [P1953].

Equations (4.12) and (4.18) together constitute the exact expression for the expectation value of $\widehat{H}_e$ in the state $|\psi_0\rangle$. The total Hartree-Fock energy is

$$E_{\mathrm{HF}} = \langle \widehat{H}_1 \rangle + \frac{1}{2} \sum_{\lambda,\nu \; (\mathrm{occ.})} [U_{\nu\lambda} - J_{\nu\lambda}]. \quad (4.21)$$

The only detail missing is the prescription for finding the orbitals ϕ_ν and ϕ_λ. To determine them, we minimize the ground-state energy E_{HF} with respect to the ϕ's, subject to the condition that the ϕ's remain normalized. This latter condition can be imposed by introducing a Lagrange multiplier, ϵ_ν, in the minimization:

$$\frac{\delta E_{\mathrm{HF}}}{\delta \phi_\nu^*(\mathbf{r})} = \epsilon_\nu \frac{\delta}{\delta \phi_\nu^*(\mathbf{r})} \int d\mathbf{r}' |\phi_\nu(\mathbf{r}')|^2. \quad (4.22)$$

We discuss below the physical significance of energies ϵ_ν. Performing the variation in Eq. (4.22) results in the celebrated Hartree-Fock equations

$$\left[\frac{-\hbar^2}{2m}\nabla^2 + \widehat{V}_{\mathrm{ion}}(\mathbf{r}) + \sum_\lambda \int d\mathbf{r}' n_\lambda(\mathbf{r}') \frac{e^2}{|\mathbf{r} - \mathbf{r}'|} \right] \phi_\nu(\mathbf{r})$$

$$-\sum_\lambda \int d\mathbf{r}' \phi_\lambda^*(\mathbf{r}')\phi_\nu(\mathbf{r}') \frac{e^2}{|\mathbf{r} - \mathbf{r}'|} \phi_\lambda(\mathbf{r}) = \epsilon_\nu \phi_\nu(\mathbf{r}) \quad (4.23)$$

for the single-particle orbitals $\phi_\nu(\mathbf{r})$. In the absence of translational invariance, these variational equations can be solved iteratively until a self-consistent set of orbitals with corresponding energies, ϵ_ν, are determined.

Multiplying Eq. (4.23) by $\phi_\nu^*(\mathbf{r})$ and integrating over $\mathbf{r}$, we derive the result for the single-particle energy ϵ_ν,

$$\epsilon_\nu = \langle \nu | h_1 | \nu \rangle + \sum_{\lambda \, (\text{occ.})} (U_{\nu\lambda} - J_{\nu\lambda}). \qquad (4.24)$$

Because these energies involve the interaction with electrons in all other occupied orbitals, ϵ_ν cannot be interpreted simply as the energy of the electron in orbital ν. In fact, the total energy is not simply the sum of these single-particle energies, but, as follows from Eq. (4.21), is rather

$$E_{\text{HF}} = \sum_{\nu \, (\text{occ.})} \epsilon_\nu - \frac{1}{2} \sum_{\lambda, \nu \, (\text{occ.})} (U_{\nu\lambda} - J_{\nu\lambda}). \qquad (4.25)$$

Koopmans [K1934] observed that the Hartree-Fock single-particle energy, ϵ_ν, is the energy required to add a particle in the previously unoccupied orbital ν. Consider adding an extra electron to a set of filled orbitals in a previously unoccupied orbital ν. The change in the Hartree-Fock energy due to adding the particle is

$$\delta E_{\text{HF}} = E_{\text{HF}}^{\text{new}} - E_{\text{HF}} = \langle \nu | h(1) | \nu \rangle + \sum_{\alpha \neq \nu} (U_{\nu\alpha} - J_{\nu\alpha}) = \epsilon_\nu. \qquad (4.26)$$

Likewise, had we removed an electron, the resultant Hartree-Fock energy would simply decrease by the energy of the orbital from which the electron was taken. Quite generally, in going from an N to an $N-1$ particle system, $\delta E_{\text{HF}} = \epsilon_N = E_N^{\text{HF}} - E_{N-1}^{\text{HF}}$. Koopmans' theorem states that the negative of ϵ_N may be taken as a variational approximation to the system's first ionization potential. For atoms, the ionization potentials predicted by Koopmans' theorem are in reasonable agreement with experiment [W1964].

4.3 DIAGRAMS

Throughout this book, we will use diagrams as a heuristic tool to represent processes involving interactions among particles. Fig. (4.1) illustrates the two-body interaction in Eq. (4.13). A forward-going arrow indicates the propagation of a particle of given momentum, and the dashed line represents the two-electron Coulomb interaction. Fig. (4.2)

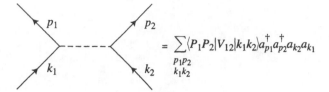

Figure 4.1 General two-body interaction.

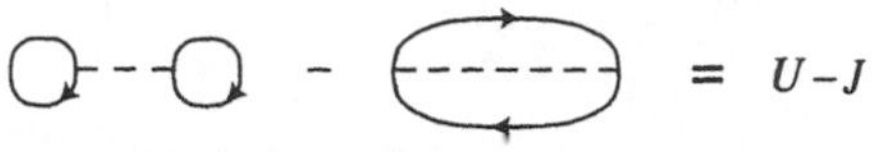

$$\bigcirc\!-\!-\!\bigcirc \;-\; \left(\!-\!-\!-\!-\!-\!\right) \;=\; U-J$$

Figure 4.2 Direct and exchange interactions.

shows diagrammatically the terms in the interaction energy, Eq. (4.18). Here, the clockwise arrow represents the propagation of a hole.

PROBLEMS

1. Show that the Hartree-Fock energy can be written as

$$E_{\mathrm{HF}} = \sum_{n=1}^{N} \epsilon_n, \tag{4.27}$$

where N is the number of single-particle orbitals and

$$\epsilon_n = \langle n|\widehat{H}_1|n\rangle + \sum_{\lambda=1}^{n} [U_{n\lambda} - J_{n\lambda}]. \tag{4.28}$$

Interpret this result in light of the fact that the true Hartree-Fock energy cannot be written as a sum of the single-particle energies.

REFERENCES

[K1934] T. Koopmans, *Physica* **1**, 104 (1934). Interestingly, this theorem is Koopmans' only physics contribution. Immediately after publishing this work, he switched to economics, a field in which he later won the Nobel Prize.

[P1953] D. Pines, *Phys. Rev.* **92**, 636 (1953).

[W1964] See A. C. Wahl, *J. Chem. Phys.* **41**, 2600 (1964) for a failure of Koopmans' theorem in the molecule F_2.

[W1950] G.-C. Wick, *Phys. Rev.* **80**, 268 (1950).

– 5 –

Interacting Electron Gas

We now turn to describing the properties of interacting electrons in metals. As a first approximation, we regard the ions as providing only a uniform background of positive charge to compensate the overall charge of the electrons. Such an approximation constitutes the Jellium model, first introduced by Herring and characterized extensively by Pines [P1955]. For ions of charge Z and density n_i, the electron density n_e equals Zn_i. We start by considering the Hartree-Fock approximation to the uniform interacting electron system, which applies at high densities, and later introduce further approximations to treat the dilute regime of the interacting electron gas.

Typically, we associate the eigenstates

$$\phi_{\mathbf{p}}(\mathbf{r}) = \frac{e^{i\mathbf{p}\cdot\mathbf{r}/\hbar}}{\sqrt{V}}, \tag{5.1}$$

with free electrons moving in a box of volume V with periodic boundary conditions. It turns out that such free-particle states also satisfy the Hartree-Fock equations for the Jellium model of an electron gas. However, as we derive below, the corresponding single-particle energies are no longer $p^2/2m$, but rather

$$\epsilon(\mathbf{p}) = \frac{p^2}{2m} - \frac{e^2 p_F}{\pi\hbar}\left(1 + \frac{p_F^2 - p^2}{2pp_F}\ln\left|\frac{p + p_F}{p - p_F}\right|\right). \tag{5.2}$$

The energies of the occupied energy levels are lowered below the free-particle result as a consequence of the exchange correlation among the electrons, which keeps particles of the same spin from lying atop one

35

another. Hence, the Hartree-Fock approximation produces a dressed electronic excitation with a well-defined momentum. The fact that such excitations are described by plane-wave eigenstates suggests that the Hartree-Fock approximation is essentially a noninteracting treatment of the electron gas. Indeed this is true. Hartree-Fock produces a set of noninteracting quasi particles that attempt to mimic the interacting system. We will see, however, that their imitation of the properties of the original system does not fare well. For example, the Hartree-Fock treatment fails to recover the linear temperature dependence of the heat capacity. It predicts instead a $\ln T$ divergence of the specific heat.

5.1 UNIFORM ELECTRON GAS

The starting point for analyzing the interacting electron gas is the Hartree-Fock equation

$$
\epsilon_\nu \phi_\nu(\mathbf{r}) = \left[\frac{-\hbar^2}{2m} \nabla_r^2 + V_{\text{ion}}(r) + e^2 \sum_\lambda \int d\mathbf{r}' n_{\phi_\lambda}(r') \frac{1}{|\mathbf{r} - \mathbf{r}'|} \right] \phi_\nu(\mathbf{r})
$$

$$
- e^2 \sum_\lambda \int d\mathbf{r}' \phi_\nu(\mathbf{r}') \phi_\lambda^*(\mathbf{r}') \phi_\lambda(\mathbf{r}) \frac{1}{|\mathbf{r} - \mathbf{r}'|}. \tag{5.3}
$$

Assuming at the outset that plane-wave states of the form of Eq. (5.1) satisfy the Hartree-Fock equations, converting sums over momentum to integrals and changing variables of integration in the exchange term to $\mathbf{x} = \mathbf{r}' - \mathbf{r}$, we obtain the eigenvalue equation

$$
\epsilon(\mathbf{p}) \phi_\mathbf{p}(\mathbf{r}) = \left[\frac{p^2}{2m} + V_{\text{ion}}(r) + e^2 n_e \int d\mathbf{r}' \frac{1}{|\mathbf{r} - \mathbf{r}'|} \right] \phi_\mathbf{p}(\mathbf{r})
$$

$$
- e^2 \int \frac{d\mathbf{p}' d\mathbf{x}}{(2\pi\hbar)^3} \frac{e^{i(\mathbf{p} - \mathbf{p}') \cdot \mathbf{x}/\hbar}}{|\mathbf{x}|} \phi_\mathbf{p}(\mathbf{r}). \tag{5.4}
$$

In deriving this equation, we first include the term with $\lambda = \nu$ in both the direct Coulomb and exchange terms in Eq. (5.3) and then let

$$
\sum_\lambda n_{\phi_\lambda(\mathbf{r}')} = n_e(\mathbf{r}') \equiv n_e \tag{5.5}
$$

in the direct term. Although the direct Coulomb term explicitly excludes the interaction of an electron with itself, inclusion of the self-interaction in the exchange term as well cancels this overestimation of the Coulomb repulsion. As we see, plane-wave states do satisfy the Hartree-Fock equations for a uniform electron gas. As a consequence of the charge neutrality constraint

$$V_{\text{ion}}(\mathbf{r}) = \sum_i \frac{-Ze^2}{|\mathbf{r} - \mathbf{R}_i^0|} \rightarrow -n_e \int \frac{d\mathbf{R}}{|\mathbf{r} - \mathbf{R}|}, \qquad (5.6)$$

the second and third terms in the eigenvalue equation cancel. The energy of a particle in a momentum state $\mathbf{p}$ is then

$$\epsilon(\mathbf{p}) = \frac{p^2}{2m} - e^2 \int_0^{PF} \frac{d\mathbf{p}'}{(2\pi\hbar)^3} \int d\mathbf{x} \frac{e^{i(\mathbf{p}-\mathbf{p}')\cdot\mathbf{x}/\hbar}}{|\mathbf{x}|}. \qquad (5.7)$$

The second term in Eq. (5.7) is the single-particle exchange energy,

$$\epsilon_{\text{exch}}(\mathbf{p}) = -e^2 \int \frac{d\mathbf{p}'}{(2\pi\hbar)^3} \int d\mathbf{x} \frac{e^{i(\mathbf{p}'-\mathbf{p})\cdot\mathbf{x}/\hbar}}{|\mathbf{x}|}. \qquad (5.8)$$

Using the integral

$$\int d\mathbf{x} \frac{e^{i(\mathbf{p}'-\mathbf{p})\cdot\mathbf{x}/\hbar}}{|\mathbf{x}|} = \frac{4\pi\hbar^2}{|\mathbf{p} - \mathbf{p}'|^2}, \qquad (5.9)$$

we find

$$\begin{aligned}
\epsilon_{\text{exch}}(\mathbf{p}) &= -\frac{e^2}{\hbar} \int_0^{PF} \frac{d\mathbf{p}'}{(2\pi)^3} \frac{4\pi}{|\mathbf{p} - \mathbf{p}'|^2} \\
&= -\frac{e^2}{\pi\hbar} \int_0^{PF} \int_{-1}^1 \frac{p'^2 \, dp' \, d\cos\theta}{p^2 + p'^2 - 2pp'\cos\theta} \\
&= \frac{e^2}{\pi p\hbar} \int_0^{PF} p' \, dp' [\ln|p - p'| - \ln|p + p'|].
\end{aligned}$$

Then, since

$$\int x \ln(x + a) \, dx = \frac{x^2 - a^2}{2} \ln(x + a) - \frac{1}{4}(x - a)^2, \qquad (5.10)$$

we derive the result

$$\epsilon_{\text{exch}}(\mathbf{p}) = \frac{e^2}{\pi p \hbar}\left[\frac{(p'^2 - p^2)}{2}\ln\left|\frac{p - p'}{p + p'}\right| - \frac{1}{4}(p + p')^2 + \frac{(p - p')^2}{4}\right]_0^{p_F}$$

$$= -\frac{e^2 p_F}{\pi \hbar}\left(1 + \frac{(p_F^2 - p^2)}{2pp_F}\ln\left|\frac{p + p_F}{p - p_F}\right|\right). \tag{5.11}$$

The single-particle energy is then as given in Eq. (5.2).

Note that the exchange energy is negative, since $p < p_F$ and $|\mathbf{p} - \mathbf{p}_F| < |\mathbf{p} + \mathbf{p}_F|$. The reason is the following. The exchange energy accounts for the fact that electrons of like spin avoid one another as a consequence of the exclusion principle, and, hence, the energy of the electron is lower than it would be were it moving in a uniform electron cloud of density n_e, as described by the direct term. Each electron develops an exchange cloud around it, in which the density of same-spin electrons is severely reduced. This space is the *exchange hole*. We will return to the energy spectrum after we evaluate the total energy.

The total Hartree-Fock energy is given as a sum over the single-particle energies of the occupied levels as

$$E_{\text{HF}} = 2V\int_0^{p_F}\frac{d\mathbf{p}'}{(2\pi\hbar)^3}\left(\frac{p^2}{2m} + \frac{1}{2}\epsilon_{\text{exch}}(\mathbf{p})\right). \tag{5.12}$$

Note that the exchange energy enters with a factor 1/2 in the total energy. Simply, it is

$$\int_0^{p_F}\frac{d\mathbf{p}}{(2\pi\hbar)^3}\epsilon_{\text{exch}}(\mathbf{p}) = -\frac{e^2 p_F^4}{4\pi^3\hbar^4}. \tag{5.13}$$

(See Problem 5.1.) Using this result, we may write the total Hartree-Fock energy per particle,

$$\frac{E_{\text{HF}}}{N} = \frac{e^2}{2a_0}\left[\frac{3}{5}\left(\frac{p_F a_0}{\hbar}\right)^2 - \frac{3}{2\pi}\left(\frac{p_F a_0}{\hbar}\right)\right]$$

$$= \left(\frac{2.21}{r_s^2} - \frac{0.916}{r_s}\right)\text{Ry}, \tag{5.14}$$

in terms of $r_s = (9\pi/4)^{1/3}e^2/\hbar v_F$, where Ry is the Rydberg, $e^2/2a_0 = 13.6$ eV. The exchange energy lowers the total energy significantly, indicating that electron correlations are important and that they must be treated correctly to describe the properties of real materials.

5.2 HARTREE-FOCK EXCITATION SPECTRUM

To delineate the limitations of the Hartree-Fock approximation, we focus on the excitation spectrum. We start by rewriting the single-particle exchange energy as $\epsilon_{\text{exch}} = \epsilon_F^0 F(x)$, where $x = p/p_F$,

$$F(x) = \frac{1}{2} + \frac{1 - x^2}{4x} \ln \left| \frac{1 + x}{1 - x} \right|, \tag{5.15}$$

and ϵ_F^0 is the unperturbed Fermi energy $p_F^2/2m$. Then,

$$\epsilon(\mathbf{p})/\epsilon_F^0 = [x^2 - 0.663 r_s F(x)]; \tag{5.16}$$

$F(x)$ is plotted in Fig. (5.1) for $r_s = 3.0$. In a free-particle system, $\partial^2 \epsilon / \partial p^2 = 1/m$. More generally, one defines the effective mass, m^*, of an electron at the Fermi surface by

$$\frac{1}{m^*} = \frac{1}{p_F} \left(\frac{\partial \epsilon(\mathbf{p})}{\partial p} \right)_{p = p_F}. \tag{5.17}$$

As $x \to 1$, the slope of $F(x)$ diverges logarithmically,

$$\frac{\partial F}{\partial x} \approx \ln \left| \frac{1 - x}{2} \right| \to -\infty. \tag{5.18}$$

Thus, as one approaches the Fermi surface, the effective mass in the Hartree-Fock approximation from Eq. (5.17) behaves as

$$m^* \propto \frac{1}{\ln |1 - x|}, \tag{5.19}$$

thereby vanishing at the Fermi surface. This result is not consistent with experimental data.

As Bardeen has shown [B1936], the vanishing of the effective mass in Hartree-Fock at the Fermi surface leads to an unphysical behavior of the low-temperature heat capacity. Quite generally, we may

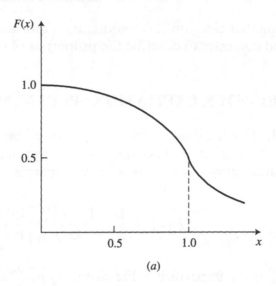

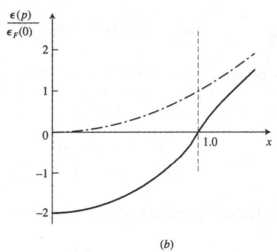

Figure 5.1 (*a*) A plot of the function $F(x)$ defined by Eq. (5.15) for $r_s = 3.0$. The slope of $F(x)$ diverges as $x \to 1$ and $F(x = 1) = \frac{1}{2}$. For large x, $F(x)$ approaches $1/3x^2$. (*b*) The scaled Hartree-Fock single-particle energy dispersion (solid line) defined by Eq. (5.16). Hartree-Fock overestimates the bandwidth by the additive constant $0.331 r_s \epsilon_F^0$. This increase in the bandwidth arises entirely from the exchange interaction. As a guide to the eye, the free-particle dispersion is plotted as a dash-dot line.

calculate the low-temperature heat capacity from Eq. (1.25). The density of states near $p = p_F$ is given in Hartree-Fock by

$$N(\epsilon) = \frac{p_F^2}{\pi^2 \hbar^3} \frac{\partial p}{\partial \epsilon} \approx \frac{m p_F / \pi^2 \hbar^3}{1 - (m e^2 / \pi \hbar p_F) \ln |(p - p_F)/2 p_F|}. \qquad (5.20)$$

The characteristic value of $p - p_F$ at low temperature, T, is given by $|p^2/2m - p_F^2/2m| \sim |p - p_F|v_F \sim T$, where v_F is the bare Fermi velocity, p_F/m; thus, carrying out the integral in Eq. (1.25), one eventually finds

$$C_V \sim \frac{T}{|\ln T|}, \tag{5.21}$$

as T goes to zero. The unphysical logarithm arises from the long range of the Coulomb potential, which is reflected in the divergent behavior of the Fourier transform of $1/r \sim 1/k^2$, as $k \to 0$. This divergence is removed in fact by screening of the interaction between electrons, where the long-ranged e^2/r becomes modified to a short-range interaction, roughly of the form $e^2 \exp(-k_{TF}r)/r$, where k_{TF} is the inverse screening length. With screening, which we discuss at length in Chapter 8, the correct linear temperature dependence of the heat capacity, Eq. (1.38), is recovered.

Another difficulty with the Hartree-Fock approximation is that it overestimates the energy bandwidth, defined as $\Delta = \epsilon(x = 1) - \epsilon(x = 0)$. For a wide range of metals, the noninteracting value of ϵ_F^0 well approximates the experimentally observed bandwidth. At the Hartree-Fock level, the bandwidth expands to

$$\Delta = \epsilon_F^0 (1 + 0.331 r_s). \tag{5.22}$$

The increase in the bandwidth, $\sim 0.331 r_s$, is illustrated in Fig. (5.1b).

5.3   COHESIVE ENERGY OF METALS

One of the key problems to which the Hartree-Fock approximation has been applied is the cohesive energy of a metal. When free atoms are brought together to form a metal, their energy decreases. The cohesive energy is defined as the difference

$$\epsilon_{coh} = (\epsilon/\text{atom})_{metal} - (\epsilon/\text{atom})_{free\,atom},$$

and it is negative in a bound metal. The measured cohesive energies in the alkalis are:

eV	Li	Na	K	Rb	Cs
$\epsilon_{coh,\,expt}$	−1.58	−1.13	−0.98	−0.82	−0.815

To calculate cohesive energies quantitatively, two improvements beyond Hartree-Fock need to be made. In addition to better describing the correlations among electrons, it is necessary to go beyond the approximation that the ions form a uniform background and take into account more precisely the interaction of the electrons with the ions and the average Coulomb energy of the ions. We follow here the strategy of Wigner and Seitz [WS1933] to calculate the cohesive energy of a metal, a scheme that enabled them to carry out the first quantitative application of quantum mechanics to calculating realistic properties of solids.

5.3.1   Wigner-Seitz Method

The main idea of the Wigner-Seitz method is to divide the crystal into (Wigner-Seitz) cells, each containing a single ion and Z electrons, and to treat the interactions within each cell reasonably accurately. Since each cell is electrically neutral, the interaction energy between cells can be ignored to a first approximation. The cohesive energy is then given by

$$\epsilon_{coh} = \epsilon_0 - \epsilon_{atom} + \epsilon_{kin} + \epsilon_{coul}, \tag{5.23}$$

where ϵ_0 is the lowest energy of a conduction electron in the cell (physically, the energy of the bottom of the conduction band), ϵ_{kin} is the average electron kinetic energy, and ϵ_{coul} is the average electron-electron Coulomb energy.

At this point, we restrict our attention to the alkali metals, in each cell of which a single conduction electron and an ion reside. One can initially ignore electron-electron interactions within each cell. Most simply, the Wigner-Seitz cell can be taken to be a sphere about the ion. To calculate ϵ_0, we solve the Schrödinger equation

$$\left(\frac{\hat{p}^2}{2m} + V_{ion}\right)\varphi_0(\mathbf{r}) = \epsilon_0 \varphi_0(\mathbf{r}) \tag{5.24}$$

for the lowest-conduction electron level in the presence of the ion potential. The boundary condition at the edge of each cell is that the normal derivative $\varphi_0'(\mathbf{r}) = 0$.

The resulting ϵ_0's are shown in the table below:

eV	Li	Na	K	Rb	Cs
ϵ_{atom}	−5.37	−5.16	−4.34	−4.17	−3.89
ϵ_0	−9.15	−8.25	−6.58	−6.18	−5.85
$\epsilon_0 - \epsilon_{atom}$	−3.78	−3.09	−2.24	−2.01	−1.95

As is evident, $\epsilon_0 - \epsilon_{atom} < 0$, resulting from interactions between neighboring ions in a crystal that delocalize the electrons, an effect that ultimately lowers their energy.

The simplest estimate of ϵ_0 assumes that the ion provides a central $1/r$ potential and that, as a consequence of the exclusion principle with the core electrons, the electron cannot get closer than a_0 to the central potential. Consequently,

$$\epsilon_0 \simeq -n_e \int_{a_0}^{r_e} d\mathbf{r} \frac{e^2}{r} \tag{5.25}$$

$$\simeq \frac{3e^2}{2a_0 r_e^3}(r_e^2 - a_0^2) = -\frac{40.82\text{eV}}{r_s}\left(1 - \frac{1}{r_s^2}\right). \tag{5.26}$$

We next include the kinetic energy of the conduction electrons. The kinetic energy per electron in a free gas is given by

$$\epsilon_{kin}^0 = \frac{3}{5}\frac{p_F^2}{2m} = \frac{2.21}{r_s^2}\text{Ry.} \tag{5.27}$$

Here is a list of r_s values and corresponding kinetic energies for the alkalis:

eV	Li	Na	K	Rb	Cs
r_s	3.22	3.96	4.87	5.18	5.57
ϵ_{kin}	2.90	1.92	1.26	1.12	0.97

To set the scale, we ignore Coulomb interactions and compute the cohesive energy from just ϵ_0, Eq. (5.26), and the kinetic energy as

$\epsilon_{coh} \approx \epsilon_0 - \epsilon_{atom} + \epsilon_{kin}$ and find:

eV	Li	Na	K	Rb	Cs
$\epsilon_{coh,NI}$	−0.89	−1.17	−0.97	−0.89	−0.98
$\epsilon_{coh,expt}$	−1.58	−1.13	−0.98	−0.82	−0.815

As is evident, our estimates of the cohesive energy at this level of theory are in fair-to-poor agreement with experimental data.

We now include the direct Coulomb energy, treating the electron density as uniform and replacing each cell by a sphere of radius r_e. In such a sphere, the number of electrons enclosed in a sphere of radius $r < r_e$ is $n(r) = 4\pi r^3 n_e / 3$. The direct Coulomb energy is

$$\epsilon_{direct} = \int_0^{r_e} d\mathbf{r} \frac{e^2}{r} n_e n(r). \tag{5.28}$$

Evaluating the integral, we obtain

$$\epsilon_{direct} = \frac{3}{5} \frac{e^2}{r_e} = \frac{6}{5 r_s} \text{Ry}. \tag{5.29}$$

The interaction energy between neutral Wigner-Seitz cells is a higher-order correction, which can be included in a more precise calculation.

We next include the exchange Coulomb energy at the Hartree-Fock level, assuming, again, that the electron density is uniform. From Eq. (5.14), the exchange energy is

$$\epsilon_{exch} = -\frac{0.916}{r_s} \text{Ry}. \tag{5.30}$$

The total electron-electron Coulomb energy is then

$$\epsilon_{coul} = \epsilon_{exch} + \epsilon_{direct} = \frac{0.294}{r_s} \text{Ry}, \tag{5.31}$$

and the cohesive energy becomes

$$\epsilon_{coh,HF} = \epsilon_0 - \epsilon_{atom} + \epsilon_{kin} + \epsilon_{coul}. \tag{5.32}$$

The Coulomb energy, together with a comparison of the new cohesive energies with the experimental values [P1955], is shown below:

eV	Li	Na	K	Rb	Cs
ϵ_{coul}	1.24	1.01	0.82	0.77	0.72
$\epsilon_{\text{coh,HF}}$	0.32	−0.16	−0.15	−0.12	−0.26
$\epsilon_{\text{coh,expt}}$	−1.58	−1.13	−0.98	−0.82	−0.82

As is evident, whereas the Hartree-Fock approximation predicts that atoms in a metal are bound, it does not give accurate quantitative predictions. In fact, the noninteracting picture, $\epsilon_{\text{coh,NI}}$, did much better. To improve the theory, we need to include higher-order electron-electron interactions; that is, we need to determine more precisely the effect of correlations on the energy for a many-body system.

5.3.2   Wigner Solid

Wigner's interpolation scheme [W1934], in which he treats the system at high densities ($r_s \rightarrow 0$) perturbatively and at low densities ($r_s \rightarrow \infty$) as a solid, is more directly applicable to calculating the effect of correlations on the energy of the electron gas. Wigner's crucial observation was that, in a low-density electron gas in a uniform ion background, the electrons should form an ordered array. The 2d analog of a Wigner crystal is shown in Fig. (5.2). The basic idea is that, in this limit, the energy of the ground state, as sum of kinetic and Coulomb terms, is dominated by the Coulomb repulsion, since $\epsilon_{\text{kin}} \sim 1/r_s^2$ while $\epsilon_{\text{coul}} \sim 1/r_s$, so that, as $r_s \rightarrow \infty$, $\epsilon_{\text{kin}} < \epsilon_{\text{coul}}$. To minimize the Coulomb repulsion, the electrons find it energetically favorable to organize themselves in an ordered array. In three dimensions, the most favorable array is a body-centered cubic lattice, although the face-centered array is quite close in energy; the simple cubic array is less favorable. An antiferromagnetic bcc lattice is stable for $r_s > 93$ [CA1999]. A transition to a ferromagnet appears to occur around $r_s \approx 65$ [OHB1999].

The energy of a Wigner solid is determined entirely by the Coulomb energy. To estimate the energy, we calculate the energy of a neutral Wigner-Seitz cell, this time containing an electron in its center with a uniform ion background. As in the earlier case, the interaction energy between neutral Wigner-Seitz cells is a higher-order correction. In a spherical cell, the interaction energy between the uniform positive background and an electron localized at the center

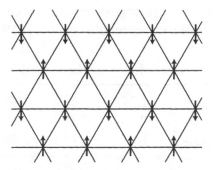

Figure 5.2 A two-dimensional (2d) Wigner lattice of electrons. For d = 2, a triangular lattice minimizes the $1/r$ Coulomb repulsion between the electrons. The arrows on each lattice site reflect the orientation of the electron spin. Because a triangular lattice is not bipartite, that is, it cannot be split into two equivalent sublattices, it does not support long-range antiferromagnetic order. Monte Carlo calculations [TC1989] indicate that for $r_s > 37$, a transition to a Wigner solid occurs with the spin configuration shown here. Wigner solid formation has been observed for electrons on the surface of liquid helium [G79], as well as in a dilute hole gas [Y1999] confined to move at the interface between GaAs and AlGaAs. In the presence of an external magnetic field, the Wigner crystal should be stable at higher densities, since a magnetic field tends to freeze out the electron zero-point motion. Experiments on a dilute 2d electron gas in a large perpendicular magnetic field also confirm the formation of a Wigner crystal [S1992].

is given by

$$\epsilon_{ei} \approx -n_e e^2 \int_0^{r_e} \frac{d\,\mathbf{r}}{r} = -\frac{3}{2}\frac{e^2}{r_e}. \tag{5.33}$$

The compensating positive background provides a uniform electrostatic energy that is the analog of the direct Coulomb interaction determined in the earlier discussion of the Wigner-Seitz method (Eq. 5.29), $\epsilon_{direct} \simeq 3e^2/5r_e$. The energy of the Wigner solid is then

$$\epsilon_{WS} = \left(\frac{3}{5} - \frac{3}{2}\right)\frac{e^2}{r_e} = -\frac{9}{10}\frac{e^2}{r_e} = -\frac{1.8}{r_s}\text{Ry}. \tag{5.34}$$

The contribution to the energy of a Wigner crystal from zero-point fluctuations of the electrons around their equilibrium positions falls off as $1/r_s^{3/2}$ (see Problem 5.5) and need not be included in extracting a first approximation to the correlation energy.

To identify the correlation energy in the Wigner solid, we write

$$\epsilon_{WS} = \epsilon_{exch} + \epsilon_{corr} = \left(\frac{-0.916}{r_s} - \frac{0.884}{r_s}\right)\text{Ry.} \qquad (5.35)$$

For large r_s, the correlation energy is thus

$$\epsilon_{corr} \approx -\frac{0.884}{r_s}\text{Ry}, \; r_s \to \infty. \qquad (5.36)$$

In the low-density regime, one can resort to the perturbative treatment of Gell-Mann and Brueckner [GB1957]. In perturbation theory, the second-order term in the electron-electron interaction is of the form

$$E_{coul}^{(2)} = \sum_{\mu \neq 0} \frac{\langle p|V_{ee}|\mu\rangle\langle\mu|V_{ee}|0\rangle}{E_0 - E_\mu}, \qquad (5.37)$$

where $|\mu\rangle$ is an intermediate excited state. All terms in the perturbative expansion can be represented diagrammatically, as can be shown using Wick's theorem. The second-order diagrams that enter are of the form shown in Fig. (5.3).

The last diagram in Fig. (5.3), the exchange term involving just a single electron loop, is logarithmically divergent. Gell-Mann and Brueckner [GB1957] recognized that, by clever resummation of a whole class of such divergent terms, they could get a finite but non-analytic result. Indeed, they find an expansion for the energy of the

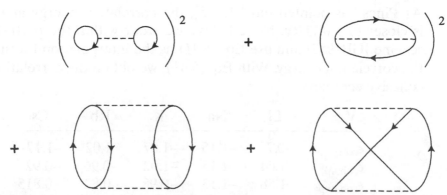

Figure 5.3 Second-order correlation diagrams in a perturbation expansion for the ground state energy of the uniform electron gas. The solid lines indicate electrons, and the dashed lines represent the two-body Coulomb interaction.

electron gas,

$$\epsilon_{GB} = \left(\frac{2.21}{r_s^2} - \frac{0.916}{r_s} + 0.062 \ln r_s - 0.096 + \cdots\right)\text{Ry.} \quad (5.38)$$

Note the presence of the nonanalytic term $\ln r_s$. This term reflects the fact that the energy of the system cannot be analytic in e^2 about $e^2 = 0$; indeed, the physics of a system of electrons with $e^2 < 0$ is that of a self-gravitating cloud and is qualitatively different from that with $e^2 > 0$. The problem with applying perturbation theory, even in its re-summed form, is that real metals lie in the region of $r_s > 1$, and, hence, one would have to sum the entire series to have quantitatively accurate results. It is customary to define the *correlation energy* by writing the energy of the electron system in the form

$$\epsilon = \epsilon_{\text{kin}} + \epsilon_{\text{exch}} + \epsilon_{\text{corr}}; \quad (5.39)$$

thus, the Gell-Mann and Brueckner result for the correlation energy in the high-density limit is

$$\epsilon_{\text{corr}} \approx 0.062 \ln r_s - 0.096, \; r_s \to 0. \quad (5.40)$$

Taking $\epsilon_{\text{corr}} = -0.096\text{Ry}$ at $r_s = 1$, we may write, in the spirit of Wigner, a simple approximate interpolation between the high- and low-density results:

$$\epsilon_{\text{corr}} = -\frac{0.884}{r_s + 8.21}\text{Ry} = -\frac{12.03}{r_s + 8.21}\text{eV.} \quad (5.41)$$

As Pines has pointed out [P1955], the correlation energy at $r_s = 1$ is closer to -0.11Ry. Nonetheless, we work within the perturbative scheme [GB1957] and use Eq. (5.41) as the interpolation formula for the correlation energy. With Eq. (5.41), we obtain the correlation and cohesive energies:

eV	Li	Na	K	Rb	Cs
ϵ_{corr}	−0.7	−1.15	−1.07	−1.02	−1.17
$\epsilon_{\text{coh,W}}$	−1.64	−1.15	−1.02	−0.96	−0.92
$\epsilon_{\text{coh,expt}}$	−1.58	−1.13	−0.98	−0.82	−0.815

The improvement with the use of Wigner interpolation is marked.

However, whereas Eq. (5.41) is a useful interpolation formula for the cohesive energy, the Wigner scheme implicitly assumes that no new phases of the electron gas appear between the dense perturbative limit and the crystalline regime. There is really no basis for this assumption. In fact, one of the unsolved problems in solid state physics is which phase of matter arises when a Wigner crystal melts. As perturbation theory cannot be used at the Wigner melting boundary, this problem has no easy resolution. At the writing of this chapter, experiments on a dilute two-dimensional electron gas indicate that this phase is exotic [K1999].

5.4 SUMMARY

We have shown in this chapter how the Hartree-Fock procedure can be implemented in the context of the electron gas in a compensating positive background. The key result is that the eigenfunctions are plane waves, but the single-particle energy levels are lowered by the exchange interaction. The exchange interaction produces a diminished electron density around each electron. The Hartree-Fock description, however, does not describe accurately the bandwidth or the specific heat, leading to a $T / \ln T$ behavior in the latter quantity. In Chapter 8, we show how inclusion of electron screening remedies some of the failures of Hartree-Fock. To make accurate estimates of the cohesive energy of crystals, the Wigner interpolation scheme is quite successful. The underlying physics in this scheme is that, at sufficiently low densities, an ordered electron lattice minimizes the energy of the electron gas. In d $=$ 3, a bcc lattice is favored, whereas in d $=$ 2, a triangular lattice of electrons forms.

PROBLEMS

1. Show that

$$\int_0^{p_F} \epsilon_{\text{exch}}(\mathbf{p}) \frac{d\mathbf{p}}{(2\pi\hbar)^3} = -\frac{e^2 p_F^4}{4\pi^3 \hbar^4}.$$

2. Assuming the Hartree-Fock expression for the specific heat of an electron gas, $\sim T/|\ln T|$, determine the temperature for Na below which corrections to the linear specific heat would become significant (10%, say).

3. Consider a uniform electron gas that interacts via a potential of the form $V(r) = V_0 e^{-r/a}/r$.
   (a) Solve the Hartree-Fock equations for this system for the eigenfunctions and excitation spectrum $\epsilon(\mathbf{p})$. Evaluate the Fermi energy $\epsilon_F = \mu$.

(b) At the Hartree-Fock level, show that the effective mass m^* is determined solely by the exchange contribution. Compute explicitly m^* in the limits $k_F a \ll 1$ and $k_F a \gg 1$.

(c) Show that the exchange interaction contribution to ϵ_F is negligible, when $k_F a \gg 1$, and that the direct and exchange terms are comparable for a short-range interaction with $k_F a \ll 1$.

4. Estimate the lowest energy ϵ_0 of an electron in a Wigner-Seitz cell by the following variational calculation. Assume that the potential due to the ion is Coulombic for $r > r_c$ and a repulsive hard core for $r \le r_c$, where r_c is the radius of the outermost occupied Bohr orbit in the ionic core. Take as a trial wave function the correct ground state (boundary conditions and all) for an electron feeling the hard core repulsion but no Coulomb attraction. Express the result in eV and compare with more refined results.

5. Assuming a spherical cell, determine the potential felt by an electron in a Wigner lattice as it moves away from its equilibrium position. Calculate the zero-point energy of the electron in this potential and show that the first correction to the energy of a low-density electron gas is $3/r_s^{3/2}$ Ry per electron.

6. A two-dimensional electron gas in a compensating background of positive charge exhibits a net magnetization, that is, a difference in the population between up and down spins. The electrons interact via a $1/r$ interaction. Define the relative magnetization in the electron gas as

$$\xi = \frac{n_\uparrow - n_\downarrow}{n_\uparrow + n_\downarrow}. \tag{5.42}$$

Show first that the Fermi momentum for up and down spins can be written as

$$p_{F\uparrow} = p_F \sqrt{1 + \xi}, \qquad p_{F\downarrow} = p_F \sqrt{1 - \xi}. \tag{5.43}$$

Calculate the ground state energy, at the level of Hartree-Fock, of the electron gas as a function of ξ. For what values of ξ is the ground state stable? Estimate the value of r_s at which the ground state energy of the fully polarized (ferromagnetic) electron gas is lower in energy than the completely unpolarized ($\xi = 0$) gas.

REFERENCES

[B1936] J. Bardeen. *Phys. Rev.* **50**, 1098 (1936); see also E. P. Wohlfarth, *Phil. Mag.* **41**, 534 (1950).

[CA1999] D. M. Ceperley, B. J. Alder. *Phys. Rev. Lett.* **45**, 566 (1980); see also D. Ceperley, *Nature* (London) **397**, 386 (1999).

[GB1957] M. Gell-Mann, K. A. Brueckner. *Phys. Rev.* **106**, 364 (1957).

[G79] C. C. Grimes, G. Adams. *Phys. Rev. Lett.* **42**, 795 (1979).

[K1999] S. V. Kravchenko, D. Simonian, M. P. Sarachik, W. Mason, J. E. Furneaux. *Phys. Rev. Lett.* **77**, 4938 (1996).

[OHB1999] G. Ortiz, M. Harris, P. Ballone. *Phys. Rev. Lett.* **82**, 5317 (1999).

[S1992] M. B. Santos, Y. W. Suen, M. Shayegan, Y. P. Li, L. W. Engel, D. C. Tsui. *Phys. Rev. Lett.* **68**, 1188 (1992). For a theoretical prediction for the onset of a Wigner crystal in the quantum Hall regime, see P. K. Lam, S. M. Girvin, *Phys. Rev.* B **30**, 473 (1984).

[P1955] D. Pines. *Advances in Solid State Physics*, ed. F. Seitz and D. Turnbull, Academic Press, New York, 1955, Vol. I, p. 373.

[TC1989] B. Tanatar, D. M. Ceperley. *Phys. Rev.* B **39**, 5005 (1989).

[W1934] E. Wigner. *Phys. Rev.* **46**, 1002 (1934).

[WS1933] E. Wigner, F. Seitz. *Phys. Rev.* **43**, 804 (1933); **46**, 509 (1933).

[Y1999] J. Yoon, C. C. Li, D. Shahar, D. C. Tsui, M. Shayegan. *Phys. Rev. Lett.* **82**, 1748 (1999).

– 6 –

Local Magnetic Moments in Metals

Many years ago, Matthias and co-workers, in a series of electron spin resonance (ESR) and nuclear magnetic resonance (NMR) experiments on nonmagnetic metals—metals with no permanent magnetic moment—observed surprising evidence for long-lived local spin packets in the ESR lineshape [M1960]. These data indicated the persistence of local magnetic moments. The magnetic moment was quickly traced to the presence of small amounts of magnetic impurities. While various systems, for example, Mn, Fe, and other iron group impurities in host materials, such as Cu, Ag, and Au, were studied, the common ingredient shared by all the impurity ions is that they possessed one or more vacant inner-shell orbitals. In addition, the experiments demonstrated that varying the kind and amount of the magnetic impurities did not always result in the formation of local magnetic moments in nonmagnetic metals. This finding added to the intrigue and established the question of the formation of local magnetic moments as central to understanding magnetism and transport in solids. In this chapter, we describe the origin of local moments, focusing primarily on Anderson's model [A1961], the model that rose to the fore as the standard microscopic view of local magnetic moment formation in metals.

6.1 LOCAL MOMENTS: PHENOMENOLOGY

An impurity in a nonmagnetic metal can give rise to a local moment if an electronic state on the impurity is singly occupied, at least on the time scale of the experiment. Friedel [F1958] was the first to introduce a phenomenological model to explain the onset of local moments. His model describes the effect of the impurity as generating a potential that is effectively a deep Coulombic core plus an angular momentum-dependent centrifugal barrier $\sim \ell(\ell+1)/r^2$, where the distance r is measured from the center of the impurity. Such a potential can support bound states. Those occurring well below the Fermi level will be doubly occupied and have no magnetic moment. A local magnetic moment can form, however, in a bound or resonant state of the impurity potential near the Fermi level if the levels corresponding to up and down spins are nondegenerate.

Anderson [A1961], adopting the basic ideas of Friedel, developed a one-band model of local moment formation, the essence of which is the following. The band in the nonmagnetic metallic host, typically a transition metal, such as titanium, is represented by a set of Bloch states. Measuring single-particle energies from the top of the Fermi sea, equivalent to choosing the Fermi energy $\epsilon_F = 0$, we take the band energies to be $\epsilon_{\mathbf{k}} = \hbar^2 k^2/2m - \hbar^2 k_F^2/2m$. The impurity is treated simply as a local site on which a single electronic orbital is placed. While this simplification does not capture the five-fold degeneracy of the d-orbitals of typical impurities, such as Co, Fe, and Ni, this deficiency is not crucial. We denote the wave function of the orbital by $\phi_d(\mathbf{r})$. The energy required to place an electron on the impurity with spin either up or down is ϵ_d. The orbital can be either singly occupied or occupied by two electrons of opposite spin. The latter case costs an energy, U, which physically arises from the Coulomb repulsion between the electrons and is thus given by

$$U = \langle dd|v_{ee}|dd\rangle = \int d\mathbf{r}_1 d\mathbf{r}_2 |\phi_d(\mathbf{r}_1)|^2 \frac{e^2}{|\mathbf{r}_1 - \mathbf{r}_2|}|\phi_d(\mathbf{r}_2)|^2. \quad (6.1)$$

The Coulomb repulsion has the effect of favoring single occupation of the impurity level. The final ingredient in the Anderson model is a spin-conserving coupling between the impurity level and the $\mathbf{k}$-states of the band in the metal, described by a matrix element $V_{\mathbf{k}d}$. This interaction causes a hybridization of the band states and the impurity level. The basic Anderson Hamiltonian is, then,

$$H^A = \sum_{\mathbf{k}\sigma} \epsilon_{\mathbf{k}} a^\dagger_{\mathbf{k}\sigma} a_{\mathbf{k}\sigma} + \sum_\sigma \epsilon_d n_{d\sigma}$$
$$+ \sum_{\mathbf{k}\sigma} V_{\mathbf{k}d}(a^\dagger_{\mathbf{k}\sigma} a_{d\sigma} + a^\dagger_{d\sigma} a_{\mathbf{k}\sigma}) + U n_{d\uparrow} n_{d\downarrow}, \quad (6.2)$$

where the $a^\dagger_{\mathbf{k}\sigma}$ create electrons in band states, $a^\dagger_{d\sigma}$ creates an electron with spin σ on the impurity, and $n_{d\sigma} = a^\dagger_{d\sigma} a_{d\sigma}$ is the number operator for a localized electron of spin σ.

The Anderson model includes only the on-site Coulomb repulsion between two localized electrons. In addition, an electron localized on a d-level interacts with localized electrons on other sites. These other terms are generally small, however. To estimate the size of these additional interactions, Hubbard [H1964] assumed a lattice model of 3d electrons in transition metals, in which a localized state is placed at each lattice site, denoted by $i, j, \ldots$. He then calculated a series of matrix elements of the Coulomb potential of the form

$$\langle ij | v_{ee} | kl \rangle = \int d\mathbf{r}_1 d\mathbf{r}_2 \phi_i^*(\mathbf{r}_1) \phi_j^*(\mathbf{r}_2) \frac{e^2}{|\mathbf{r}_1 - \mathbf{r}_2|} \phi_l(\mathbf{r}_2) \phi_k(\mathbf{r}_2). \quad (6.3)$$

The magnitudes of the various Coulomb matrix elements are listed below:

Symbol	Integral	Description	Magnitude
U	$\langle ii \| v_{ee} \| ii \rangle$	on-site	20 eV
V	$\langle ij \| v_{ee} \| ij \rangle$	nearest neighbor	2–3 eV
Y	$\langle ij \| v_{ee} \| ji \rangle$	exchange	1/40 eV

In this table, i and j ($i \neq j$) denote "nearest neighbor" lattice sites. The nearest-neighbor term is the direct Coulomb energy between densities of electrons at neighboring sites, and the exchange term is the corresponding exchange energy, as discussed in Chapter 4. As is evident, U is the largest contributor to the Coulomb interactions, and in many solid-state problems involving strongly-correlated electrons, for example, the Hubbard model [H1964], only the U-term is retained. Generally, inclusion of U does not guarantee that all the relevant physics will be described. Polyacetylene is a famous case in point [BC1989], in which V, as well as the other Coulomb matrix elements, play a crucial role in the physics of dimerization and the optical properties. Here the additional density-bond matrix element, $X = \langle ii | v_{ee} | ij \rangle$, and the bond-bond matrix element, $\tilde{Y} = \langle ii | v_{ee} | jj \rangle$—of order of $1/2$ eV and $1/40$ eV in the transition metals—are significant. The notation *density* refers to the case $k = i$, where the corresponding wave functions give the electron density at site i; the *bond* contributions are so-named because they are nonzero only if an electron has nonzero overlap in two neighboring localized states.

The Coulomb interaction favors the formation of local moments because it tends to inhibit double occupation of a site. On the other hand, the charge fluctuations on the impurity caused by strong hybridization of the impurity level with the band states tend to wash out local moments. The energy scale measuring this coupling is, as we shall see, essentially the rate at which the coupling causes transitions between electrons initially in the impurity state and a **k**-state in the band, and vice versa. From Fermi's golden rule, the rate of hopping of an electron from the impurity to a state in the band of the metal with momentum **k** is given by $1/\tau = 2\pi|V_{\mathbf{k}d}|^2 N(\epsilon_d)V/\hbar \equiv 2\Delta/\hbar$, where $N(\epsilon_d)$ is the density of electron k states of a particular spin at the impurity energy ϵ_d and V is the volume. As a result of such virtual transitions, the energy levels on the impurity will be broadened; the width of the broadening is governed by the energy scale Δ. We refer to Δ as the *hybridization energy*, as this energy arises solely from the overlap between the band and impurity states.

The physics of the Anderson model is governed by several parameters: the magnitude of the impurity energy ϵ_d and its location relative to the Fermi level, the on-site repulsion energy U, and the hybridization energy Δ. As illustrative cases, let us consider the limits of strong and weak hybridization. In the case of weak hybridization, $U \gg \epsilon_d \gg \Delta$, that is, the cost of doubly occupying the d-level on the impurity far exceeds ϵ_d and the hybridization energy Δ. In this limit, the system generally supports local moment formation. We assume generally that the impurity level lies below the Fermi sea, that is, $\epsilon_d < 0$. In thermal equilibrium, the state with energy ϵ_d will be at least singly occupied. The energy cost of putting two electrons on the impurity is $2\epsilon_d + U$. Should $2\epsilon_d + U$ exceed the Fermi energy, that is, $2\epsilon_d + U > 0$, the upper state of the impurity would be unoccupied and the impurity would be magnetic. The energy level diagram for this case is shown in Fig. (6.1a). However, in certain experimental situations, for example, rare earth compounds, such as SmB_6 and heavy fermion systems, the impurity levels are close to the Fermi level. In these "mixed valence" situations, slight changes in the hybridization energy lead to a loss of the local moment. Consider the case in which $U \gg \Delta \gg \epsilon_d$, so that the upper level remains well above the Fermi level. Then the absolute energy of the lower impurity state is smaller than the width of the state. As a consequence, the occupation of this state undergoes rapid fluctuations, as illustrated in Fig. (6.1b). This regime is nonmagnetic.

Consider next the regime in which U is small compared with Δ. In this case, Δ determines the physics, not U. Because the broadening of each level exceeds the energy cost for double occupancy, the impurity could with equal probability be partially occupied with a spin-up or a spin-down electron. This state is not magnetic because the occupancy in the up and down spin levels is equal. This state of affairs is termed

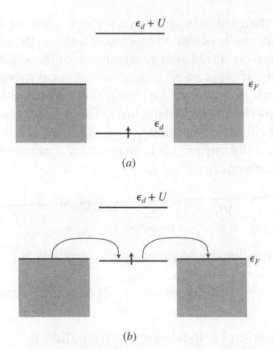

Figure 6.1 Energy level diagrams for the Anderson model in the local weakly coupled regime, $U \gg \Delta$ (Δ is the hybridization energy). The left and right shaded regions show the same electronic states in the band of the host metal—as is customary in this problem to emphasize virtual second-order transitions from the continuum to the localized state to the continuum. The Fermi level is fixed at $\epsilon_F = 0$. The energies of the available single-particle states on the impurity are ϵ_d if the level is unoccupied and $\epsilon_d + U$ if the level is already singly occupied. (*a*) When ϵ_d is sufficiently below the Fermi surface, local moment formation is favored. (*b*) Shown here is the energy level diagram in the mixed valence regime for $U \gg \Delta \gg \epsilon_d$. The lower impurity level is very close to the Fermi surface. Charge fluctuations on the impurity are rapid and lead to a local spin fluctuation on the impurity. This case is not magnetic.

a *localized spin fluctuation*. Since, by adjusting U and Δ, we can tune continuously from the magnetic to the localized spin fluctuation state, the model is thus a good starting point to discuss the physics of local moment formation.

6.2   MEAN-FIELD SOLUTION

The central quantity required to understand local moment formation is the occupancy of the states of the impurity. If the states have a greater probability of being occupied by electrons of a given spin than the other, then the site will have a local moment. The occupancy of the impurity site is obtained by calculating the probability that an electron in a given occupied energy eigenstate of the system will be found on the impurity. If $\langle n|$ is an eigenstate of the local moment Hamiltonian, H^A, with energy $\epsilon_{n\sigma}$, then $|\langle n\sigma|d\sigma\rangle|^2$ is the probability that the state $|n\sigma\rangle$ overlaps the impurity site, $|d\sigma\rangle$. This overlap is nonzero as a result of the hybridization coupling, $V_{\mathbf{k}d}$. The net occupancy of an electron of spin σ on the impurity is then given by a sum over all occupied electron states with energy $\epsilon \leq \epsilon_F$:

$$\langle n_{d\sigma}\rangle = \sum_{n,\epsilon_{n\sigma} \leq \epsilon_F} |\langle n\sigma|d\sigma\rangle|^2 = \int_{-\infty}^{\epsilon_F} d\epsilon\, \rho_{d\sigma}(\epsilon), \qquad (6.4)$$

where the density of states on the impurity is given by

$$\rho_{d\sigma}(\epsilon) = \sum_{n} \delta(\epsilon_{n\sigma} - \epsilon)|\langle n\sigma|d\sigma\rangle|^2. \qquad (6.5)$$

The criterion for local moment formation is that $\langle n_{d\sigma}\rangle \neq \langle n_{d-\sigma}\rangle$.

The interaction term on the impurity makes constructing the single-particle energy levels and the impurity density of states nontrivial. To solve this problem, we adopt a mean-field or Hartree-Fock approximation to the interacting problem, within which the single-particle energy levels are well defined. The criterion for local moment formation resulting from this procedure has in fact had spectacular success in predicting the onset of local moment formation in magnetic alloys [GZ1974]. In Hartree-Fock, the ground state of the system has the form

$$|\Phi_0\rangle = \prod_{\epsilon_n < \epsilon_F} a_{n\sigma}^\dagger |0\rangle, \qquad (6.6)$$

where

$$a_{n\sigma}^\dagger = \sum_{\mathbf{k}} \langle n\sigma|\mathbf{k}\sigma\rangle a_{\mathbf{k}\sigma}^\dagger + \langle n\sigma|d\sigma\rangle a_{d\sigma}^\dagger \qquad (6.7)$$

is a linear combination of band and impurity states. In the Hartree-Fock approximation in terms of the new states (Eq. 6.7), the one-body Hamiltonian is

$$H_{HF}^A = \sum_{n\sigma} \epsilon_{n\sigma} a_{n\sigma}^\dagger a_{n\sigma}. \qquad (6.8)$$

The Hartree-Fock approximation to the Hamiltonian amounts to replacing the interaction term, $U n_{d\uparrow} n_{d\downarrow}$, by $U\langle n_{d\uparrow}\rangle n_{d\downarrow} + U\langle n_{d\downarrow}\rangle n_{d\uparrow} -$

$U\langle n_{d\uparrow}\rangle\langle n_{d\downarrow}\rangle$, where the averages are in the state $|\Phi_0\rangle$. The last term provides an overall shift in the zero of energy; since it cannot affect the physics of the local moment, we drop it. With a redefinition of the defect site energies as

$$E_{d\sigma} = \epsilon_d + U\langle n_{d-\sigma}\rangle, \tag{6.9}$$

the Hartree-Fock approximation to the Anderson Hamiltonian is

$$H_{HF}^A = \sum_\sigma E_{d\sigma} n_{d\sigma} + \sum_{\mathbf{k}\sigma} \epsilon_{\mathbf{k}} n_{\mathbf{k}\sigma} + \sum_{\mathbf{k}\sigma} V_{\mathbf{k}d}(a_{\mathbf{k}\sigma}^\dagger a_{d\sigma} + a_{d\sigma}^\dagger a_{\mathbf{k}\sigma}). \tag{6.10}$$

As we see, the Hartree-Fock approximation of the Hamiltonian with on-site Coulomb repulsion amounts to a simple renormalization of the site energies; that is, $\epsilon_{d\sigma} \rightarrow E_{d\sigma}$. The criterion for a local moment, that $\langle n_{d\sigma}\rangle \neq \langle n_{d-\sigma}\rangle$, is equivalent to the condition that $E_{d\sigma} \neq E_{d-\sigma}$. Once we define the single-particle levels, $\epsilon_{n\sigma}$, it is straightforward to show that Eqs. (6.8) and (6.10) are equivalent (see Problem 6.3).

The single-particle energies, $\epsilon_{n\sigma}$, are defined through the operator equations of motion,

$$[H_{HF}^A, a_{n\sigma}^\dagger] = \epsilon_{n\sigma} a_{n\sigma}^\dagger. \tag{6.11}$$

To evaluate the commutator in Eq.(6.11), we use the commutators,

$$
\begin{aligned}
[H_{HF}^A, a_{\mathbf{k}\sigma}^\dagger] &= \sum_{\mathbf{k}'\sigma} \epsilon_{\mathbf{k}'}[a_{\mathbf{k}'\sigma}^\dagger a_{\mathbf{k}'\sigma}, a_{\mathbf{k}\sigma}^\dagger] \\
&+ \sum_{\mathbf{k}'\sigma} V_{\mathbf{k}'d}[(a_{\mathbf{k}'\sigma}^\dagger a_{d\sigma} + a_{d\sigma}^\dagger a_{\mathbf{k}'\sigma}), a_{\mathbf{k}\sigma}^\dagger] \\
&= \epsilon_{\mathbf{k}} a_{\mathbf{k}\sigma}^\dagger + V_{\mathbf{k}d} a_{d\sigma}^\dagger
\end{aligned} \tag{6.12}
$$

and

$$[H_{HF}^A, a_{d\sigma}^\dagger] = E_{d\sigma} a_{d\sigma}^\dagger + \sum_{\mathbf{k}} V_{\mathbf{k}d} a_{\mathbf{k}\sigma}^\dagger, \tag{6.13}$$

and find

$$
\begin{aligned}
[H_{HF}^A, a_{n\sigma}^\dagger] &= \sum_{\mathbf{k}} \langle n\sigma|\mathbf{k}\sigma\rangle(\epsilon_{\mathbf{k}} a_{\mathbf{k}\sigma}^\dagger + V_{\mathbf{k}d} a_{d\sigma}^\dagger) \\
&+ \langle n\sigma|d\sigma\rangle\left(E_{d\sigma} a_{d\sigma}^\dagger + \sum_{\mathbf{k}} V_{\mathbf{k}d} a_{\mathbf{k}\sigma}^\dagger\right) \\
&= \epsilon_{n\sigma} \sum_{\mathbf{k}} \langle n\sigma|\mathbf{k}\sigma\rangle a_{\mathbf{k}\sigma}^\dagger + \langle n\sigma|d\sigma\rangle a_{d\sigma}^\dagger.
\end{aligned} \tag{6.14}
$$

Thus,

$$\epsilon_{n\sigma}\langle n\sigma|\mathbf{k}\sigma\rangle = \epsilon_{\mathbf{k}}\langle n\sigma|\mathbf{k}\sigma\rangle + V_{\mathbf{k}d}\langle n\sigma|d\sigma\rangle \tag{6.15}$$

and

$$\epsilon_{n\sigma}\langle n\sigma|d\sigma\rangle = E_{d\sigma}\langle n\sigma|d\sigma\rangle_{\sigma} + \sum_{\mathbf{k}}\langle n\sigma|\mathbf{k}\sigma\rangle V_{\mathbf{k}d}. \tag{6.16}$$

Equations (6.15) and (6.15) define the single-particle levels, $\epsilon_{n\sigma}$. The explicit solution, which is left as a homework exercise (see Problem 6.2), requires eliminating the states $\langle n\sigma|d\sigma\rangle$ and $\langle n\sigma|\mathbf{k}\sigma\rangle$ from Eqs. (6.15) and (6.16).

Now that we have shown how to obtain the single-particle energy levels that determine the density of states, we compute this quantity explicitly. We carry out the calculation in terms of resolvents or Green functions, which allow us to include directly the widths of the impurity levels. We begin by rewriting the density of states making use of the relation,

$$\begin{aligned}
\delta(\epsilon - \epsilon_{n\sigma}) &= \frac{1}{\pi}\lim_{\Gamma\to 0}\frac{\Gamma}{(\epsilon - \epsilon_{n\sigma})^2 + \Gamma^2} \\
&= \frac{1}{2\pi i}\lim_{\Gamma\to 0}\left[\frac{1}{\epsilon - \epsilon_{n\sigma} - i\Gamma} - \frac{1}{\epsilon - \epsilon_{n\sigma} + i\Gamma}\right] \\
&= -\frac{1}{\pi}\lim_{\Gamma\to 0}\operatorname{Im}\frac{1}{\epsilon - \epsilon_{n\sigma} + i\Gamma}. \tag{6.17}
\end{aligned}$$

As a consequence,

$$\rho_{d\sigma}(\epsilon) = -\frac{1}{\pi}\lim_{\Gamma\to 0}\operatorname{Im}\sum_{n}\frac{|\langle n\sigma|d\sigma\rangle|^2}{\epsilon - \epsilon_{n\sigma} + i\Gamma}. \tag{6.18}$$

The quantity $(\epsilon - \epsilon_{n\sigma} + i\Gamma)^{-1}$ is a resolvent related to the Green function for this problem. We define the Green (operator) function $G(E + i\Gamma)$ through

$$(E + i\Gamma - H)G = 1. \tag{6.19}$$

Dividing both sides of (6.19) by $(E - H + i\Gamma)$ and expanding in a complete set of eigenstates of H^A, we obtain

$$G(E + i\Gamma) = \sum_{n}\frac{|n\rangle\langle n|}{E + i\Gamma - E_n} \tag{6.20}$$

as the spectral resolution of the Green function, where $H|n\rangle = E_n|n\rangle$; G is singular with poles at the eigenenergies E_n. The eigenfunctions of G are then $|n\rangle$.

For the single-particle Hamiltonian $H = H_{HF}^A$, we define the matrix elements of G through

$$\sum_{\beta}(\epsilon + i\Gamma - H)_{\alpha\beta}G_{\beta\mu}(\epsilon) = \delta_{\alpha\mu}, \tag{6.21}$$

where the subscripts α, β, and μ label band or impurity states with a particular spin, that is, $|k\sigma\rangle$ and $|d\sigma\rangle$, respectively. From the definition of G and Eq. (6.18), we can then rewrite $\rho_{d\sigma}(\epsilon)$ as

$$\rho_{d\sigma} = -\frac{1}{\pi}\lim_{\Gamma\to0}\text{Im}\sum_{n}\frac{|\langle n\sigma|d\sigma\rangle|^2}{\epsilon - \epsilon_{n\sigma} + i\Gamma}$$

$$= -\frac{1}{\pi}\lim_{\Gamma\to0}\text{Im}\langle d\sigma|G(\epsilon + i\Gamma)|d\sigma\rangle$$

$$= -\frac{1}{\pi}\lim_{\Gamma\to0}\text{Im}G_{dd}^{\sigma} \tag{6.22}$$

if G is expressed in terms of the exact eigenstates of H. Equation (6.22), which relates the density of states to the diagonal matrix element of G, is the principal relationship we need to formulate the local moment problem. From the equations for the reduced Hamiltonian and the Green function, we find that, for $\alpha = \mu = d\sigma$,

$$(\epsilon + i\Gamma - E_{d\sigma})G_{dd}^{\sigma} - \sum_{k}V_{dk}G_{kd}^{\sigma} = 1, \tag{6.23}$$

and, for $\alpha = k\sigma, \mu = d\sigma$,

$$(\epsilon + i\Gamma - \epsilon_k)G_{kd} - V_{kd}G_{dd}^{\sigma} = 0. \tag{6.24}$$

Eliminating G_{dk}^{σ} from Eq. (6.23) by using Eq. (6.24), we obtain

$$G_{dd}^{\sigma}(\epsilon + i\Gamma) = \left[\epsilon + i\Gamma - E_{d\sigma} - \sum_{k}\frac{|V_{kd}|^2}{\epsilon + i\Gamma - \epsilon_k}\right]^{-1}. \tag{6.25}$$

[Note that the zeroes of the denominator lie at the energies calculated in second-order perturbation Brillouin-Wigner theory, with the full ϵ in the energy denominator.] In the absence of the last term in the denominator of Eq. (6.25), the singularities of G_{dd}^{σ} lie at the renormalized site energy, $E_{d\sigma}$.

The transfer interaction $V_{\mathbf{k}d}$ broadens the impurity level. To see the broadening, we rewrite the last term of $(G^{\sigma}_{dd})^{-1}$ as

$$\sum_{\mathbf{k}} \frac{|V_{\mathbf{k}d}|^2}{\epsilon - \epsilon_{\mathbf{k}} + i\Gamma} = \sum_{\mathbf{k}} |V_{\mathbf{k}d}|^2 \frac{\epsilon - \epsilon_{\mathbf{k}} - i\Gamma}{(\epsilon - \epsilon_{\mathbf{k}})^2 + \Gamma^2}. \tag{6.26}$$

In the limit $\Gamma \to 0$, Eq. (6.26) reduces to

$$\lim_{\Gamma \to 0} \sum_{\mathbf{k}} \frac{|V_{\mathbf{k}d}|^2}{\epsilon - \epsilon_{\mathbf{k}} + i\Gamma} = P\left(\sum_{\mathbf{k}} \frac{|V_{\mathbf{k}d}|^2}{\epsilon - \epsilon_{\mathbf{k}}}\right) - i\pi \sum_{\mathbf{k}} |V_{\mathbf{k}d}|^2 \delta(\epsilon - \epsilon_{\mathbf{k}})$$

$$= P\left(\sum_{\mathbf{k}} \frac{|V_{\mathbf{k}d}|^2}{\epsilon - \epsilon_{\mathbf{k}}}\right) - i\pi |\overline{V}_{\mathbf{k}d}|^2 N(\epsilon)V, \tag{6.27}$$

where P indicates the principal value; we have used Eq. (6.17) to obtain the final result and written $|\overline{V}_{\mathbf{k}d}|^2 = |\overline{V}_{\mathbf{k}d}|^2_{\epsilon_{\mathbf{k}}=\epsilon}$. The first term in Eq. (6.26) is purely real and, hence, represents a shift of the d-impurity energy level. This term affects the physics only if it fluctuates wildly as a function of energy. The density of states of the host band, $N(\epsilon)$, is fairly constant, however, on the scale over which $E_{d\sigma}$ changes. Hence, the real part of (6.26) can be ignored, and G becomes

$$G^{\sigma}_{dd}(\epsilon + i\Gamma) = \frac{1}{\epsilon + i\Gamma - E_{d\sigma} + i\Delta}, \tag{6.28}$$

where $2\Delta = 2\pi \langle |\overline{V}_{\mathbf{k}d}|^2 \rangle N(\epsilon)V$ is the effective transition rate between the impurity and the conduction electrons.

If we interpret $E_{d\sigma} - i\Delta$ as the new site energy, then, as advertised, $2\hbar/\Delta$ is the lifetime of the impurity level. From Eqs. (6.22) and (6.28), we find that the density of states at the d-impurity reduces to the Lorentzian form

$$\rho_{d\sigma}(\epsilon) = \frac{1}{\pi} \frac{\Delta}{(\epsilon - E_{d\sigma})^2 + \Delta^2}, \tag{6.29}$$

in which, as expected, Δ appears as the half-width of the d-level. The density of states approaches a delta function as the hybridization between the d-level and the conduction states vanishes. Why is this so? The density of states is determined by the imaginary part of the Green function. If, as E approaches the real axis, $G(E)$ is purely real, then the singularities of $G(E)$ are simple poles well separated in energy. In this case, the density of states corresponds to a series of sharp (delta-

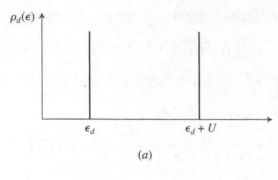

(a)

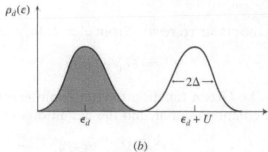

(b)

Figure 6.2 (a) Single-particle density of states, $\rho_d^{tot}(\epsilon) = \rho_{d\uparrow}(\epsilon) + \rho_{d\downarrow}(\epsilon)$, in the Anderson model when the hybridization between the d-level and the $\mathbf{k}$-states vanishes. The sharp peak at single-particle energy ϵ_d corresponds to a vacancy in the impurity level, whereas the sharp peak at $\epsilon_d + U$ corresponds to the situation in which the impurity level is already singly occupied by an electron of opposite spin. Including interactions at the Hartree-Fock level shifts these peaks to $E_{d\sigma} = \epsilon_d + U\langle n_{d-\sigma}\rangle$. (b) Density of states when the d- and $\mathbf{k}$-states are mixed. The width of each state is broadened by hybridization. As in (a), the lower level corresponds to a situation in which the impurity level is vacant, and the upper peak to the situation in which an electron of given spin already resides on the lower level. When the lower level is occupied and the upper vacant, as indicated by the unequal shading under the two peaks, the system has a localized moment.

function) peaks at the eigenenergies $E_{d\sigma} = \epsilon_d + U\langle n_{d-\sigma}\rangle$. For single occupation of the level by an electron of given spin, the poles are separated by the on-site Coulomb repulsion, U, as illustrated in Fig. (6.2a). This state of affairs persists if there is no mixing between the d-level and the $\mathbf{k}$-states in the conduction band. Once mixing is turned on, the poles move off the real axis causing the pure d-levels to broaden. The resultant energy spectrum is no longer well separated in energy (see Fig. 6.2b).

Before we derive the condition for local moment formation, it is useful to see the structure of our results in terms of the phase shifts for scattering of conduction electrons by the impurity. Using Eq. (6.21), we find that the Green function for the conduction electrons obeys

$$(\epsilon + i\Gamma - \epsilon_{\mathbf{k}})G^{\sigma}_{\mathbf{kk}}(\epsilon + i\Gamma) = \delta_{\mathbf{kk'}} + V_{\mathbf{k}d} G^{\sigma}_{d\mathbf{k'}}(\epsilon + i\Gamma). \quad (6.30)$$

Eliminating $G^{\sigma}_{d\mathbf{k'}}$ by using Eqs. (6.23) and (6.24), we obtain the closed expression

$$G^{\sigma}_{\mathbf{kk'}}(\epsilon + i\Gamma) = \frac{\delta_{\mathbf{kk'}}}{\epsilon + i\Gamma - \epsilon_{\mathbf{k}}} + \frac{V_{\mathbf{k}d}}{\epsilon + i\Gamma - \epsilon_{\mathbf{k}}} G^{\sigma}_{dd}(\epsilon + i\Gamma) \frac{V^{*}_{\mathbf{k'}d}}{\epsilon + i\Gamma - \epsilon_{\mathbf{k'}}}.$$

$$(6.31)$$

We now use the standard results from elementary scattering theory,

$$G = G_0 + G_0 V G, \quad (6.32)$$

where G_0 is the Green function for the noninteracting system and V represents the bare potential, and the definition of the T-matrix

$$T = V + VGT, \quad (6.33)$$

to derive the scattering equation

$$G = G_0 + G_0 T G_0, \quad (6.34)$$

which expresses G in terms of G_0 and the T-matrix. Comparing Eqs. (6.31) and (6.34), we identify the T-matrix for scattering of a conduction electron from momentum $\mathbf{k}$ to momentum $\mathbf{k'}$ as

$$T^{\sigma}_{\mathbf{kk'}}(\epsilon + i\Gamma) = V_{\mathbf{k}d} G^{\sigma}_{dd}(\epsilon + i\sigma) V^{*}_{\mathbf{k'}d}. \quad (6.35)$$

Note that since the interaction $V_{\mathbf{k}d}$ is, by assumption, independent of the direction of $\mathbf{k}$, the scattering occurs only in s-waves. Substituting Eq. (6.28), we find that

$$T^{\sigma}_{\mathbf{kk}}(\epsilon + i\Gamma) = \frac{|V_{\mathbf{k}d}|^2}{\epsilon + i\Gamma - E_{d\sigma} + i\pi N(\epsilon)V|\overline{V}_{\mathbf{k}d}|^2}, \quad (6.36)$$

and consequently, the imaginary part of the T-matrix

$$\mathrm{Im}\, T^{\sigma}_{\mathbf{kk}}(\epsilon + i\Gamma) = -\pi|V_{\mathbf{k}d}|^2 \rho_{d\sigma}(\epsilon) \quad (6.37)$$

is proportional to the density of states on the d-level.
From elementary scattering theory,

$$T^{\sigma}_{\mathbf{kk'}}(\epsilon + i\Gamma) \propto e^{2i\delta_{\sigma}} - 1 \propto \sin\delta_{\sigma} e^{i\delta_{\sigma}} = 1/(\cot\delta_{\sigma} - i). \quad (6.38)$$

Thus, we identify the s-wave phase shift as

$$\cot \delta_\sigma(\epsilon) = \frac{E_{d\sigma} - \epsilon}{\Delta}. \tag{6.39}$$

"On shell," that is, for $\epsilon_k = \epsilon_{k'} = \epsilon$, the T-matrix element reduces to the standard form

$$T_{kk}^\sigma(\epsilon + i\Gamma) = -\frac{1}{\pi N(\epsilon)V} \sin \delta_\sigma e^{i\delta_\sigma}. \tag{6.40}$$

As discussed above, the average occupation, $\langle n_{d\sigma} \rangle$, of the d-level is given by the density of states integrated over the filled states,

$$\langle n_{d\sigma} \rangle = \int_{-\infty}^{\epsilon_F = 0} \rho_{d\sigma}(\epsilon)\, d\epsilon$$

$$= \frac{\Delta}{\pi} \int_{-\infty}^{0} \frac{d\epsilon}{(\epsilon - E_{d\sigma})^2 + \Delta^2}$$

$$= \frac{1}{\pi} \cot^{-1}\left(\frac{E_{d\sigma}}{\Delta}\right). \tag{6.41}$$

Using Eq. (6.39), we find

$$\langle n_{d\sigma} \rangle = \frac{\delta_\sigma(0)}{\pi}, \tag{6.42}$$

which is a simplified form of Friedel's sum rule between the occupancy of the impurity site and the phase shifts of conduction electrons at the Fermi energy scattering on the impurity. We emphasize that the phase shift calculated here arises entirely from non–spin-flip scattering at the d-level. The role of spin-flip scattering will be stressed in the next chapter, when we consider explicitly the Kondo problem.

Substituting the expression $E_{d\sigma} = \epsilon_d + U\langle n_{d-\sigma} \rangle$ into Eq. (6.41), we obtain two coupled equations,

$$\langle n_{d\uparrow} \rangle = \frac{1}{\pi} \cot^{-1}\left(\frac{\epsilon_d + U\langle n_{d\downarrow} \rangle}{\Delta}\right) \tag{6.43}$$

and

$$\langle n_{d\downarrow} \rangle = \frac{1}{\pi} \cot^{-1}\left(\frac{\epsilon_d + U\langle n_{d\uparrow} \rangle}{\Delta}\right), \tag{6.44}$$

for the occupations of the ↑ and ↓ spin levels of the impurity. These are the central equations of the Hartree-Fock treatment of the Anderson Hamiltonian.

For local moment formation, we seek solutions in which $\langle n_{d\uparrow} \rangle \neq \langle n_{d\downarrow} \rangle$. Only in this case is the d-impurity magnetic. First, we see immediately that if $U = 0$, the only solution is $\langle n_{d\uparrow} \rangle = \langle n_{d\downarrow} \rangle$. Now in the limit in which $\Delta \to \infty$, we again find a nonmagnetic solution, $\langle n_{d\uparrow} \rangle = \langle n_{d\downarrow} \rangle = 1/2$. A magnetic solution exists, however, in the intermediate parameter range $\Delta/U \ll 1$. To simplify the notation, we introduce the dimensionless parameters $x = -\epsilon_d/U$ and $y = U/\Delta$. The value $x = 0$ corresponds to the d-state energy ϵ_d lying right at the Fermi level. At $x = 1$, $\epsilon_d = -U$ and the upper d-state with energy $\epsilon_d + U$ is degenerate with the Fermi level. Hence, the energies 0 and $-U$ are symmetrically located around $x = 1/2$ or $\epsilon_d = -U/2$. As we will see, $x = 1/2$ is the most favorable case for a magnetic moment to form. In fact, magnetism persists only for $0 \leq x \leq 1$. When $\epsilon_d = -U/2$, the impurity terms in the Hamiltonian $\sum_\sigma (\epsilon_d n_{d\sigma} + a_{d\sigma}^\dagger a_{k\sigma}) + U n_{d\uparrow} n_{d\downarrow}$ are particle-hole symmetric under the transformation $a_d \leftrightarrow a_d^\dagger$ (see Problem 4).

In the limit $y \gg 1$ and $1 > x > 0$, Eqs. (6.43) and (6.44) predict a magnetic solution of the form

$$\pi\langle n_{d\uparrow} \rangle \simeq \pi + \frac{1}{y(\langle n_{d\downarrow} \rangle - x)} \tag{6.45}$$

and

$$\pi\langle n_{d\downarrow} \rangle \simeq \frac{1}{y(\langle n_{d\uparrow} \rangle - x)}, \tag{6.46}$$

in which $\langle n_{d\uparrow} \rangle \sim 1$ and $\langle n_{d\downarrow} \rangle \sim 0$. In the limit of large y, the solution is

$$\langle n_{d\uparrow} \rangle = 1 - \frac{1}{\pi x y} + \cdots,$$

$$\langle n_{d\uparrow} \rangle = x + \frac{1}{\pi x y} + \cdots. \tag{6.47}$$

The phase diagram for the complete range of magnetic parameters is shown in Fig. (6.3). On the transition curve and beyond, $\langle n_{d\uparrow} \rangle = \langle n_{d\downarrow} \rangle \equiv n_d$, or, equivalently, $\cot \pi n_d = (\epsilon_d + U)/\Delta$. The mixed-valence region corresponds to the lower-left and upper-left regions of the phase diagram in which either the lower or upper levels is nearly

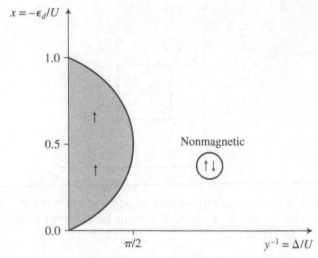

Figure 6.3 Local moment phase diagram for the Anderson model. In the shaded region, local moments form stably. The solid curve separates the magnetic and nonmagnetic regions.

degenerate with the Fermi level. In either of these regions, increasing Δ slightly results in a nonmagnetic solution. As mentioned, many electronic systems, such as the heavy fermion materials, reside in this regime. At the boundary that defines the region of local moment formation, an infinitesimal perturbation of the equal occupation probability, $n_d = n_c$, thereby $\langle n_\sigma \rangle = n_c + \sigma \delta n$, leads to unequal occupancy of the impurity levels. Hence, from Eqs. (6.43) and (6.44), we simply find that on the phase boundary, the occupation must also be a solution to the derivative of the local moment equations, so that

$$1 = \frac{U}{\Delta \pi} \frac{1}{1 + [(\epsilon_d + Un_c)/\Delta]^2}. \tag{6.48}$$

An equivalent way of deducing this form for the phase boundary is to consider solutions around the point at which the curves $\cot \pi n$ and $y(n - x)$ intersect, as shown in Fig. (6.4). A solution of Eqs. (6.43) and (6.44) is described by a rectangle of the type shown in Fig. (6.4) around such points, where for $\langle n_{d\uparrow} \rangle > \langle n_{d\downarrow} \rangle$, the vertical side on the right is at $\langle n_{d\uparrow} \rangle$, and on the left at $\langle n_{d\downarrow} \rangle$, as shown in the figure. Magnetic solutions will exist as long as the vertices of the rectangle touch the curves. An infinitesimal variation away from the intersection point where $\cot \pi n = y(n - x)$ leads to a magnetic solution, if the slopes of the two curves $\cot \pi n$ and $y(n - x)$ are equal in magnitude and opposite in sign. This is precisely the condition in Eq. (6.48).

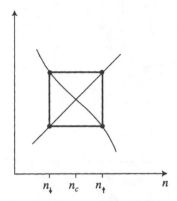

Figure 6.4 A plot of the intersection of $y(n - x)$ and $\cot \pi n$. The vertices of the rectangle enscribing the intersection point correspond to magnetic solutions of Eqs. (6.43) and (6.44).

Comparison of Eq. (6.48) with Eq. (6.29) reveals that the magnetic stability condition is equivalent to

$$U\rho_d(\epsilon = 0) = U\rho_d(\epsilon_F) = 1, \tag{6.49}$$

where $\rho_d(\epsilon) = \rho_{d\uparrow}(\epsilon)(\rho_{d\downarrow}(\epsilon))$ for equal occupancy. This equation is the principal result in Anderson's theory of local moments. Magnetic impurities with large densities of states at the Fermi level are expected to form local moments in nonmagnetic metals. In fact, we can see the local moment criterion more intuitively by noticing that local moment formation is most favorable when the impurity levels are half-filled (see Fig. 6.3). At half-filling, the local moment criterion, Eq. (6.48), becomes $U/\pi\Delta > 1$. The hybridization energy is determined by the square of the hopping matrix element, through Δ. Hence, local moment formation arises from a balancing of the on-site interaction with the hybridization energy.

A related result is Stoner's criterion for the onset of ferromagnetism in metals. Stoner [S1938] showed that within a simple model with a δ-function repulsion between electrons of the form, $U\delta(\mathbf{r}_i - \mathbf{r}_j)$, the mean-field or Hartree-Fock criterion for the onset of ferromagnetism is $UN(\epsilon_F) \geq 1$, where $N(\epsilon_F)$ is the density of states per unit energy at the Fermi level of the metal. That ferromagnetism results from an interplay between the density of states and the Coulomb interaction can be understood as follows. Hopping processes that result in double occupancy of any lattice site with two electrons of the same spin are forbidden by the Pauli principle. However, when two neighboring lattice sites are occupied by electrons of opposite spin, the hopping process that doubly occupies one of the lattice sites has finite probability. Such antiferromagnetic double occupancy increases the energy of a lattice site as a result of the Coulomb repulsion. Nonetheless, hopping processes lower the kinetic energy of the electrons. Hence, whether the system of localized electrons becomes ferromagnetic and/or antiferromagnetic is ultimately determined by the competition between electron kinetic energy and the Coulomb repulsion; if Coulomb interactions dominate, the system is ferromagnetic. Effectively, at a given electron density, the density of states varies inversely with the electron kinetic energy, and the Stoner criterion reflects this competition between the Coulomb interaction and the kinetic energy.

6.3 SUMMARY

We have seen that the Hartree-Fock solution to the Anderson model provides a simple criterion for local moment formation,

$$U\rho_d(\epsilon_F) = 1. \tag{6.50}$$

Magnetic moments form when $U\rho_d(\epsilon_F) > 1$. Hence, an ion will have a tendency to be magnetic if either its density of d-states at the Fermi level or the on-site repulsion, U, is large. Ions with vanishing densities of states at the Fermi level will be nonmagnetic. For a half-filled impurity, the local moment criterion can be written as $U/\pi\Delta > 1$. This criterion lays plain the competition between the local interaction energy and the hybridization that gives rise to local moment formation.

PROBLEMS

1. The matrix element V_{kd} takes into account the interaction between electrons on the impurity and the electrons in the band, which is omitted in the Hartree-Fock determination of the impurity states and the continuum states. The Coulomb matrix element, $\langle k, k'|v_{ee}|d, k'\rangle$, summed over other band electrons k' moves an electron in the impurity level to a continuum level k. Estimate the magnitude of V_{kd} from this residual Coulomb interaction.

2. Calculate explicitly the $\epsilon_{n\sigma}$'s in the Hartree-Fock approximation to the Anderson model.

3. Show explicitly that the Hartree-Fock approximation to the Anderson Hamiltonian can be written in the form of Eq. (6.6), with the $a_{n\sigma}^{\dagger}$ defined in Eq. (6.5).

4. Show that when $\epsilon_d = -U/2$, the impurity terms in the Anderson model, $\sum_{\sigma}(\epsilon_d n_{d\sigma} + a_{d\sigma}^{\dagger} a_{k\sigma}) + U n_{d\uparrow} n_{d\downarrow}$, are invariant under the transformation $a_d^{\dagger} \leftrightarrow a_d$.

REFERENCES

[A1961] P. W. Anderson, *Phys. Rev.* **124**, 41 (1961).

[BC1989] D. Baeriswyl, D. K. Campbell, eds. *Interacting Electrons in Reduced Dimensions* (Plenum Press, New York, 1989).

[F1958] J. Friedel, *Nuovo Cimento Suppl.* **VII**, 287 (1958).

[GZ1974] G. Grüner, A. Zawadowski, *Repts. Prog. Phys.* **37**, 1497 (1974).

[H1964] J. Hubbard, *Proc. Roy. Soc. Lond.* A **276**, 238 (1964).

[M1960] B. T. Matthias, M. Peter, H. J. Williams, A. M. Clogston, E. Corenzwit, R. C. Sherwood, *Phys. Rev. Lett.* **5**, 542 (1960).

[S1938] E. C. Stoner, *Proc. Roy. Soc. Lond.* A **165**, 372 (1938).

– 7 –

Quenching of Local Moments: The Kondo Problem

In the previous chapter, we developed a mean-field criterion for local magnetic moment formation in a metal. As mean-field theory is valid typically at high temperatures, we anticipate that at low-temperatures, significant departures from this treatment occur. The questions we focus on in this chapter are: 1) how does the presence of local magnetic moments affect the low-temperature transport and magnetic properties of the host metal, and 2) what is the fate of local magnetic moments at low temperatures in a metal? These questions are of extreme experimental importance because it has been known since the early 1930s that the resistivity of a host metal, such as Cu with trace amounts of magnetic impurities, typically Fe, reaches a minimum and then increases as $-\ln T$ as the temperature subsequently decreases. This behavior is illustrated in Fig. (7.1) for various Mo and Nb alloys [SCL1964]. A resistivity minimum and subsequent logarithmic-temperature dependence are in stark contrast to the resistivity of the pure metal that tends to zero monotonically as the temperature decreases. An additional surprise is that the $-\ln T$-dependence of the resistivity does not continue indefinitely to low temperature, but rather, below a characteristic temperature, the Kondo temperature, T_k, it phases out. Moreover, the spin properties of a magnetic impurity change fundamentally in the neighborhood of the Kondo temperature, as magnetic susceptibility measurements showed [H1969]. Well above the Kondo temperature, the magnetic susceptibility of the impurity spins obeys the Curie $1/T$ law for free magnetic moments. Below T_k, however, the susceptibility tends to a constant [H1969]. A constant susceptibility at $T = 0$ is characteristic of a singlet state polarized by a magnetic field (see Problem 7.1). Hence, in addition to being the temperature at which the $-\ln T$

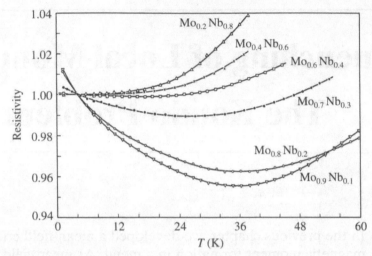

Figure 7.1 Resistance minima for Fe in a series of Mo-Nb alloys (from Sarachik et al., *Phys. Rev.* **135**, 1041 (1964)).

behavior of the resistivity ceases, T_k is the temperature at which the impurity and conduction electron spins begin to condense into singlet states. This condensation is complete at $T = 0$. The vanishing of the local moment below some characteristic temperature makes the Kondo problem fundamentally different from bulk ferromagnetism in which there is an onset of magnetism below some characteristic temperature, the Curie temperature. In the Kondo problem, just the opposite occurs; magnetism ceases at low temperatures.

Since the $-\ln T$-dependence of the resistivity vanishes once the impurities begin to condense into singlet states, the resistivity minimum must be due to the interaction of the impurity spin with those of the host metal. The inception of the first model that was capable of describing the interaction of a local spin with itinerant conduction electrons dates back to work in the 1940s and 1950s by Vousovskii [V1946], Zerner [Z1951], and Kasuya [K1956]. The essence of this model is that an exchange interaction, J, exists between a local impurity spin and the conduction electrons. This model, as we show, is related to the Anderson model in a certain parameter range.

Kondo showed that a $-\ln T$-term appears in a perturbative theory to second-order in J. Such scattering processes appear in addition to the scattering rate that arises between conduction electrons and phonons, which scales as T^5 at low temperatures. Because the resistivity is directly proportional to the total scattering probability, the logarithmic contribution from spin-flip scattering implies that a minimum occurs in the resistivity. Hence, Kondo's perturbative treatment offered a solution to the long-standing problem of the resistivity minimum in metal alloys.

However, all was not well with the Kondo solution. As $T \rightarrow 0$, the ln T-terms diverge. The temperature at which the second-order term in the perturbative treatment of the scattering probability becomes comparable in magnitude to the first-order determines when the divergence occurs. This temperature is known as T_k, and it is the temperature around which the magnetic properties of the system change. The divergence implies that perturbation theory breaks down at low temperatures. As a consequence, Kondo's solution is valid only for $T \gg T_k$; other methods must be developed to understand the ultimate fate of a magnetic impurity near and below T_k. Experiments tell us, however, that ultimately the local moment does not survive at $T = 0$. The search for the theory that removes the divergence in Kondo's perturbative treatment and accounts for the formation of a bound-singlet state at an impurity is known as the *Kondo problem*.

Anderson and Yuval's [AYH1970] scaling hypothesis was the key concept that paved the way to the solution of this quite subtle problem. They showed within a perturbative scheme that the exchange interaction in the Kondo problem increases in magnitude as the effects of more-and-more high-energy excitations on the effective integration are included. Their scheme, while perturbative, was instrumental in pinpointing a physical mechanism by which the local moment vanishes and a ground-state singlet ensues. The key breakthrough in this problem, however, took place in the late 1970s when Wilson [W1975] developed a numerical renormalization group procedure to solve the Kondo problem. His solution confirmed the scaling hypothesis of Anderson and Yuval. The fact that the resolution of the Kondo problem required one of the key ideas in modern theoretical physics was certainly not anticipated. Indeed, the simplicity of the Kondo model belied the complexity of the physics it embodied. It is partly for this reason that the origin of the bound-singlet state remained shrouded in divergent perturbation sums long after the resistivity minimum was satisfactorily explained.

In this chapter, we present the key ideas that led to the resolution of the resistivity minimum and Kondo problems. We first present the Kondo model and establish the origin of the antiferromagnetic interaction between the local moment and the conduction electrons. In so doing, we will be able to establish the relationship between the Anderson and Kondo models. With the antiferromagnetic interaction in hand, we then perform the second-order perturbative analysis to uncloak the ln T-dependence in the resistivity. It will become evident that a ln T-dependence of the conductivity is anticipated anytime a local impurity has a degree of freedom, for example, spin in the case of a local magnetic moment. We finally discuss the variational and scaling analyses of the bound-singlet state formation.

7.1   KONDO HAMILTONIAN

The Kondo Hamiltonian describes the interaction of impurity spins, assumed to be spin-$1/2$, with those of the conduction electrons. We first introduce the two-component spinor operators that remove electrons from impurity and conduction states

$$\Psi_{\mathbf{k}} = \begin{pmatrix} a_{\mathbf{k}\uparrow} \\ a_{\mathbf{k}\downarrow} \end{pmatrix}, \quad \Psi_d = \begin{pmatrix} a_{d\uparrow} \\ a_{d\downarrow} \end{pmatrix} \tag{7.1}$$

and spin matrix operators

$$\mathbf{S} = \frac{\hbar\boldsymbol{\sigma}}{2}, \tag{7.2}$$

where the σ's are the usual Pauli matrices,

$$\sigma_x = \begin{pmatrix} 0 & 1 \\ 1 & 0 \end{pmatrix}, \quad \sigma_y = \begin{pmatrix} 0 & -i \\ i & 0 \end{pmatrix}, \quad \sigma_z = \begin{pmatrix} 1 & 0 \\ 0 & -1 \end{pmatrix}. \tag{7.3}$$

We will also find useful the spin-raising, $\hbar\sigma^+ = S_x + iS_y = S^+$, and spin-lowering, $\hbar\sigma^- = S_x - iS_y = S^-$, spin operators.

Kondo physics arises from the coupling of the spin of an impurity to that of a conduction electron. Hence, the simplest Hamiltonian that represents the interaction of a local spin with a band of itinerant electrons is

$$H_K = \sum_{\mathbf{k},\sigma} \epsilon_{\mathbf{k}} n_{\mathbf{k}\sigma} - \sum_{\mathbf{k},\mathbf{k}'} \frac{J_{\mathbf{k}\mathbf{k}'}}{\hbar^2} (\Psi_{\mathbf{k}'}^\dagger \mathbf{S} \Psi_{\mathbf{k}}) \cdot (\Psi_d^\dagger \mathbf{S} \Psi_d). \tag{7.4}$$

The operator $\Psi_d^\dagger \mathbf{S} \Psi_d$ is the spin operator of electrons in impurity state d, whereas the spin operator $\Psi_{\mathbf{k}'}^\dagger \mathbf{S} \Psi_{\mathbf{k}}$ is the *transition* spin operator of electrons between conduction states $\mathbf{k}$ and $\mathbf{k}'$. As the local impurity energy simply shifts the zero of the energy, we have dropped this term from the Hamiltonian. The interaction $J_{\mathbf{k}\mathbf{k}'}$, with units of energy, is the analog of the Heisenberg spin exchange interaction between the spin of a band electron and that of a localized electron. A positive $J_{kk'} > 0$ tends to favor spin alignment (see Eq. 4.18 in Chapter 4) and thus describes a ferromagnetic interaction, whereas $J_{kk'} < 0$ corresponds to an antiferromagnetic interaction.

7.2 WHY IS J NEGATIVE?

We offer here a simple physical derivation of the exchange interaction in the Kondo Hamiltonian, appealing to the magnetic impurity model of Anderson. The complete derivation is detailed in the Appendix. The interaction in the Kondo problem is nonzero when there is a net spin on the impurity—in the local moment phase for an impurity that can at most be doubly occupied. As we discussed in the previous chapter, the local moment phase in the Anderson model persists in the limit where the on-site Coulomb interaction, U, is much larger than the hybridization energy, Δ. It is precisely in the local moment limit that the Anderson and Kondo models describe the same physics. However, a fundamental difference between the Anderson and Kondo models is that the Anderson model includes charge fluctuations, which determine the hybridization energy, but they are absent in the Kondo model, which includes only spin-spin interactions. As we show now, the effect of charge fluctuations to second order in the hybridization matrix element, V_{kd}, results in an antiferromagnetic interaction of the type in the Kondo model.

  We focus on the processes, in the Anderson model, that lead to scattering of a conduction electron with a local moment and calculate to second order in the hybridization interaction, $\propto V_{kd}$, which couples conduction electrons to the local moment. In perturbation theory, the amplitudes of such processes are of the form

$$V_{kd} \frac{1}{E_i - E_{\text{int}}} V_{dk}, \tag{7.5}$$

where E_i is the energy of the initial state and E_{int} is the energy of the intermediate. Let us first consider the scattering of a conduction electron in state $\mathbf{k}$ with spin up by a spin-down impurity to a final electron state $\mathbf{k}'$ of spin up with the impurity remaining with spin down. This process occurs by the electron $\mathbf{k}$ hopping, via the hybridization interaction, to the impurity, as illustrated in Fig. (7.2a); in this intermediate state, the impurity is doubly occupied. The energy of the intermediate state is

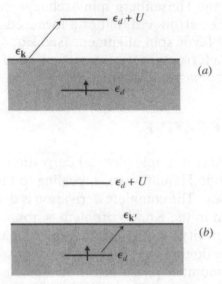

Figure 7.2 Two processes that contribute to the second-order charge fluctuations on a singly-occupied d-level. The solid line denotes the initial conduction electron state, just below the Fermi surface. The energy levels are labeled as in Chapter 6.

$2\epsilon_d + U$. Then the spin-up electron on the impurity hops to the final conduction electron state, $\mathbf{k}'$. The second-order matrix element for this process is

$$T_{(\mathbf{k}\sigma)+(d-\sigma)\rightarrow(\mathbf{k}'\sigma)+(d-\sigma)} = \frac{V_{\mathbf{k}d}V_{d\mathbf{k}'}}{\epsilon_{\mathbf{k}} - \epsilon_d - U}. \qquad (7.6)$$

On the other hand, the scattering of a conduction electron in state $\mathbf{k}$ with spin up from a spin-up impurity to a final electron state $\mathbf{k}'$ of spin up with the impurity remaining with spin up can only proceed by the electron on the impurity first hopping to $\mathbf{k}'$, as illustrated in Fig. (7.2b). In the intermediate state, both electrons are in conduction states. The matrix element is then

$$T_{(\mathbf{k}\sigma)+(d\sigma)\rightarrow(\mathbf{k}'\sigma)+(d\sigma)} = -\frac{V_{\mathbf{k}d}V_{d\mathbf{k}'}}{\epsilon_d - \epsilon_{\mathbf{k}'}}. \qquad (7.7)$$

The minus sign arises from the exchange of the conduction and impurity electrons; the initial conduction electron ends on the impurity, whereas the initial electron on the impurity ends up in the conduction band. Similarly, the spin-flip scattering of a spin-up electron in state $\mathbf{k}$ and a spin-down impurity to a final electron state $\mathbf{k}' \downarrow$ with the impurity ending with spin up proceeds either by this same process or as follows. A conduction electron first hops to the impurity, and then the down-spin electron on the impurity hops to the final state $\mathbf{k}'$. The amplitude is a linear superposition of those for the individual processes,

$$
T_{(\mathbf{k}\sigma)+(d-\sigma)\rightarrow(\mathbf{k}'-\sigma)+(d\sigma)} = -V_{\mathbf{k}d}V_{d\mathbf{k}'}\left(\frac{1}{\epsilon_{\mathbf{k}} - \epsilon_d - U} + \frac{1}{\epsilon_d - \epsilon_{\mathbf{k}'}}\right).
$$

(7.8)

The minus sign again arises from exchange of the conduction and impurity electrons.

As shown in the Appendix, the spin-spin interaction of the Kondo model, Eq. (7.4), yields the amplitude for the spin-flip process

$$
T_{(\mathbf{k}\sigma)+(d-\sigma)\rightarrow(\mathbf{k}'-\sigma)+(d\sigma)} = -\frac{1}{2}J_{\mathbf{k}\mathbf{k}'}.
$$

(7.9)

Consequently, we infer from Eqs. (7.8) and (7.9) that

$$
J_{\mathbf{k}\mathbf{k}'} = 2V_{\mathbf{k}d}V_{d\mathbf{k}'}\left(\frac{1}{\epsilon_{\mathbf{k}} - \epsilon_d - U} + \frac{1}{\epsilon_d - \epsilon_{\mathbf{k}'}}\right).
$$

(7.10)

If we assume that the relevant $\mathbf{k}$-states of the host metal are close to the Fermi level, that is, $\epsilon_{\mathbf{k}} \approx \epsilon_{\mathbf{k}'} \approx 0$, then this expression reduces to

$$
J_{\mathrm{eff}} = -|V_{\mathbf{k}d}|^2 \frac{U}{|\epsilon_d|(U - |\epsilon_d|)} < 0.
$$

(7.11)

The effective coupling between the d-impurity and the band of electrons is indeed antiferromagnetic. To determine its magnitude, we consider the particle-hole symmetric point at which $\epsilon_d = -U/2$. In this limit, the exchange interaction,

$$
J_{\mathrm{eff}} = -4\frac{|V_{\mathbf{k}d}|^2}{U},
$$

(7.12)

is inversely proportional to the on-site Coulomb repulsion. In the Anderson model, charge fluctuations between the conduction band and the impurity mediate the Kondo interaction. The coupling constant in the Kondo limit of the Anderson Hamiltonian scales as $\Delta/U \ll 1$. The interaction we derive here is the lowest term in a perturbation series in

$V_{\mathbf{k}d}$. Why does taking charge fluctuations into account to second order lead to a spin dependence of the interaction? Physically, the answer is the Pauli exclusion principle, which forbids intermediate states in which the impurity is occupied by two electrons of the same spin orientation.

In the Appendix to this chapter we derive, via the Schrieffer-Wolff transformation, the full Hamiltonian to second order in $V_{\mathbf{k}d}$. The only difference for $J_{\mathbf{kk}'}$ from Eq. (7.10) is that the full answer, Eq. (7.120), is symmetrized with respect to the two wave vectors $\mathbf{k}$ and $\mathbf{k}'$. As the reader should verify, the full amplitude for the equal spin process in Eq. (7.7) arises from the direct interaction H_{dir}, Eq. (7.123), whereas the full amplitude for the opposite spin process in Eq. (7.6) arises from both the direct and spin-spin interactions.

The equivalence of the Kondo model with the Anderson model in the local moment phase suggests that the ground state of both models should be identical. Naively, one would associate the formation of a singlet state in the Anderson model with double occupancy of the impurity levels. This interpretation is problematic because in the local moment phase of the Anderson model, the upper level is unoccupied. Further, the energy levels of the impurity remain fixed relative to the Fermi level when the temperature is lowered, the upper level (with energy $\epsilon_d + U$) remaining well above the Fermi level. Hence, it must be empty, as illustrated in Fig. (7.3).

Understanding the origin of the singlet ground state in the Anderson model at $T = 0$ is the essence of the Kondo problem. The singlet state emerges as a result of a new resonant level [GZ1974] that forms and remains pinned at the Fermi level. The change in the single-particle density of states in the Anderson model as a result of the formation of the resonant level is shown in Fig. (7.4). The peak grows logarithmically down [L1981] to the Kondo temperature; its width is proportional to T_k. Collectively, all the conduction electrons with energies within T_k of the Fermi surface contribute to the formation of the resonant level, as depicted in Fig. (7.3). For a spin-1/2 impurity, the resonant level is occupied on average by just a single conduction electron, which compensates the spin on the impurity. As we discuss at the end of Section 4, it is the logarithmic singularity in the electron-impurity scattering rate that gives rise to the sharp peak in the single-particle density of states at the Fermi level. The excess density of states at the Fermi level grows as the impurity loses its spin.

To summarize, at high temperatures the Hartree-Fock theory of local moment formation is valid, and the impurity density of states is well described by the two-peaked Lorentzian function derived in the last

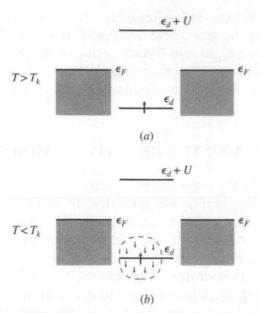

$$T > T_k$$

$\epsilon_d + U$

ϵ_F ϵ_F

ϵ_d

(a)

$\epsilon_d + U$

$$T < T_k$$

ϵ_F ϵ_F

ϵ_d

(b)

Figure 7.3 Schematic depiction of the energy levels in the Anderson model (a) above and (b) below the Kondo temperature. Even though the local moment begins to disappear as the temperature falls below T_k, the occupancy in the upper level remains unchanged from zero. This strange occurrence is the Kondo problem. Note that the Fermi sea does not develop a net magnetization below the Kondo temperature.

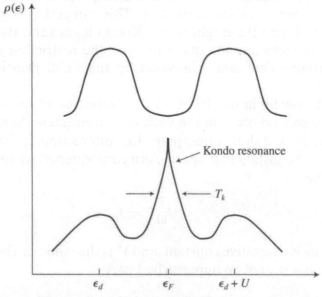

$\rho(\epsilon)$

Kondo resonance

T_k

ϵ_d ϵ_F $\epsilon_d + U$

Figure 7.4 Single-particle density of states in the Anderson model in the Kondo regime. The emergence of the sharp peak at the Fermi level is a many-body resonance effect. This peak increases logarithmically as the temperature is lowered.

chapter. However, at low temperatures, new physics arises that is not described by the mean-field description of the Anderson model. At low temperatures, rapid spin-flip scattering occurs between the conduction electrons and the local moments. We now show how such processes underlie the resistivity minimum in metallic alloys.

7.3   SCATTERING AND THE RESISTIVITY MINIMUM

The origin of the logarithmic contribution to the resistivity is the spin degree of freedom at the impurity. To see the role of the spin, we outline the key notions that underlie the structure of the scattering probability. As we will see, the scattering probability to lowest order in the exchange interaction is well behaved and temperature independent. Temperature dependence first arises in a calculation of the scattering amplitude to second order in the exchange interaction. Because the interaction depends on the spin degree of freedom, a second-order term can involve spin-flip processes on the impurity. However, the order in which the spins are flipped matters because, for example, $S^+ S^- \neq S^- S^+$. This lack of commutativity, combined with the restriction of the allowed occupancy of the intermediate states arising from the Pauli principle, leads to a nontrivial temperature dependence of the scattering amplitudes. In second order, the contribution is of the form f/ϵ. This integral is logarithmically divergent, hence the origin of the Kondo logarithm. By contrast, if the impurity lacks any internal structure, the restriction on the filling of the intermediate states imposed by the Pauli principle cancels out entirely.

We now fill in the details of the derivation of the scattering amplitude to second order in the exchange interaction. Since the exchange interaction is a short-range point-like interaction, its Fourier transform should be constant in k-space. As a consequence, we make the approximation

$$J_{\mathbf{k}\mathbf{k}'} = J_0/V, \tag{7.13}$$

where J_0 is a negative constant, and V is the volume. The Kondo Hamiltonian is a sum of an unperturbed part

$$H_0 = \sum_{\mathbf{k},\sigma} \epsilon_{\mathbf{k}} n_{\mathbf{k}\sigma} \tag{7.14}$$

and the spin-spin interaction, which we treat perturbatively,

$$H' = -\frac{J_0}{2\hbar V} \sum_{\mathbf{k}',\mathbf{k}'',\sigma} [S_d^z(a_{\mathbf{k}'\uparrow}^\dagger a_{\mathbf{k}''\uparrow} - a_{\mathbf{k}'\downarrow}^\dagger a_{\mathbf{k}''\downarrow})$$

$$+ S_d^+ a_{\mathbf{k}'\downarrow}^\dagger a_{\mathbf{k}''\uparrow} + S_d^- a_{\mathbf{k}'\uparrow}^\dagger a_{\mathbf{k}''\downarrow}]. \tag{7.15}$$

Since we are interested only in the coupling to a single impurity, we write the interaction here in terms of the single impurity spin operator, S_d, rather than the second-quantized description of the impurity in the Kondo Hamiltonian, Eq. (7.4). We assume quite generally that the impurity has total spin $S = \hbar s$; the eigenstates of S_z are denoted by the eigenvalue $\hbar m_s$.

Let us first calculate in lowest order the total rate of scattering by an impurity of a conduction electron initially of momentum $\mathbf{k}$ to momentum $\mathbf{k}'$; we let the impurity have initial spin orientation m_s. The many-particle electron wave function is a Slater determinant of single-particle states in the conduction band. Because H' involves one pair of creation and annihilation operators for the conduction band, it can only connect states that differ by a single orbital. Hence, the energy differences entering the calculation of scattering rates reduce to differences of single orbital energies.

The total scattering rate is the sum of the rates of two incoherent processes: the first, in which the initial electron, assumed to be spin up, does not flip its spin and remains with spin up, and the second, in which the electron flips from spin up to spin down, whereas the impurity changes from m_s to $m_s + 1$. Consider first the non–spin-flip process, for which the bare matrix element is

$$T_{\mathbf{k}\uparrow \to \mathbf{k}'\uparrow;m_s}^{(0)} = -\frac{J_0}{2V} m_s. \tag{7.16}$$

The lowest-order rate of scattering is

$$\Gamma_{\mathbf{k}\to\mathbf{k}'}^0(m_s) = \frac{2\pi}{\hbar} \sum_{\mathbf{k}'} \delta(\epsilon_{\mathbf{k}} - \epsilon_{\mathbf{k}'}) \left(\frac{J_0}{2V}\right)^2 m_s^2, \tag{7.17}$$

where the sum is over a limited group of final states $\mathbf{k}'$. Note that the scattering is isotropic (s-wave). To compute the total rate of this process, we sum $\mathbf{k}'$ over all final states, which gives a factor $N(0)V$ for $\mathbf{k}$ close to the Fermi surface, where $N(0) = mk_F/2\pi^2\hbar^3 = 3n_e/4\epsilon_F$ is the single-spin electron density of states at the Fermi surface. The total scattering rate from a single impurity with no spin-flip is thus

$$\Gamma^0_{\text{non-flip}}(m_s) = \frac{\pi}{2\hbar V} N(0) J_0^2 m_s^2. \tag{7.18}$$

In the spin-flip process, the matrix element is

$$T^{(0)}_{\mathbf{k}\uparrow \to \mathbf{k'}\downarrow;m_s} = -\frac{J_0}{2\hbar V} \langle s, m_s + 1 | S_d^+ | s, m_s \rangle. \tag{7.19}$$

From the quantum theory of angular momentum,

$$\langle s, m_s \pm 1 | S^{\pm} | s, m_s \rangle = \hbar \sqrt{s(s+1) - m_s(m_s \pm 1)}, \tag{7.20}$$

so that

$$\Gamma^0_{\text{flip}}(m_s) = \frac{\pi}{2\hbar V} N(0) J_0^2 (s(s+1) - m_s(m_s + 1)). \tag{7.21}$$

The total rate of scattering is the sum of Eqs. (7.18) and (7.21),

$$\Gamma^0(m_s) = \frac{\pi}{2\hbar V} N(0) J_0^2 (s(s+1) - m_s). \tag{7.22}$$

If we average over all initial spin orientations of the impurity, the m_s-term vanishes. Summing over all impurities, we derive the spin-averaged scattering rate,

$$\Gamma^0 = \frac{\pi}{2\hbar} s(s+1) J_0^2 n_{\text{imp}} N(0)$$

$$= \frac{8\pi}{3} \frac{\epsilon_F}{\hbar} s(s+1) c (N(0) J_0)^2, \tag{7.23}$$

where $c = n_{\text{imp}}/n_e$ is the fractional concentration and n_{imp} the density of impurities. The non-flip contribution is $1/3$ of the total, as one sees from the relation,

$$\frac{1}{2s+1} \sum_{m_s} m_s^2 = \frac{1}{3} s(s+1) \tag{7.24}$$

The scattering rate (Eq. 7.23) is well behaved and completely independent of temperature. All things being equal, each order of perturbation theory is smaller by a factor of $J_0 N(0)$. Hence, it might seem ill-advised to explore the second-order term. However, all things are not equal and a crucial divergence lurks at second order.

To determine the scattering rate to next order in H', we calculate the scattering amplitude to second order. In general the second-order amplitude is given by

$$T^{(2)}_{a \to b} = \sum_{c \neq a} \frac{\langle b|H'|c\rangle\langle c|H'|a\rangle}{E_a - E_c}. \tag{7.25}$$

As in lowest order, the scattering rate is a sum of processes in which the initial and final electron state have the same spin and in which the final electron spin is opposite to the initial spin. Figure (7.5) shows the second-order contributions to the amplitude in which the final electron has the same spin as the initial, and Fig. (7.6) shows the processes in which the final spin has flipped. In each of these processes, two types of contributions enter the intermediate states: non–spin-flip terms, in which the electron spin remains unchanged and which involve only S_z, and spin-flip terms involving the operators $S^{\pm}$.

Let us first consider the upper two processes in Fig. (7.5), in which an electron of initial momentum $\mathbf{k}$ is scattered to final momentum $\mathbf{k}'$. On the one hand, the electron can scatter from momentum $\mathbf{k}$ to an un-occupied state with momentum $\mathbf{q}$ and subsequently to the final state

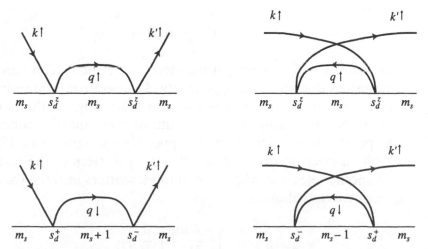

Figure 7.5 Second-order processes between initial and final electron states $\mathbf{k} \uparrow$ and $\mathbf{k}' \uparrow$. The straight lines depict the spin of the impurity and the curved lines the electron. The line with a backward arrow in the intermediate state represents a hole, that is, a particle that is present in the initial state and final state but is absent in the intermediate state. The label on the hole is the momentum and spin of that electron in the initial state.

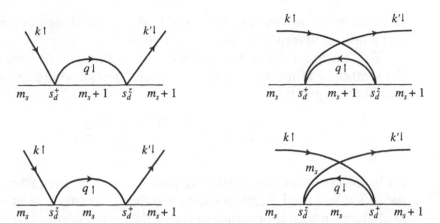

Figure 7.6 Second-order spin-flip processes between initial and final electron states $\mathbf{k}\uparrow$ and $\mathbf{k}'\downarrow$.

with momentum $\mathbf{k}'$, as shown in the upper-left diagram. The contribution of this process to $T^{(2)}_{\text{non-flip}}(\mathbf{k} \to \mathbf{k}'; m_s)$ is

$$\left(\frac{-J_0 m_s}{2}\right)^2 \frac{1}{V}\alpha, \tag{7.26}$$

where the sum over intermediate particle states is

$$\alpha = \frac{1}{V}\sum_{\mathbf{q}}\frac{1-f_{\mathbf{q}}}{\epsilon_{\mathbf{k}}+i\eta-\epsilon_{\mathbf{q}}}. \tag{7.27}$$

The factor $1-f_{\mathbf{q}}$, where $f_{\mathbf{q}}$ is the electron Fermi distribution function, guarantees that the state $\mathbf{q}$ is empty. The positive infinitesimal imaginary number $i\eta$ guarantees that the scattering produces outgoing, rather than incoming, waves. Quantum mechanically coherent with this process is that shown in the upper-right diagram (see Fig. 7.5), in which an electron in an occupied state $\mathbf{q}$ scatters to the final state $\mathbf{k}'$ and another electron with momentum $\mathbf{k}$ scatters into the vacated state $\mathbf{q}$, giving a contribution

$$\left(\frac{-J_0 m_s}{2V}\right)^2 \frac{1}{V}\gamma, \tag{7.28}$$

where

$$\gamma = -\frac{1}{V}\sum_{\mathbf{q}}\frac{f_{\mathbf{q}}}{\epsilon_{\mathbf{q}}+i\eta-\epsilon_{\mathbf{k}'}}. \tag{7.29}$$

In this latter process, the intermediate momentum state must be occupied; the rate is thus proportional to f_q. The minus sign arises from
the interchange of the operations of removing the initial electron and
creating the final electron, $a_k a_{k'}^\dagger = -a_{k'}^\dagger a_k$ for $k \neq k'$. In an energy-
conserving scattering ($\epsilon_k = \epsilon_{k'}$) process, the distribution functions
cancel in the sum of the amplitudes for the two processes, and the
amplitude for non–spin-flip processes is

$$T^{(2)}_{\text{non flip}}(k \uparrow \to k' \uparrow; m_s) = \left(\frac{-J_0 m_s}{2V}\right)^2 \sum_q \frac{1}{\epsilon_k - \epsilon_q}. \qquad (7.30)$$

Because the f's cancel, the non–spin-flip amplitudes are independent
of temperature. The summation is formally divergent because we have
ignored the momentum dependence of the matrix elements; once taken
into consideration, this term is finite.

The lower two processes shown in Fig. (7.5) involve a single-
electron spin-flip in the intermediate state. The same two intermediate momentum states enter as in the non–spin-flip processes discussed above; however, they do not have the same weights. In the
lower-left diagram, the matrix elements give a factor $(-J_0/2\hbar V)^2 P_{+-}$,
where

$$P_{+-} = \frac{1}{\hbar^2}|\langle s, m_s + 1|S_d^+|s, m_s\rangle|^2 = s(s+1) - m_s(m_s + 1). \qquad (7.31)$$

In the lower-right diagram, where the intermediate state involves
a filled state, the matrix elements yield instead $(-J_0/2\hbar V)^2 P_{-+}$,
where

$$P_{-+} = \frac{1}{\hbar^2}|\langle s, m_s - 1|S_d^-|s, m_s\rangle|^2 = s(s+1) - m_s(m_s - 1). \qquad (7.32)$$

The sum of the matrix elements for these processes, for given m_s, is

$$\left(\frac{J_0}{2V}\right)^2 (P_{+-}\alpha + P_{-+}\gamma) = \left(\frac{J_0}{2V}\right)^2 (P_{+-}(\alpha + \gamma) + 2m_s\gamma). \qquad (7.33)$$

Because $P_{+-} \neq P_{-+}$, the Fermi distributions do not cancel. The presence of the Fermi distribution function in the spin-flip transition amplitude, which introduces a temperature dependence into the transition
rate, is the fundamental difference between the cases involving a spin-
flip and a non–spin-flip in the intermediate states. The temperature dependence is intrinsically tied to the dynamics between the degenerate

spin levels of the conduction electrons and the impurity. The sum over hole states γ is also generally dependent on the energy of the incoming particle. As in the non–spin-flip case, the part of the above expression proportional to $\mathcal{P} + \gamma$ is divergent because we have ignored the momentum dependence of the matrix elements. With such dependence, this term becomes finite but remains independent of temperature. The sum of the contributions of all four terms in Fig. (7.5) is:

$$T^{(2)}(\mathbf{k}\uparrow \to \mathbf{k'}\uparrow; m_s) = \left(\frac{J_0}{2}\right)^2 \frac{1}{V}(2m_s\,\gamma + (s(s+1) - m_s)(\alpha + \gamma)).$$

(7.34)

The crucial temperature-dependent term in the second-order amplitude, which comes from the terms with an intermediate spin-flip, is the first on the right side of Eq. (7.34). We drop the second term, so that

$$T^{(2)}(\mathbf{k}\uparrow \to \mathbf{k'}\uparrow; m_s) \approx \frac{2m_s}{V}\left(\frac{J_0}{2}\right)^2 \gamma.$$

(7.35)

Converting the sum over $\mathbf{q}$ in γ to an integral and working in the $T = 0$ limit with the free-particle dispersion relation $\epsilon_{\mathbf{k}} = k^2/2m$, we have

$$\gamma = \frac{m}{\pi^2\hbar^3}\int_0^{k_F}\frac{q^2 dq}{k^2 - q^2 - i\eta}.$$

(7.36)

The real part of the integral over momentum is of the type encountered in Chapter 5 in the context of the interacting electron gas:

$$\int_0^{k_F}\frac{q^2 dq}{k^2 - q^2} = -k_F - \frac{k}{2}\ln\left|\frac{k - k_F}{k + k_F}\right|.$$

(7.37)

For $T \neq 0$ but $\ll T_F$, the range of k for thermally excited electrons is given by $|k^2/2m - k_F^2/2m| \lesssim k_B T$; we thus approximate $|k - k_F|$ in the numerator of the logarithm by $mk_B T/k_F$ and let $k = k_F$ everywhere else. The leading $-k_F$-term on the right is a constant, which we drop. Thus,

$$\gamma \approx N(0)\left(\ln\frac{T_F}{T} + \cdots\right).$$

(7.38)

Substituting this expression into Eq. (7.35) and adding in the lowest-order scattering amplitude, we find the amplitude, to within constant terms, in second order to be

$$T(\mathbf{k}\uparrow \rightarrow \mathbf{k}'\uparrow; m_s) = -\frac{J_0 m_s}{2V}\left(1 - J_0 N(0)\ln\frac{T_F}{T} + \cdots\right). \quad (7.39)$$

Note that the second-order correction increases the magnitude of the scattering amplitude for $J_0 < 0$.

By a similar calculation, which is left as an exercise, one can derive the total amplitude,

$$T^{(2)}(\mathbf{k}\uparrow \rightarrow \mathbf{k}'\downarrow; m_s) = -\left(\frac{J_0}{2}\right)^2 \frac{1}{\hbar V}\langle s, m_s + 1|S_d^+|s, m_s\rangle(\alpha - \gamma),$$

$$(7.40)$$

for the four processes shown in Fig (7.6). Adding in the lowest-order amplitude, we find the electron spin-flip amplitude, including the temperature-dependent part in second order:

$$T(\mathbf{k}\uparrow \rightarrow \mathbf{k}'\downarrow; m_s)$$

$$\approx -\frac{J_0}{2\hbar V}\langle s, m_s + 1|S_d^+|s, m_s\rangle\left(1 - J_0 N(0)\ln\frac{T_F}{T} + \cdots\right). \quad (7.41)$$

Note that both the non–spin-flip and spin-flip amplitudes are modified by the same temperature-dependent factor, $(1 - J_0 N(0)\ln(T_F/T))$.

To calculate the scattering rate to third order in J_0, we need simply therefore replace J_0 in Eq. (7.23) by $J_0(1 - J_0 N(0)\ln(T_F/T))$. The scattering rate to next order is thus

$$\Gamma = \Gamma^0\left(1 - 2J_0 N(0)\ln\frac{T_F}{T} + \cdots\right). \quad (7.42)$$

Note that the temperature-dependent correction to the scattering rate is independent of the total spin of the impurity. The logarithmic behavior in the scattering rate gives rise to the logarithmic increase in the resistivity. Resistance to electrical flow in a metal arises from electron scattering processes taking momentum from the electrons; the resistivity is inversely proportional to the time between scattering events. The transport equation that captures this result is the Boltzmann equation, as we discuss in Chapter 10. As we show there, the contribution from phonon scattering scales as T^5. Note that although the Kondo contribution to the scattering time is positive, it *decreases* with increasing temperature. Combining the electron-phonon scattering result with the logarithmic Kondo contribution, we see that the resistivity has the form

$$\rho(T) = aT^5 - bc \ln \frac{T}{T_F}, \tag{7.43}$$

where a and b are positive constants. The result, Eq. (7.43), has a minimum at temperature

$$T_{\min} = \left(\frac{b}{5ac}\right)^{1/5}. \tag{7.44}$$

This expression for the temperature at the minimum, which depends only weakly on the concentration of magnetic impurities, is in excellent agreement with numerous Kondo alloys, most notably Fe in Au [K1964].

Although the scattering rate looks as if it can grow indefinitely as the temperature is lowered, the scattering amplitude is limited by unitarity. We will see this limit once we develop a little more technology to deal with the scattering amplitudes.

7.4   ELECTRON-IMPURITY SCATTERING AMPLITUDES

Since the overall spin of the impurity has little effect on the temperature-dependent terms in the scattering amplitude, we assume from here on that the impurity has spin-$1/2$. Since the interaction Hamiltonian is rotationally invariant and thus conserves total angular momentum, it is often most convenient to analyze the scattering in terms of the total spin of the electron plus impurity. Two spin-$1/2$ particles can have total spin $S = \hbar$ or 0, corresponding to spin triplet or spin singlet, respectively. The spin eigenstates are:

$$|s = 1, m_s = 1\rangle = |e_\uparrow d_\uparrow\rangle,$$

$$|s = 1, m_s = 0\rangle = \frac{1}{\sqrt{2}}(|e_\uparrow d_\downarrow\rangle + |e_\downarrow d_\uparrow\rangle),$$

$$|s = 1, m_s = -1\rangle = |e_\downarrow d_\downarrow\rangle,$$

$$|s = 0, m_s = 0\rangle = \frac{1}{\sqrt{2}}(|e_\uparrow d_\downarrow\rangle - |e_\downarrow d_\uparrow\rangle). \tag{7.45}$$

The two $S = 0$ terms can be written equivalently as

$$|e_\uparrow d_\downarrow\rangle = \frac{1}{\sqrt{2}}(|s = 1, m_s = 0\rangle + |s = 0, m_s = 0\rangle)$$

$$|e_\downarrow d_\uparrow\rangle = \frac{1}{\sqrt{2}}(|s = 1, m_s = 0\rangle - |s = 0, m_s = 0\rangle). \tag{7.46}$$

Rotational invariance (for two spin-1/2 particles) has the lovely consequence that all scatterings are described by just two amplitudes, T_0, the spin-singlet amplitude, and T_1, the spin-triplet amplitude. The scattering amplitude from electron-impurity state $|a\rangle$ to $|b\rangle$ is

$$T_{a \to b} = \sum_{s, m_s} |\langle s, m_s | a \rangle|^2 T_s. \tag{7.47}$$

The scattering $(\mathbf{k}\sigma) + (d\sigma) \to (\mathbf{k}'\sigma) + (d\sigma)$ is purely spin triplet, and thus

$$T_{(\mathbf{k}\sigma)+(d-\sigma) \to (\mathbf{k}'\sigma)+(d-\sigma)} = T_1. \tag{7.48}$$

On the other hand, the scattering $(\mathbf{k}\sigma) + (d - \sigma) \to (\mathbf{k}'\sigma) + (d - \sigma)$ proceeds through the singlet and triplet channels with equal weights. From Eqs. (7.46) and (7.47),

$$T_{(\mathbf{k}\sigma)+(d-\sigma) \to (\mathbf{k}'\sigma)+(d-\sigma)} = \frac{1}{2}(T_1 + T_0). \tag{7.49}$$

Similarly,

$$T_{(\mathbf{k}\sigma)+(d-\sigma) \to (\mathbf{k}'-\sigma)+(d\sigma)} = \frac{1}{2}(T_1 - T_0). \tag{7.50}$$

The minus sign on the right results from the opposite signs of the singlet state in the two parts of Eq. (7.46).

The results of the previous subsection for the scattering amplitudes to second order are

$$T(\mathbf{k}\uparrow \to \mathbf{k}'\uparrow; m_s) = -\frac{J_0}{2V}\left(m_s - \frac{J_0}{8}[3(\alpha + \gamma) - 4m_s(\alpha - \gamma)]\right) \tag{7.51}$$

and

$$T(\mathbf{k}\uparrow \to \mathbf{k}'\uparrow; m_s) = -\frac{J_0}{2V}\left(1 + \frac{1}{2}(\alpha - \gamma)\right). \qquad (7.52)$$

Thus, from Eqs. (7.49) and (7.50), we derive the results for the singlet

$$T_0 = \frac{3J_0}{4V}\left[1 + \frac{3J_0}{4}\left(\alpha - \frac{1}{3}\gamma\right)\right] \qquad (7.53)$$

and triplet

$$T_1 = -\frac{J_0}{4V}\left[1 - \frac{J_0}{4}(\alpha + 5\gamma)\right] \qquad (7.54)$$

amplitudes to second order.

The lowest-order amplitudes in these results can be readily understood by rewriting the interaction term, $-J_0/\hbar^2 V\,\mathbf{S}_e \cdot \mathbf{S}_d$, in terms of the total spin of the electron and the impurity. Using the identity for spin-$1/2$ particles,

$$\mathbf{S}^2 = (\mathbf{S}_e + \mathbf{S}_d)^2 = \frac{3\hbar^2}{2} + 2\mathbf{S}_e \cdot \mathbf{S}_d, \qquad (7.55)$$

we see immediately that

$$\mathbf{S}_e \cdot \mathbf{S}_d = -\frac{3\hbar^2}{4} + \frac{1}{2}\mathbf{S}^2. \qquad (7.56)$$

In a singlet state, $\mathbf{S}_e \cdot \mathbf{S}_d = -3\hbar^2/4$, and thus the bare interaction is $3J_0/4V$, as in Eq. (7.53); similarly, in a triplet state, $\mathbf{S}_e \cdot \mathbf{S}_d = \hbar^2/4$, so the bare interaction is $-J_0/4V$, as in Eq. (7.54).

At this stage we can replace the bare interactions in the second-order terms in Eqs. (7.54) and (7.53) by the corresponding full-scattering amplitudes to derive the T-matrix equations,

$$T_1 = -\frac{J_0}{4V}[1 + VT_1(\alpha + 5\gamma)],$$

$$T_0 = \frac{3J_0}{4V}\left[1 + VT_0\left(\alpha - \frac{1}{3}\gamma\right)\right], \qquad (7.57)$$

which give the total spin-scattering amplitudes to all orders in the coupling. These equations are algebraic; we write their solutions in the forms

$$T_1 = -\frac{J_0/4V}{1 + J_0\gamma + (J_0/4)(\alpha + \gamma)} \tag{7.58}$$

and

$$T_0 = \frac{3J_0/4V}{1 + J_0\gamma - (3J_0/4)(\alpha + \gamma)}. \tag{7.59}$$

We can now see how unitarity limits the total scattering amplitude and prevents the logarithmic result, Eq. (7.42), from growing indefinitely. Let us consider $\mathbf{k}$ just outside the Fermi surface. We use Dirac's identity to do integrals with a small imaginary part in the denominator

$$\frac{1}{x + i\eta} = P\frac{1}{x} - i\pi\delta(x), \tag{7.60}$$

where P indicates that the principal value of the integral must be taken. Then

$$\alpha = P\int \frac{d\mathbf{k}}{(2\pi)^3} \frac{1 - f_\mathbf{q}}{\epsilon_\mathbf{k} - \epsilon_\mathbf{q}} - i\pi N(0). \tag{7.61}$$

On the other hand, since the denominator of the integral for γ does not vanish for $\mathbf{k}$ outside the Fermi surface, γ has no imaginary part. By dropping the real part of the denominators in Eqs. (7.58) and (7.59), we readily derive the unitarity bound,

$$|T_s| \le \frac{1}{\pi N(0)V}, \tag{7.62}$$

where $s = 0, 1$. Using the techniques of elementary scattering theory, one can write the singlet- and triplet-scattering amplitudes, in the case of isotropic scattering, in terms of the s-wave phase shifts δ_s as

$$T_s = -\frac{1}{\pi N(0)V} e^{i\delta_s} \sin \delta_s. \tag{7.63}$$

From this form, we see that the unitary bound is realized for $\delta_s = (n + 1/2)\pi$, that is, a scattering resonance.

The important terms in the Kondo problem are the temperature-dependent terms in the denominators of the scattering amplitudes,

Eqs. (7.58) and (7.59). If we define the effective coupling J_{eff} by

$$J_{eff} = \frac{J_0}{1 + J_0\gamma} \approx \frac{J_0}{1 + J_0 N(0)\ln(T_F/T)}, \qquad (7.64)$$

we see that the singlet and triplet amplitudes have the structures

$$T_0 = \frac{3J_{eff}/4V}{1 - (3J_{eff}/4)(\alpha + \gamma)} \approx \frac{3J_0/4V}{1 + J_0 N(0)\ln(T_F/T)} \qquad (7.65)$$

and

$$T_1 = -\frac{J_{eff}/V}{1 + (J_{eff}/4)(\alpha + \gamma)} \approx -\frac{J_0/V}{1 + J_0 N(0)\ln(T_F/T)}. \qquad (7.66)$$

The amplitudes are those for lowest-order scattering, with the effective coupling J_{eff} replacing the bare coupling J_0. Comparing with Eq. (7.63), we see that the singlet and triplet phase shifts are given by

$$\tan \delta_0 = -\frac{3\pi}{4}J_{eff}N(0), \quad \tan \delta_1 = \frac{\pi}{4}J_{eff}N(0). \qquad (7.67)$$

Since $J_{eff} < 0$, the singlet phase shift is positive, and, hence, this channel is attractive, with a tendency toward formation of a bound state. The triplet phase shift is negative, indicating repulsion and no bound state formation. With increasing magnitude of J_{eff}, the singlet phase shift approaches the resonance point $\pi/2$.

This value of the singlet phase shift is particularly illuminating. From the simplified version of Friedel's sum-rule that we derived in Chapter 6, Eq. (6.42), a value of $\pi/2$ for the singlet phase shift at the Fermi level implies that the occupancy on the impurity is $1/2$. To put the Kondo singlet state in the context of the Anderson impurity, we focus on an impurity ion with ionic charge Z. Hence, Z is the total charge that must be screened by the conduction electrons. If the impurity is a transition metal ion with d-orbital symmetry, the condition for complete screening (the charge neutrality condition) is

$$Z = 5(\langle n_{d\sigma} \rangle + \langle n_{d\bar{\sigma}} \rangle) = \frac{5}{\pi}(\delta_{d\sigma} + \delta_{d-\sigma}). \qquad (7.68)$$

In the absence of a magnetic field, the magnetic moment of the impurity spin can be oriented parallel or antiparallel to the conduction electron spin. Hence, there is no uniquely defined phase shift for the conduction electron scattering. We are free to choose $\delta_{d\sigma}$ or $\delta_{d-\sigma}$ but with equal probability. At $T = 0$, however, a well-defined single phase shift exists, $\delta = (\delta_{d\sigma} + \delta_{d-\sigma})/2$. Consequently, we can rewrite the charge neutrality condition as

$$\delta = \frac{\pi Z}{10}. \tag{7.69}$$

The resultant value for the T-matrix at the Fermi level,

$$T_{\mathbf{kk}}(\epsilon = 0 + i\delta) = \frac{5}{\pi N(0)V}\left[\exp\left(i\,\frac{\pi Z}{10}\right)\sin\left(\frac{\pi Z}{10}\right)\right], \tag{7.70}$$

follows directly from Eq. (6.38). Taking the imaginary part of this expression results in the density of states,

$$\rho_{d\sigma}(0) = \frac{1}{\pi\Delta}\sin^2\left(\frac{\pi Z}{10}\right), \tag{7.71}$$

at the impurity (see Eq. 6.37) at the Fermi level. This value of the density of states arises from the charge neutrality condition. Comparing this value with that arising from the Hartree-Fock approximation (Eq. 6.48),

$$\rho_{d\sigma}^{HF}(0) = \frac{1}{\pi\Delta}\frac{1}{1 + [(\epsilon_d + Un_c)/\Delta]^2}, \tag{7.72}$$

we find complete disagreement. In fact, the density of states at the Fermi level arising from the charge neutrality condition exceeds the Hartree-Fock value. To illustrate, for $Z = 5$, the charge neutrality condition leads to an impurity density of states at the Fermi level of $1/(\pi\Delta)$. In the Hartree-Fock approximation, the impurity density of states is always less than $1/(\pi\Delta)$ except at the impurity energy levels where it identically acquires this value. The discrepancy between Eqs. (7.71) and (7.72) arises because spin-flip scattering is absent in the Hartree-Fock procedure, and, hence, it is incapable of describing the formation of the singlet ground state at $T = 0$.

As Grüner and Zawadowski [GZ1974] have argued, there are two ways to resolve the large value of the density of states at the Fermi level arising from the charge neutrality condition. Either a single maximum occurs at the Fermi level or at low temperatures an additional peak appears (superimposed on the Hartree-Fock density of states) as a result of the correlation effects arising from multiple scattering events. The first option would require a reorganization of the impurity levels on an energy scale on the order of the on-site Coulomb repulsion, U. As U is typically of order 1eV, such a reorganization would occur at a temperature comparable to 10^4 K. This option can be ruled out on experimental grounds; no reorganization of the impurity density of states occurs at such energy scales. The only option remaining is that a modification occurs at an energy scale on the order of T_k. Consequently, superimposed on the Hartree-Fock density of states is an additional resonance peak whose width must be on the order of T_k. As a result, the Kondo resonance causes a modification of the density of states at temperatures distinct from the Hartree-Fock energy scales. The precise form of the density of states in response to the formation of the Kondo resonance is shown in Fig. (7.3). The pinning of the Kondo resonance at the Fermi energy arises entirely from the fact that spin-flip scattering costs zero energy. It is for this reason that the Kondo resonance peak is sometimes called the *zero-bias anomaly*. Extensive calculations [L1981] reveal that the Kondo resonance grows logarithmically down to the Kondo temperature. On average, for a spin-$1/2$ impurity, the Kondo resonance is occupied by a single electron of opposite spin to account for the spin singlet state at the impurity. However, the multiple powers of the density of states that appear in the T-matrices indicate that all the electrons within T_k of the Fermi energy contribute to the resonance.

The effective coupling J_{eff} sums the leading logarithms in all orders [AYH1970]. Note that J_{eff} diverges at the temperature

$$T_k = T_F e^{1/N(0)J_0}. \tag{7.73}$$

As a consequence, the present treatment is valid only above T_k. To investigate the physical consequences of this singularity, we look first at a simple variational approach to the ground state of the Kondo problem.

7.5   KONDO TEMPERATURE

In the presence of an antiferromagnetic interaction, a bound singlet state forms between the impurity spin and electrons in the conduction band. The simplest model of this phenomena, first suggested by Yosida [Y1966], is to assume that only a single conduction electron forms the total spin singlet state on the impurity, with the other electrons remaining in the Fermi sea, and to retain only the interactions between the impurity and this conduction electron. The role of the Fermi sea in the model is only to block states below the Fermi sea to this electron.

A singlet, $S = 0$, state d_σ, is given by the linear combination of spins,

$$|S = 0\rangle = \frac{1}{\sqrt{2}}(|e_\uparrow, d_\downarrow\rangle - |e_\downarrow, d_\uparrow\rangle). \qquad (7.74)$$

The interaction term between the spins is proportional to $\mathbf{S}_e \cdot \mathbf{S}_d$. In a singlet state of the conduction electron and impurity spins, $\mathbf{S}_e \cdot \mathbf{S}_d = 3\hbar^2/4$, and the term in the interaction $\Psi_{\mathbf{k}'}^\dagger \mathbf{S} \Psi_{\mathbf{k}} \cdot \Psi_d^\dagger \mathbf{S} \Psi_d$ simplifies to $-(3/4)\Psi_{\mathbf{k}'}^\dagger \Psi_{\mathbf{k}} \Psi_d^\dagger \Psi_d$; for a single impurity $\Psi_d^\dagger \Psi_d = 1$. The effective Hamiltonian for this problem is thus

$$H_K^{\text{eff}} = \sum_\sigma \left(\sum_k \epsilon_k a_{\mathbf{k}\sigma}^\dagger a_{\mathbf{k}\sigma} + \sum_{k,k'} \frac{3}{4} J_{\mathbf{k}\mathbf{k}'} a_{\mathbf{k}'\sigma}^\dagger a_{\mathbf{k}\sigma} \right), \qquad (7.75)$$

where $\mathbf{k}$ and $\mathbf{k}'$ are restricted to be outside the Fermi sea. As a consequence of this restriction, the effective Hamiltonian does not contain the constant energy of the electrons in the Fermi sea, a term with no effect here.

To construct the full state of the system, we note that the operator

$$b_{\mathbf{k}}^\dagger = \frac{1}{\sqrt{2}}(a_{\mathbf{k}\uparrow}^\dagger \Psi_{d\downarrow}^\dagger - a_{\mathbf{k}\downarrow}^\dagger \Psi_{d\uparrow}^\dagger) \qquad (7.76)$$

creates a spin singlet state of the conduction electron of momentum $\mathbf{k}$ ($k > k_F$) with the impurity spin. The wave function of the singlet state plus the Fermi sea has the general form

$$|\Phi\rangle = \sum_{k>k_F} \alpha_{\mathbf{k}} b_{\mathbf{k}}^\dagger |F\rangle, \qquad (7.77)$$

where $\alpha_{\mathbf{k}}$ is the amplitude for the conduction electron in the singlet state to have momentum $\mathbf{k}$, and the Fermi sea is given by $|F\rangle = \prod_{\mathbf{k}<k_F} a_{\mathbf{k}\uparrow}^\dagger a_{\mathbf{k}\downarrow}^\dagger |0\rangle$, where $|0\rangle$ is the vacuum. The amplitude for forming a singlet state is given by the overlap of the filled Fermi sea with the state $b_{\mathbf{k}}|\Phi\rangle$:

$$\langle F|b_{\mathbf{k}}|\Phi\rangle = \sum_{k'>k_F} \alpha_{\mathbf{k}'} \langle F|b_{\mathbf{k}} b_{\mathbf{k}'}^\dagger |F\rangle = \alpha_{\mathbf{k}}, \qquad (7.78)$$

where $k > k_F$. The state $|\Phi\rangle$, while approximate, captures key features of the interaction of the spin moment with the conduction electrons.

To determine the energy eigenstates of this system, we construct the eigenvalue equation $H_K^{\text{eff}}|\Phi\rangle = E|\Phi\rangle$. From Eqs. (7.75) and (7.77), we have

$$H_K^{\text{eff}}|\Phi\rangle = \sum_{\mathbf{k}} \epsilon_{\mathbf{k}} \alpha_{\mathbf{k}} b_{\mathbf{k}}^\dagger |F\rangle + \sum_{\mathbf{k},\mathbf{k}'} \frac{3}{4} J_{\mathbf{k}\mathbf{k}'} \alpha_{\mathbf{k}'} b_{\mathbf{k}'}^\dagger |F\rangle = E|\Phi\rangle. \quad (7.79)$$

Using Eq. (7.78), we then obtain the eigenvalue equation

$$\epsilon_{\mathbf{k}} \alpha_{\mathbf{k}} + \frac{3}{4} \sum_{k'>k_F} J_{\mathbf{k}\mathbf{k}'} \alpha_{\mathbf{k}'} = E\alpha_{\mathbf{k}}, \quad (7.80)$$

where $k > k_F$.

As in the previous subsection, we write $J_{\mathbf{k}\mathbf{k}'} \approx J_0/V$. With this approximation, the eigenvalue equation becomes separable. We find

$$\alpha_{\mathbf{k}} = \frac{3J_0}{4V} \frac{1}{E - \epsilon_{\mathbf{k}}} \sum_{\mathbf{k}'} \alpha_{\mathbf{k}'}, \quad (7.81)$$

from which a sum over $\mathbf{k}$ ($k > k_F$) yields the equation for E:

$$1 = \frac{3J_0}{4V} \sum_{k>k_F} \frac{1}{E - \epsilon_{\mathbf{k}}}. \quad (7.82)$$

This type of equation is characteristic of the formation of a bound state. We will meet a similar equation in Chapter 11 in the Cooper pair problem, a prelude to understanding superconductivity. For $J_0 < 0$, the equation must admit a solution with $E < \epsilon_F$. To solve for E, we convert the sum to an integral, assuming that the density of states is a constant, $N(0)$, in an energy interval, $\epsilon_F < \epsilon < D$, where D is the upper energy in the conduction band. Then, the energy of the singlet state is

$$E = \epsilon_F - (D - \epsilon_F) e^{4/3J_0 N(0)} \simeq \epsilon_F - D e^{4/3J_0 N(0)}. \quad (7.83)$$

Because E is less than the Fermi energy, we interpret

$$E_b = D e^{4/3J_0 N(0)} \quad (7.84)$$

as the binding energy of the singlet state. As we see, in the presence of a weak antiferromagnetic interaction, the Fermi sea is unstable to the formation of a singlet state with a localized impurity.

The singlet state breaks up for temperatures $T \gtrsim E_b / k_B$. The Kondo temperature in this picture emerges as the temperature scale

$$T_k = \frac{D}{k_B} e^{4/3J_0 N(0)}, \tag{7.85}$$

below which the singlet state is stable. Although this derivation of T_k is not exact, Eq. (7.85) provides an excellent approximation to the experimentally observed Kondo temperature. In general, the Kondo temperature is only a function of the density of states of the host metal and the exchange interaction with the metal ion. For Fe-impurities in Cu, the Kondo temperature is 6 K, considerably smaller than the bandwidth.

We can, in fact, understand the Yosida result directly in terms of the singlet channel-scattering amplitude, Eq. (7.59), derived in the previous subsection. The assumption that the only function of the electron sea is to block states below the Fermi surface is equivalent to neglecting the term γ in the denominator of Eq. (7.59). For E below the Fermi sea, α has no imaginary part, and we see that the scattering amplitude has a pole when $(3J_0/4)\alpha = 1$, which is precisely Yosida's equation (7.82).

Experimentally, the Kondo effect is observed only for an antiferromagnetic interaction. The analogous triplet trial wave function in the ferromagnetic case corresponding to Eq. (7.77) also leads to a bound state. This unphysical result is an indication that the simple variational approach is insufficient to describe the true ground state of the Kondo model. A key piece of physics that is missing is the collective many-body quenching of the local moment by all the electrons within T_k of the Fermi surface. This quenching can be thought of as the formation of a cloud of electrons around the impurity that screens out the spin moment. The existence of the low-energy scale, T_k, implies that a macroscopically large length scale, $\xi_k = \hbar v_F / T_k$, might describe the screening cloud. For Fe in Cu, ξ_k is on the order of 2000Å. The first correct attempt to include the many-body effects that lead to the ground state of the Kondo model is the "poor man's scaling" approach of Anderson, to which we now turn.

7.6   POOR MAN'S SCALING

The behavior of a many-body system at long wavelengths depends on interactions of the microscopic degrees of freedom on all length scales.

In such a system, the short-wavelength degrees of freedom determine the effective macroscopic parameters of the long-wavelength theory. A very familiar example is hydrodynamics, which describes fluids on length and time scales that are large compared with particle mean-free paths and scattering times, and in which only effective parameters, such as the sound velocity and viscosities, enter; these parameters are determined by the microscopic physics. The goal of a scaling analysis is to formulate a series of descriptions of a system valid on longer and longer wavelength scales with more and more of the higher momentum scales integrated over. Once integrated over, the degrees of freedom on higher momentum scales no longer appear explicitly in the problem. Thinning the degrees of freedom results in a simplified effective model from which the low-energy physics can be extracted. In this way, one formulates recursive equations for the physically relevant coupling constants as various degrees of freedom are integrated out.

The simplest scaling analysis of the Kondo problem successively eliminates high-energy excitations as intermediate states in perturbation theory. At each level, the spin-spin coupling constant, $J_{k,k'}$, is modified. To understand this method, we work only to second-order in the effective spin-spin coupling, whence the name *poor man's scaling*. While higher-order terms in the perturbation theory do contribute to scaling, they do not lead to qualitatively new features [SZ1974].

Before presenting Anderson's scaling analysis of the isotropic Kondo model, let us first quote its key result. As the temperature is varied, the number of states explored by the electrons increases, and the effective coupling, which we denote in this subsection as J, obeys the equation

$$\frac{dJ}{d\ln(T_F/T)} = -J^2 N(0), \qquad (7.86)$$

with solution Eq. (7.64). This expression correctly predicts the essential physics of the Kondo problem, that smaller and smaller energy excitations (dominant at low temperatures) lead, through a bootstrapping process, to a divergence of the exchange interaction and the binding of an antiparallel electron at the local moment [A1978]. Although Eq. (7.86) does not remove the divergence found earlier by Kondo, it does serve to illuminate how the exchange coupling increases in magnitude. The process by which this increase occurs is analogous to the confinement of color degrees of freedom in quark physics [A1978].

There are many ways to develop the scaling ideas, some simple and some complicated. We will present a streamlined version of the simple

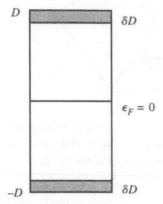

Figure 7.7 The particle and hole states to be eliminated in the "poor man's scaling" approach. The width of states to be removed is δD.

approach [H1993]. We consider an electronic band with $-D < \epsilon_k < D$. As illustrated in Figure (7.7), we proceed by eliminating a narrow band of particle and hole states from the lower and upper regions of the band, respectively. The width of the states to be eliminated is δD. The basic idea will be to consider the scattering processes that enter the second-order transition amplitude and to determine how they change when the bandwidth is decreased by an infinitesimal amount, δD. Our hope is that as the bandwidth is reduced, we will be able to find a set of equations that relate the new exchange couplings to the original ones.

For simplicity, we will treat an $S = 1/2$ impurity but, for the sake of generality, we consider an anisotropic spin-flip Hamiltonian

$$
H' = -\frac{1}{2\hbar V} \sum_{k',k'',\sigma} [J_z S_d^z (a_{k'\uparrow}^\dagger a_{k''\uparrow} - a_{k'\downarrow}^\dagger a_{k''\downarrow})
$$
$$
+ J_+ S_d^+ a_{k'\downarrow}^\dagger a_{k''\uparrow} + J_- S_d^- a_{k'\uparrow}^\dagger a_{k''\downarrow}],
$$

in which the spin-flip ($J_\pm$) and longitudinal (J_z) parts have different interaction strengths. Our goal is to simplify the second-order perturbative terms and show that they lead to a renormalization of the bare values of J_z and $J_\pm$. We consider first the two spin-flip events shown in Fig. (7.8). Because a two-spin-flip process leaves the impurity spin unchanged, we anticipate that the two-spin-flip terms will lead to a renormalization of the longitudinal coupling J_z. The precise form of the correction to J_z can be obtained by performing the scaling analysis on each of the two-spin-flip diagrams. Consider the first

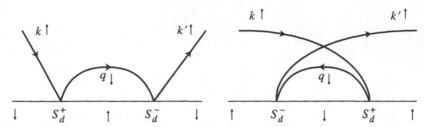

Figure 7.8 Second-order two-spin-flip diagrams in which the intermediate state involves either a particle or a hole. The solid straight line indicates the spin of the impurity and the curved lines the electron.

diagram,

$$J_+ J_- \sum_{\mathbf{q}} \frac{S_d^-}{2\hbar V} a_{\mathbf{k}'\uparrow}^\dagger a_{\mathbf{q}\downarrow} (E - \widehat{H}_0)^{-1} \sum_{\mathbf{q}'} \frac{S_d^+}{2\hbar V} a_{\mathbf{q}'\downarrow}^\dagger a_{\mathbf{k}\uparrow}. \qquad (7.87)$$

To implement the scaling procedure, we restrict the sum on $\mathbf{q}'$ over the states within δD from the top of the band. These states are created with the excitation $a_{\mathbf{q}'\downarrow}^\dagger$. Because the band edge states are originally unoccupied, we must set $a_{\mathbf{q}} a_{\mathbf{q}'}^\dagger = \delta_{\mathbf{q},\mathbf{q}'}$.

In simplifying Eq. (12.1), we are not free to move the rightmost creation and annihilation operators through the operator $1/(E - \widehat{H}_0)$, because they do not commute with $\widehat{H}_0$. For any operator $\widehat{A}$, such that $[\widehat{H}_0, \widehat{A}] = b\widehat{A}$, where b is a c-number, then $\widehat{H}_0^n \widehat{A} = \widehat{A}(b + \widehat{H}_0)^n$. As a consequence,

$$\frac{1}{E - \widehat{H}_0} \widehat{A} = \frac{\widehat{A}}{E} \sum_{n=0}^{\infty} \frac{(\widehat{H}_0 + b)^n}{E^n} \qquad (7.88)$$

$$= \widehat{A} \frac{1}{E - b - \widehat{H}_0}. \qquad (7.89)$$

Using the fact that $[\widehat{H}_0, a_{\mathbf{q}'\downarrow}^\dagger a_{\mathbf{k}\uparrow}] = a_{\mathbf{q}'\downarrow}^\dagger a_{\mathbf{k}\uparrow}(\epsilon_{\mathbf{q}'} - \epsilon_{\mathbf{k}})$, we simplify the matrix element in Eq. (12.1) to

$$J_+ J_- N(0) V |\delta D| \frac{S_d^- S_d^+}{4V\hbar^2} a_{\mathbf{k}'\uparrow}^\dagger a_{\mathbf{k}\uparrow} (E - \epsilon_{\mathbf{q}} + \epsilon_{\mathbf{k}} - \widehat{H}_0)^{-1}. \qquad (7.90)$$

Because of the $a_{\mathbf{k}'\uparrow}^\dagger a_{\mathbf{k}\uparrow}$-dependence, we see clearly that the two-spin-flip term changes the value of the bare transverse exchange interaction. To extract the exact coefficient, we use $S_d^- S_d^+ = \hbar^2/2 - \hbar S_d^z$ and set

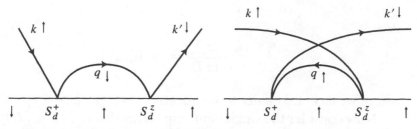

Figure 7.9 Second-order one-spin-flip diagrams in which the intermediate state involves either a particle or a hole. The solid straight line indicates the spin of the impurity and the curved lines the electron.

$\epsilon_{\mathbf{q}} = D$. Further, if we measure E from the bottom of the band, then we can ignore the $\widehat{H}_0$ dependence in the denominator of Eq. (8.2). The matrix element simplifies to

$$J_+J_-N(0)\frac{|\delta D|}{4\hbar V}\left(\frac{1}{2} - \frac{S_d^z}{\hbar}\right)a^\dagger_{\mathbf{k'}\uparrow}a_{\mathbf{k}\uparrow}(E - D + \epsilon_{\mathbf{k}})^{-1}. \qquad (7.91)$$

We can now evaluate the second diagram in Fig. (7.9) in the same way. Here the intermediate states are particle states; hence, we set $\epsilon_{\mathbf{q}} = -D$ and obtain

$$J_+J_-N(0)\frac{|\delta D|}{4\hbar V}\left(\frac{1}{2} + \frac{S_d^z}{\hbar}\right)a_{\mathbf{k}\uparrow}a^\dagger_{\mathbf{k'}\uparrow}(E - D - \epsilon_{\mathbf{k'}})^{-1}. \qquad (7.92)$$

In deriving this result, we used the relationship $S_d^+ S_d^- = \hbar^2/2 + \hbar S_d^z$.

The remaining scaling equations involving two-spin-flips we obtain by including the contribution from $\mathbf{k}\downarrow$ and $\mathbf{k'}\downarrow$. Scattering between down-spin states will generate matrix elements proportional to $a_{\mathbf{k}\downarrow}a^\dagger_{\mathbf{k'}\downarrow}$. This term will contribute to the ($m_s = -1$)-spin component of the S_z operator. We now define a new longitudinal interaction $J_z \to J_z + \delta J_z$. Noting that the coefficient of the bare interaction in Eq. (7.87) is $1/2\hbar V$, we find that Eqs. (7.91) and (7.92), combined with the down-spin equations, lead to

$$\delta J_z = \frac{1}{2}J_+J_-N(0)|\delta D|\left[\frac{1}{E - D + \epsilon_{\mathbf{k}}} + \frac{1}{E - D - \epsilon_{\mathbf{k'}}}\right] \qquad (7.93)$$

as the correction to J_z. The renormalization of J_z is energy-dependent. Consider now excitations low in energy relative to D. We also evaluate $\epsilon_{\mathbf{k}}$ and $\epsilon_{\mathbf{k'}}$ at the Fermi level, $\epsilon_F = 0$. The J_z scaling equation simplifies to

$$\frac{dJ_z}{d\ln D} = N(0)J_{\pm}^2, \tag{7.94}$$

where we have taken δD to be negative and set $J_+ = J_- = J_{\pm}$.

Now consider the single-spin-flip terms shown in Fig. (7.9). The first diagram is of the form

$$-J_+J_z \sum_{\mathbf{q}} \frac{S_d^z}{2\hbar V} a_{\mathbf{k}'\downarrow}^\dagger a_{\mathbf{q}\downarrow}(E - \widehat{H}_0)^{-1} \sum_{\mathbf{q}'} \frac{S_d^+}{2\hbar V} a_{\mathbf{q}'\downarrow}^\dagger a_{\mathbf{k}'\uparrow}. \tag{7.95}$$

The minus sign in front of this equation arises from the minus sign that accompanies the down-spin component of S_d^z. Using the same techniques as in the two-spin-flip case, we reduce this matrix element to

$$\frac{-J_+J_z N(0)|\delta D|S_d^+ a_{\mathbf{k}'\downarrow}^\dagger a_{\mathbf{k}\uparrow}}{8\hbar V(E - D + \epsilon_{\mathbf{k}})}, \tag{7.96}$$

where we have used the fact that $S_d^z S_d^+ = \hbar S_d^+/2$. Hence, the renormalization of the spin-flip interaction involves the longitudinal component of J. Analogously, the other diagram in Fig. (7.9) has a value of

$$\frac{J_+J_z N(0)|\delta D|S_d^+ a_{\mathbf{k}\uparrow} a_{\mathbf{k}'\downarrow}^\dagger}{8\hbar V(E - D - \epsilon_{\mathbf{k}'})}. \tag{7.97}$$

Additional contributions of this form can be obtained by interchanging the order of S_d^z and S_d^+. In this case, the identity $S_d^+ S_d^z = -\hbar S_d^+/2$ should be used. The contribution here will be identical to that in Eqs. (7.96) and (7.97).

Similarly, we combine the single-spin-flip results with those for scattering of a spin-down electron. Defining a new transverse interaction $J_{\pm}/2 \to (J_{\pm} + \delta J_{\pm})/2$ and using the results of Eqs. (7.96) and (7.97) produces the scaling equation

$$\delta J_{\pm} = \frac{1}{2}J_{\pm}J_z N(0)|\delta D|\left[\frac{1}{E - D + \epsilon_{\mathbf{k}}} + \frac{1}{E - D - \epsilon_{\mathbf{k}'}}\right] \tag{7.98}$$

for the transverse components of the exchange interaction. Here again, the scaling of the transverse coupling constant is energy-dependent. As in the longitudinal case, we consider excitations low in energy relative to D and evaluate $\epsilon_{\mathbf{k}}$ and $\epsilon_{\mathbf{k}'}$ at the Fermi level $\epsilon_F = 0$. The scaling equation simplifies to

$$\frac{dJ_\pm}{d \ln D} = N(0)J_z J_\pm. \tag{7.99}$$

Taking the ratio of Eqs. (7.94) and (7.99) and integrating, we see that the solutions are a family of hyperbolae determined by

$$J_z^2 - J_\pm^2 = \kappa, \tag{7.100}$$

where κ is a constant. To extract the physical consequences of these scaling equations, it is advantageous to substitute Eq. (7.100) into the scaling equation for $J_\pm$. For the ferromagnetic and antiferromagnetic cases, we find that

$$\frac{dJ_\pm}{d|\ln D|} = \begin{cases} -2\rho J_\pm \sqrt{\kappa + J_\pm^2} & J_z > 0 \quad \text{ferromagnetic} \\ 2\rho J_\pm \sqrt{\kappa + J_\pm^2} & J_z < 0 \quad \text{antiferromagnetic}. \end{cases} \tag{7.101}$$

The physical consequences of the scaling analysis are now clear. Because the right-hand side of Eq. (7.101) is negative for $J_z > 0$, the scaling equations dictate that $J_\pm \to 0$ in this case. As a result, in the ferromagnetic case, the transverse exchange coupling becomes weaker and weaker as the bandwidth is reduced. However, in the antiferromagnetic case, the right-hand side of Eq. (7.101) is positive and the scaling equations dictate that $|J_\pm| \to \infty$. This is the strong coupling regime in which the Kondo effect occurs. The exchange coupling grows in magnitude upon a reduction in the bandwidth. It is this bootstrapping of the exchange interaction to larger values as the bandwidth decreases that leads to the formation of a singlet state at a Kondo impurity.

The flow diagrams shown in Fig. (7.10) lay plain the scaling analysis. The hyperbola is the result for anisotropic coupling with $|J_+| > |J_z|$, and the straight lines are in the limit of isotropic coupling, when the solution collapses to $|J_+|^2 = J_z^2$. The arrow points in the direction of decreasing D. The curves for positive J_z (for a ferromagnetic coupling) show that J_z and $|J_+|$ flow as D goes to zero, whereas the curves for negative J_z (for antiferromagnetic coupling) show that both couplings diverge as D goes to zero. Both the isotropic and anisotropic limits exhibit the same physics.

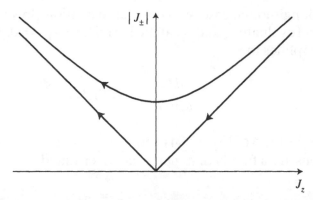

Figure 7.10 Scaling trajectories for the Kondo model using Anderson's second-order perturbation theory argument. The hyperbola is for anisotropic coupling, and the straight lines are for isotropic coupling. The region with $J_z > 0$ corresponds to ferromagnetic coupling and the region where $J_z < 0$ to antiferromagnetic coupling.

In the isotropic limit, $J_{\pm}(D) = J_z(D) = J(D)$, and the scaling equations reduce to the single equation

$$\frac{dJ}{d \ln D} = J^2 N(0), \qquad (7.102)$$

which has the solution

$$J(D) = \frac{J(D_0)}{1 - J(D_0)N(0) \ln (D/D_0)}, \qquad (7.103)$$

where D_0 is the initial bandwidth. Physically, we are interested in the situation that the effective bandwidth is in the range of states that are explored at finite temperature, of width $k_B T$ about the Fermi surface. Thus, we may set, to logarithmic accuracy, $D/D_0 = k_B T/\epsilon_F = T/T_F$. With this identification, Eq. (7.103) is completely equivalent to Eq. (7.64).

Let us turn to extracting the Kondo temperature from the scaling equations. In doing so, it is important to keep in mind that the Kondo temperature is a physical quantity and, hence, should not depend on the bandwidth D that we choose in our model description; that is, T_K should be scale invariant,

$$D \frac{\partial T_K}{\partial D} = 0. \qquad (7.104)$$

As a result, all physical quantities, such as the resistivity and the susceptibility, will depend on the ratio T/T_K (as well as have less singular behavior on T/T_F), but not on D.

If we define, in the isotropic case, the dimensionless coupling constant $g = N(0)J$, the scaling equation becomes

$$\frac{dg}{d \ln D} \equiv \beta(g) = g^2. \tag{7.105}$$

The logarithmic derivative of the coupling constant is generally referred to as the β-*function* after Gell-Mann and Low [GL1954]; we will meet this concept again in Chapter 12 when we discuss localization. On dimensional grounds, the Kondo temperature must have the form $k_B T_K = D y(g)$, where D sets the energy scale. To determine y, we substitute this form for T_K into the scale-invariant condition, Eq. (7.104), and obtain the linear differential equation,

$$y(g) + \frac{\partial y}{\partial g} \frac{\partial g}{\partial \ln D} = 0. \tag{7.106}$$

The solution is simply

$$y(g) \sim e^{1/g}. \tag{7.107}$$

Hence, the Kondo temperature is given by $k_B T_K = D \exp(1/(N(0)J))$, consistent with the structure that we saw earlier (Eq. 7.85).

The complete scaling structure of the Kondo problem was determined by Wilson in his key renormalization group analysis [W1975]. Wilson showed that as the energy scale goes to zero, the distribution of eigenstates becomes roughly equivalent to those of a system in which $J \to -\infty$. Hence, a system initially having a small exchange interaction crosses over smoothly, as the energy scale decreases, to one with a negatively diverging interaction. It is the divergence of the interaction that leads to the crossover to the singlet ground state in the Kondo problem. The divergence in the Wilson renormalization group approach occurs, however, at $T = 0$, not at a finite T_K. The divergence here at T_K is an artifact of the leading logarithm summation in perturbation theory.

This pretty much completes our story on the Kondo problem. However, more recent theoretical work includes: the Fermi liquid picture of Nozières [N1974] below T_K; the exact Bethe ansatz solution [TW1983, AFL1983] to the Anderson and Kondo models; and the conformal field-theoretic approach [AL1991, BA1996]. The inclusion of electron

correlations [LT1992] leads to a Kondo temperature that scales algebraically with the exchange coupling, as opposed to exponentially, as
discussed here; the power is proportional to the strength of the interactions among the conduction electrons. In addition, a Kondo singlet state
is predicted to be stable even for a ferromagnetic interaction [FN1994],
a result established within the framework of the Luttinger liquid, which
we discuss in Chapter 9.

7.7   SUMMARY

The quenching of local moments in metals at low temperatures is
the physics behind the Kondo problem. This effect arises from the
many-body resonance that is pinned at the Fermi level. In this chapter we focused on extracting this picture from quite simple considerations. The Kondo temperature is a scale-invariant quantity given by
$T_K = D \exp(1/g)$, where $g = N(0)J$ is the effective bare Kondo
exchange interaction. Perturbatively, the Kondo temperature can be
identified as the energy scale at which the second-order terms in the
perturbation theory become on the order of the terms linear in J, a
signature that perturbation theory is breaking down and higher-order
terms are important, as well. The resonance at the Fermi level is a true
multiparticle effect; nth-order terms in perturbation theory contain the
interactions of n electrons with the local moment. It is the summation
of all such terms that creates the resonance.

7.8   APPENDIX: THE SCHRIEFFER-WOLFF TRANSFORMATION

In this Appendix, we present the complete derivation of the exchange
interaction in the Kondo Hamiltonian from the Anderson Hamiltonian. Because the Anderson Hamiltonian contains empty as well as
doubly occupied impurity states, a transformation that generates the
Kondo Hamiltonian is equivalent to a diagonalization of the Anderson Hamiltonian in the subspace of the singly occupied impurity states.
This diagonalization is carried out by the Schrieffer-Wolff transformation [SW1966].

We first separate the Anderson Hamiltonian into a *zero*th-order
part,

$$H_0 = \sum_{\mathbf{k},\sigma} \epsilon_{\mathbf{k}} n_{\mathbf{k}\sigma} + \epsilon_d \sum_{\sigma} n_{d\sigma} + U n_{d\uparrow} n_{d\downarrow}, \tag{7.108}$$

and a perturbed part,

$$H_1 = \sum_{\mathbf{k}} V_{\mathbf{k}d} (a_{\mathbf{k}\sigma}^{\dagger} a_{d\sigma} + a_{d\sigma}^{\dagger} a_{\mathbf{k}\sigma}). \tag{7.109}$$

To proceed, we perform a canonical or similarity transformation, S, on the original Hamiltonian:

$$\tilde{H} = e^{S} H e^{-S}$$

$$= H + [S,H] + \frac{1}{2}[S,[S,H]] + \cdots.$$

Note that since e^{S} is unitary, S must be antihermitian. If we choose S so as to cancel the linear dependence

$$H_1 + [S,H_0] = 0 \tag{7.110}$$

on the perturbation H_1, the new Hamiltonian to lowest order becomes

$$\tilde{H} = H_0 + \frac{1}{2}[S,H_1], \tag{7.111}$$

and, hence, incorporates charge fluctuations to second order in $V_{\mathbf{k}d}$. The similarity transformation method is a general way of performing perturbative analyses, once a small quantity has been identified.

The explicit form of S can be constructed by noting that because $[S,H_0] = -H_1$, the operator S must contain terms $\propto a_{\mathbf{k}\sigma}^{\dagger} a_{d\sigma}$; furthermore, its commutator with $Un_{d\uparrow}n_{d\downarrow}$ yields $\propto n_{d-\sigma} a_{\mathbf{k}\sigma}^{\dagger} a_{d\sigma}$. This suggests that we try a transformation of the form

$$S = \sum_{\mathbf{k},\sigma} (A_{\mathbf{k}} + B_{\mathbf{k}} n_{d-\sigma}) a_{\mathbf{k}\sigma}^{\dagger} a_{d\sigma} - h.c., \tag{7.112}$$

where $A_{\mathbf{k}}$ and $B_{\mathbf{k}}$ are c-numbers to be determined by Eq. (7.110).

Using the commutators

$$[n_{d\sigma}, a_{d\sigma'}] = -\delta_{\sigma\sigma'} a_{d\sigma},$$

$$[n_{d\sigma}, n_{d\sigma'}] = 0$$

$$[n_{d\sigma} n_{d-\sigma}, a_{d\sigma'}] = -\delta_{\sigma\sigma'} n_{d-\sigma} a_{d\sigma} - \delta_{-\sigma\sigma'} n_{d\sigma} a_{d\sigma'}, \tag{7.113}$$

together with the relation $[A, B^\dagger] = -[A, B]^\dagger$ provided $A = A^\dagger$, we straightforwardly evaluate the commutator of H_0 with S and find

$$[H_0, S] = \sum_{\mathbf{k},\sigma} \epsilon_\mathbf{k} (A_\mathbf{k} + B_\mathbf{k} n_{d-\sigma}) a^\dagger_{\mathbf{k}\sigma} a_{d\sigma} + h.c.$$

$$+ \sum_{\mathbf{k},\sigma} \epsilon_d (-A_\mathbf{k} - B_\mathbf{k} n_{d-\sigma}) a^\dagger_{\mathbf{k}\sigma} a_{d\sigma} + h.c.$$

$$+ U \sum_{\mathbf{k},\sigma} (-A_\mathbf{k} n_{d-\sigma} - B_\mathbf{k} n_{d-\sigma}) a^\dagger_{\mathbf{k}\sigma} a_{d\sigma} + h.c.$$

$$= \sum_{\mathbf{k},\sigma} [(\epsilon_\mathbf{k} - \epsilon_d) A_\mathbf{k}$$

$$+ (\epsilon_\mathbf{k} - \epsilon_d - U) n_{d-\sigma} B_\mathbf{k} - A_\mathbf{k} U n_{d-\sigma}] a^\dagger_{\mathbf{k}\sigma} a_{d\sigma} + h.c., \quad (7.114)$$

where $h.c.$ denotes the hermitian conjugate. In order to satisfy Eq. (7.110), we require that $(\epsilon_\mathbf{k} - \epsilon_d) A_\mathbf{k} = V_{\mathbf{k}d}$ and $(\epsilon_\mathbf{k} - \epsilon_d - U) B_\mathbf{k} - A_\mathbf{k} U = 0$, so that

$$A_\mathbf{k} = \frac{V_{\mathbf{k}d}}{\epsilon_\mathbf{k} - \epsilon_d}$$

and

$$B_\mathbf{k} = V_{\mathbf{k}d} \left[\frac{1}{\epsilon_\mathbf{k} - (\epsilon_d + U)} - \frac{1}{\epsilon_\mathbf{k} - \epsilon_d} \right].$$

Equation (7.112), coupled with the definitions of the constants $A_\mathbf{k}$ and $B_\mathbf{k}$, specifies the Schrieffer-Wolff transformation.

To find the new Hamiltonian, we need to evaluate the commutator $[S, H_1]$:

$$[S, H_1] = \sum_{\mathbf{k},\sigma,\mathbf{k}',\sigma'} \{ A_\mathbf{k} V_{\mathbf{k}'d} [\rho_{\mathbf{k}d\sigma}, (\rho_{\mathbf{k}'d\sigma'} + \rho^\dagger_{\mathbf{k}'d\sigma'})]$$

$$- B_\mathbf{k}^* V_{\mathbf{k}'d} [n_{d-\sigma} \rho^\dagger_{\mathbf{k}d\sigma}, (\rho_{\mathbf{k}'d\sigma'} + \rho^\dagger_{\mathbf{k}'d\sigma'})]$$

$$+ B_\mathbf{k} V_{\mathbf{k}'d} [n_{d-\sigma} \rho_{\mathbf{k}d\sigma}, (\rho_{\mathbf{k}'d\sigma'} + \rho^\dagger_{\mathbf{k}'d\sigma'})]$$

$$- A_\mathbf{k}^* V_{\mathbf{k}'d} [\rho^\dagger_{\mathbf{k}d\sigma}, (\rho_{\mathbf{k}'d\sigma'} + \rho^\dagger_{\mathbf{k}'\sigma'})] \}, \quad (7.115)$$

where we have simplified the notation by defining $\rho_{\mathbf{k}d\sigma} = a^\dagger_{\mathbf{k}\sigma} a_{d\sigma}$. At this stage, these commutators are useful:

$$[\rho_{\mathbf{k}\,d\sigma}, \rho_{\mathbf{k}'d\sigma'}] = 0$$

$$[\rho_{\mathbf{k}\,d\sigma}, \rho_{\mathbf{k}'d\sigma'}^{\dagger}] = \delta_{\sigma\sigma'}[-\delta_{\mathbf{k}\mathbf{k}'}n_{d\sigma} + a_{\mathbf{k}\sigma}^{\dagger}a_{\mathbf{k}'\sigma}]$$

$$[n_{d-\sigma}\rho_{\mathbf{k}\,d\sigma}, \rho_{\mathbf{k}'d\sigma'}] = -\delta_{\sigma'-\sigma}\rho_{\mathbf{k}\,d\sigma}\rho_{\mathbf{k}'d-\sigma}$$

$$[n_{d-\sigma}\rho_{\mathbf{k}\,d\sigma}, \rho_{\mathbf{k}'d\sigma'}^{\dagger}] = \delta_{-\sigma\sigma'}\rho_{\mathbf{k}\,d\sigma}\rho_{\mathbf{k}'d-\sigma}^{\dagger}$$

$$+ \delta_{\sigma\sigma'}(a_{\mathbf{k}\sigma}^{\dagger}a_{\mathbf{k}'\sigma}n_{d-\sigma} - \delta_{\mathbf{k}\mathbf{k}'}n_{d-\sigma}n_{d\sigma}).$$

Substituting these relationships into Eq. (7.115), we find

$$[S, H_1] = -\sum_{\mathbf{k},\sigma}(A_{\mathbf{k}}V_{\mathbf{k}d} + B_{\mathbf{k}}V_{\mathbf{k}d}n_{d-\sigma})n_{d\sigma} + h.c.$$

$$+ \sum_{\mathbf{k},\mathbf{k}',\sigma}A_{\mathbf{k}}V_{\mathbf{k}'d}\,a_{\mathbf{k}\sigma}^{\dagger}a_{\mathbf{k}'\sigma} + h.c.$$

$$- \sum_{\mathbf{k},\mathbf{k}',\sigma}B_{\mathbf{k}}V_{\mathbf{k}'d}\,\rho_{\mathbf{k}\sigma}\rho_{\mathbf{k}'-\sigma} + h.c.$$

$$+ \sum_{\mathbf{k},\mathbf{k}',\sigma}B_{\mathbf{k}}V_{\mathbf{k}'d}\,[a_{\mathbf{k}\sigma}^{\dagger}a_{\mathbf{k}'\sigma}n_{d-\sigma} + \rho_{\mathbf{k}\sigma}\rho_{\mathbf{k}'-\sigma}^{\dagger}] + h.c.. \quad (7.116)$$

Let us write the operators in the last term as

$$[a_{\mathbf{k}\sigma}^{\dagger}a_{\mathbf{k}'\sigma}n_{d-\sigma} + \rho_{\mathbf{k}\,d\sigma}\rho_{\mathbf{k}'d-\sigma}^{\dagger}] = \frac{1}{2}(n_{d\sigma} + n_{d-\sigma})\,a_{\mathbf{k}\sigma}^{\dagger}a_{\mathbf{k}'\sigma}$$

$$- \frac{1}{2}[(n_{d\sigma} - n_{d-\sigma})\,a_{\mathbf{k}\sigma}^{\dagger}a_{\mathbf{k}'\sigma} - 2\rho_{\mathbf{k}\,d\sigma}\rho_{\mathbf{k}'d-\sigma}^{\dagger}]. \quad (7.117)$$

The second part of this term gives rise to the Kondo exchange interaction. To see how, we note that the product

$$\frac{4}{\hbar^2}(\Psi_{\mathbf{k}'}^{\dagger}\mathbf{S}\Psi_{\mathbf{k}})\cdot(\Psi_d^{\dagger}\mathbf{S}\Psi_d) = (\Psi_{\mathbf{k}'}^{\dagger}\sigma_z\Psi_{\mathbf{k}})\cdot(\Psi_d^{\dagger}\sigma_z\Psi_d)$$

$$+ 2(\Psi_{\mathbf{k}'}^{\dagger}\sigma^{+}\Psi_{\mathbf{k}})(\Psi_d^{\dagger}\sigma^{-}\Psi_d)$$

$$+ 2(\Psi_{\mathbf{k}'}^{\dagger}\sigma^{-}\Psi_{\mathbf{k}})(\Psi_d^{\dagger}\sigma^{+}\Psi_d)$$

$$= \sum_{\sigma}[a_{\mathbf{k}'\sigma}^{\dagger}a_{\mathbf{k}\sigma}(n_{d\sigma} - n_{d-\sigma}) - 2\rho_{\mathbf{k}'d\sigma}\rho_{\mathbf{k}d-\sigma}^{\dagger}]$$

$$(7.118)$$

is precisely of the form of the last term in Eq. (7.117). The resulting contribution to $\tilde{H}$ is

$$H_{\text{exch}} = -\frac{1}{\hbar^2}\sum_{\mathbf{k},\mathbf{k}'}J_{\mathbf{k}\mathbf{k}'}(\Psi_{\mathbf{k}'}^{\dagger}\mathbf{S}\Psi_{\mathbf{k}})\cdot(\Psi_d^{\dagger}\mathbf{S}\Psi_d), \quad (7.119)$$

where

$$J_{\mathbf{kk'}} = (B_{\mathbf{k'}}V_{\mathbf{k}d} + B_{\mathbf{k}}^*V_{\mathbf{k'}d})$$

$$= V_{\mathbf{k'}d}V_{\mathbf{k}d}\left[\frac{1}{\epsilon_{\mathbf{k'}} - (\epsilon_d + U)} + \frac{1}{\epsilon_{\mathbf{k}} - (\epsilon_d + U)}\right.$$

$$\left. - \frac{1}{\epsilon_{\mathbf{k}} - \epsilon_d} - \frac{1}{\epsilon_{\mathbf{k'}} - \epsilon_d}\right]. \tag{7.120}$$

This term, which arises from the last term in Eq. (7.117), is the Kondo interaction term describing the spin-spin interaction between a conduction electron and an impurity spin. Note that this result is the form (7.10), deduced from second-order perturbation theory but symmetrized in $\mathbf{k}$ and $\mathbf{k'}$. We see here a particular advantage of the similarity transformation method; it generates correctly the interaction matrix elements when the initial and final states do not have the same energy. By contrast, we can identify the second-order scattering amplitude with the interaction matrix element only when the initial and final states have the same energy, in which case Eqs. (7.10) and (7.120) agree. In the vicinity of the Fermi level, $k, k' \sim k_F$, the spin-exchange amplitude reduces to

$$J_{k_F k_F} \equiv J_o = -2|V_{\mathbf{k}d}|^2 \frac{U}{|\epsilon_d|(|\epsilon_d| - U)} < 0, \tag{7.121}$$

as advertised.

All together we can write $\tilde{H}$ as

$$\tilde{H} = H_0 + \frac{1}{2}[S, H_1]$$

$$= H_0 + H_0' + H_0'' + H_{\text{exch}} + H_{\text{dir}} + H_{\text{ch}}. \tag{7.122}$$

The direct term

$$H_{\text{dir}} = \sum_{\mathbf{k},\mathbf{k'}} \frac{1}{4}J_{\mathbf{kk'}}(\Psi_d^\dagger \Psi_d)(\Psi_{\mathbf{k'}}^\dagger \Psi_{\mathbf{k}}) \tag{7.123}$$

results from the first term in Eq. (7.117). Here

$$W_{\mathbf{kk'}} = \frac{1}{2}(A_{\mathbf{k'}}V_{\mathbf{k}d} + A_{\mathbf{k}}^*V_{\mathbf{k'}d})$$

$$= \frac{1}{2}V_{\mathbf{k'}d}V_{\mathbf{k}d}^*\left[\frac{1}{\epsilon_{\mathbf{k'}} - \epsilon_d} + \frac{1}{\epsilon_{\mathbf{k}} - \epsilon_d}\right]. \tag{7.124}$$

The energy denominators occurring in $W_{kk'}$ involve only the excitation process involving only the lowest state on the impurity, as shown in Fig. (7.2b), whereas $J_{kk'}$ includes both processes in Fig. (7.2). Because $J_{kk'}$ and $W_{k'k}$ result from successive excitation and de-excitation processes, the impurity remains singly occupied in the process they describe.

The term

$$H_0' = -\sum_{k,k'\sigma} \left(W_{kk'} + \frac{1}{2}J_{kk'}n_{d-\sigma}\right)n_{d\sigma}, \qquad (7.125)$$

which emerges from the first term in Eq. (7.116), essentially renormalizes the coefficients in H_0. Similarly, the term

$$H_0'' = \sum_{k,k'} W_{kk'}\Psi_k^\dagger \Psi_{k'}, \qquad (7.126)$$

which arises from the first term in Eq. (7.116) and the second term in Eq. (7.117), represents a renormalization of the potential felt by single electrons. Both H_0' and H_0'' are unimportant for understanding the Kondo problem. The final term

$$H_{ch} = -\frac{1}{2}\sum_{k,k'\sigma} \left(B_k V_{k'd}\, \rho_{k\sigma}\rho_{k'-\sigma} + h.c.\right), \qquad (7.127)$$

which arises from the third term in Eq. (7.116), changes the occupancy of the d-impurity by two and thus is also not important for the Kondo problem. The two important terms in the interaction are the spin-exchange process and the direct impurity-electron scattering term.

Let us isolate the terms that correspond to single occupancy on the impurity. We note first that, since H_{ch} eliminates both electrons on the d-impurity, it cannot connect to the one-electron states in the Hilbert space; we drop this term. In addition, in the one-electron subspace, $\Psi_d^\dagger \Psi_d = 1$. As a consequence, H_{dir} is a one-electron term, as are H_0' and H_0''. In this subspace, H_{exch} is the only important term. To second order in V_{kd}, the Anderson model yields the Kondo model with an antiferromagnetic exchange coupling. It is the antiferromagnetic nature of this interaction that leads to condensation into a singlet state at the d-impurity.

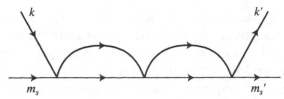

Figure 7.11 Third-order Kondo diagram that contributes a linear correction to scaling (see Problem 7.4).

PROBLEMS

1. **(a)** The energy of a magnetic moment of fixed magnitude is linear in the magnetic field. Show that the high temperature (Curie) magnetic susceptibility of a noninteracting collection of such moments is inversely proportional to T.

    (b) On the other hand, the energy of a polarizable singlet state is quadratic in the magnetic field. Show that the magnetic susceptibility of a noninteracting collection of such moments becomes constant at low temperatures.

2. Repeat Yosida's single-electron variational calculation to derive a triplet bound state in the case of a ferromagnetic spin-spin interaction between an impurity and the conduction electrons. Solve the corresponding triplet stability equation and compute the triplet binding energy. Compared to the singlet, which is lower in energy for the same $|J_0|$?

3. Solve the anisotropic scaling equations (analytically or numerically) and verify that they satisfy the flow diagram shown in Fig. (7.10.)

4. Consider the third-order Kondo diagram shown in Fig. (7.11). Explain how this diagram can give rise to a correction to scaling linear in δD. Assume that $\mathbf{q}$ and $\mathbf{q}'$ are independent. Evaluate the scaling coefficient explicitly.

5. Use the full Hamiltonian, Eq. (7.112), derived from the Schrieffer-Wolff transformation to evaluate the lowest-order scattering amplitudes: $T[(\mathbf{k}\sigma)+(d-\sigma)] \rightarrow (\mathbf{k}'\sigma)+(d-\sigma)]$, $T[(\mathbf{k}\sigma)+(d\sigma)] \rightarrow (\mathbf{k}'\sigma)+(d\sigma)]$, and $T[(\mathbf{k}\sigma)+(d-\sigma)] \rightarrow (\mathbf{k}'-\sigma)+(d\sigma)]$. Compare with the results given in Subsection 7.2.

REFERENCES

[AL1991] I. Affleck, A. W. W. Ludwig, *Nucl. Phys.* B **352**, 849 (1991).

[AYH1970] P. W. Anderson, G. Yuval, D. R. Hamann, *Phys. Rev.* B **1**, 4464 (1970).

[A1978] P. W. Anderson, *Rev. Mod. Phys.* **50**, 191 (1978).

[AFL1983] N. Andrei, K. Furuya, J. H. Lowenstein, *Rev. Mod. Phys.* **55**, 331 (1983).

[BA1996] V. Barzykin, I. Affleck, *Phys. Rev. Lett.* **76**, 4959 (1996).

[FN1994] A. Furusaki, N. Nagaosa, *Phys. Rev. Lett.* **72**, 892 (1994).

[GL1954] M. Gell-Mann, F. Low, *Phys. Rev.* **111**, 582 (1954).

[GZ1974] G. Grüner, A. Zawadowski, *Solid St. Comm.* **11**, 663 (1972).

[H1969] A. J. Heeger in *Solid State Physics*, ed. F. Seitz, D. Turnbull, and H. Ehrenreich (Academic Press, NY, 1969) **23**, 283.

[H1993] A. C. Hewson, *The Kondo Problem to Heavy Fermions* (Cambridge Univ. Press, NY, 1993).

[K1956] T. Kasuya, *Prog. Theor. Phys.* **16**, 45 (1956).

[K1964] J. Kondo, *Prog. Theor. Phys.* **32**, 37 (1964).

[L1981] C. Lacroix, *J. Phys. F: Metal Phys.* **11**, 2389 (1981).

[LT1992] D. H. Lee, J. Toner, *Phys. Rev. Lett.* **72**, 892 (1992).

[N1974] P. Noziéres, *J. Low Temp. Phys.* **17**, 31 (1974).

[SCL1964] M. Sarachik, E. Corenzwit, L. D. Longinotti, *Phys. Rev.* A **135**, 1041 (1964).

[SW1966] J. R. Schrieffer, P. A. Wolff, *Phys. Rev.* **149**, 491 (1966).

[SZ1974] J. Solyom, A. Zawadowski, *J. Phys. F: Metal Phys.* **4**, 80 (1974).

[TW1983] A. M. Tsvelick, P. B. Wiegmann. *Adv. in Phys.* **32**, 453 (1983).

[V1946] S. Vousovskii, *Zh. Eksp. Teor. Fiz.* **16**, 981 (1946).

[W1975] K. Wilson, *Rev. Mod. Phys.* **47**, 773 (1975).

[Y1966] K. Yosida, *Phys. Rev.* **147**, 223 (1966).

[Z1951] C. Zener, *Phys. Rev.* **81**, 440 (1951).

– 8 –

Screening and Plasmons

Screening of the local moment below T_k is the essence of the Kondo problem. A much less exotic example of screening takes place when a positive charge Q is introduced into an electron gas. To restore charge neutrality at the site of the positive charge, the electrons will screen the charge, thereby decreasing its net electric field. In fact, screening occurs regardless of the sign of the test charge. This state of affairs arises because as an electron moves through the electron gas, it does not have to push all the other electrons out of the way. The mutual repulsions among all the electrons in the electron gas help to clear a path for an electron to move. As a result, the effective interaction between the electrons is diminished from the long-range Coulomb $1/r$ to a much more short-ranged interaction. At the level of Thomas-Fermi, the new interaction falls off exponentially. While this interaction overestimates the effect of screening, it does illustrate how efficiently repulsive interactions screen electrons in an electron gas.

8.1 THOMAS-FERMI SCREENING

Let us assume a charge Q is at the origin. In the absence of the electron gas, an electron a distance r away from the origin feels a potential $\phi = Q/r$, such that the potential energy is given by

$$U = -e\phi = -e\frac{Q}{r}. \tag{8.1}$$

In the presence of the electron gas, the potential around the charge Q will be screened. Physically, the screening charge is determined by the difference between the charge density of the electron gas in the presence of the charge, Q, and the charge density in the absence of the external charge. Let $-e\delta n(\mathbf{r})$ represent this difference. Poisson's equation for the total charge distribution,

$$-\nabla^2 \phi_{\text{eff}}(\mathbf{r}) = 4\pi[Q\delta(\mathbf{r}) - e\delta n(\mathbf{r})], \tag{8.2}$$

depends on both Q and the screening charge, $-e\delta n(\mathbf{r})$, with ϕ_{eff}, the true potential of the charge Q in the electron gas. The easiest way to solve Eq. (8.2) is by Fourier transform. We define

$$f(\mathbf{k}) = \int d\mathbf{r} \, e^{-i\mathbf{k}\cdot\mathbf{r}} f(\mathbf{r}). \tag{8.3}$$

As a consequence,

$$k^2 U_{\text{eff}}(\mathbf{k}) = -4\pi e Q + 4\pi e^2 \delta n(\mathbf{k}). \tag{8.4}$$

Solving this exact equation requires some ansatz for the screening charge. The simplest approximation is that of Thomas [T1927] and Fermi [F1928] in which the electron gas is assumed to respond to the charge Q as if it were locally a free-electron gas. Physically, this assumption implies that the potential is a slowly varying function on a distance scale set by the Fermi wavelength. The net effect, then, is a shift in the chemical potential of the form $\mu \rightarrow \mu + U_{\text{eff}}(r)$. Let us define a new single-particle energy level,

$$\epsilon_{\mathbf{k}}(\mathbf{r}) = \frac{\hbar^2 k^2}{2m} + U_{\text{eff}}(\mathbf{r}). \tag{8.5}$$

The effective Fermi-Dirac distribution function is

$$n_{\mathbf{k}}(\mathbf{r}) = \frac{1}{1 + e^{\beta(\epsilon_{\mathbf{k}}(\mathbf{r}) - \mu)}}, \tag{8.6}$$

so that the effective density at $\mathbf{r}$ is

$$\langle n(\mathbf{r}) \rangle = 2 \int \frac{d\mathbf{k}}{(2\pi)^3} n_{\mathbf{k}}(\mathbf{r})$$

$$= n_e - 2\beta \int \frac{d\mathbf{k}}{(2\pi)^3} U_{\text{eff}}(\mathbf{r}) \frac{e^{\beta(\epsilon_{\mathbf{k}} - \mu)}}{(1 + e^{\beta(\epsilon_{\mathbf{k}} - \mu)})^2} + \cdots$$

$$\simeq n_e - \frac{\partial n_e}{\partial \mu} U_{\text{eff}}(\mathbf{r}) + \cdots, \tag{8.7}$$

where we expanded the distribution function and retained only the linear term in U_{eff}. This expansion is valid only if $\epsilon_F \gg U_{\text{eff}}(r)$. We can now calculate $\delta n(\mathbf{r})$ because

$$\delta n(\mathbf{r}) = \langle n(\mathbf{r}) \rangle - n_e = -\frac{\partial n_e}{\partial \mu} U_{\text{eff}}(\mathbf{r}), \tag{8.8}$$

which implies that

$$k^2 U_{\text{eff}}(\mathbf{k}) = -4\pi e \left(Q + e \frac{\partial n_e}{\partial \mu} U_{\text{eff}}(k) \right), \tag{8.9}$$

or, equivalently,

$$U_{\text{eff}}(\mathbf{k}) = \frac{-4\pi e Q}{k^2 + 4\pi e^2 \partial n_e / \partial \mu}. \tag{8.10}$$

The inverse Fourier transform of $U_{\text{eff}}(\mathbf{k})$,

$$
\begin{aligned}
U_{\text{eff}}(\mathbf{r}) &= -\frac{4\pi e Q}{(2\pi)^3} \int \frac{e^{i\mathbf{k}\cdot\mathbf{r}} d\mathbf{k}}{k^2 + 4\pi e^2 \partial n_e / \partial \mu} \\
&= -\frac{eQ}{\pi r} \int_{-\infty}^{\infty} dk \frac{k \sin kr}{k^2 + 4\pi e^2 \partial n_e / \partial \mu} \\
&= -eQ \frac{e^{-\kappa_{\text{TF}} r}}{r},
\end{aligned}
\tag{8.11}
$$

demonstrates that the electrostatic potential arising from the electrons around the charge Q,

$$\phi_{\text{eff}}(\mathbf{r}) = -e \frac{e^{-\kappa_{\text{TF}} r}}{r}, \tag{8.12}$$

decays exponentially. The electron gas attenuates the bare Coulomb field by screening the charge on a length scale determined by

$$\kappa_{\text{TF}}^2 = 4\pi e^2 \frac{\partial n_e}{\partial \mu}. \tag{8.13}$$

To reiterate, this result is valid in the limit of a slowly varying field induced by the electron gas. We estimate κ_{TF} by recalling from Chapter 1 that

$$n_e = \left(\frac{2m\mu_0}{\hbar^2} \right)^{3/2} \frac{1}{3\pi^2}. \tag{8.14}$$

As a consequence,

$$\frac{\partial n_e}{\partial \mu_0} = \frac{3}{2}\frac{n_e}{\mu_0} = \frac{3n_e}{mv_F^2} \tag{8.15}$$

and

$$\kappa_{TF}^2 = \frac{12\pi e^2 n_e}{mv_F^2} = \frac{4}{\pi}\left(\frac{9\pi}{4}\right)^{1/3}\frac{1}{a_o r_e} = \frac{2.434}{a_o^2}\frac{1}{r_s}. \tag{8.16}$$

Recall that $r_s \sim 2 - 6$. Hence, $\kappa_{TF}^{-1} = 0.34\sqrt{r_s}\text{Å} \approx 0.45\text{Å} - 0.9\text{Å}$. While Thomas-Fermi theory predicts a screening length of electrons in metals generally shorter than the interparticle spacing, the result only describes the long-distance falloff of the potential surrounding a given charge, as in Eq. (8.11). The theory does not correctly account for screening at distances comparable to the interparticle spacing. To see the behavior at short distances requires a more rigorous treatment, which we develop in a subsequent section.

8.2 PLASMA OSCILLATIONS

We have developed a theory for the screening effects in an electron gas that assumes slow motion of the electrons. In general this is not true, especially at large-wave vector comparable to the inverse interparticle spacing. In this limit, collective excitations of the electron gas become accessible. Such collective excitations are termed *plasmons*.

The existence of plasma excitations can be established straight-forwardly from the equations of motion for the electron density in momentum space. To proceed, we recall the form of the Coulomb interaction

$$\sum_{\mathbf{k}} \frac{4\pi}{k^2}e^{i\mathbf{k}\cdot(\mathbf{r}_i - \mathbf{r}_j)} = \frac{1}{|\mathbf{r}_i - \mathbf{r}_j|} \tag{8.17}$$

in momentum space, which implies that we can rewrite the Hamiltonian for our electron gas as

$$H = \sum_i \frac{p_i^2}{2m} + \frac{1}{2}\sum_{i \neq j}\sum_{\mathbf{k}} \frac{4\pi e^2}{k^2}e^{i\mathbf{k}\cdot(\mathbf{r}_i - \mathbf{r}_j)}$$

$$= \sum_i \frac{p_i^2}{2m} + \sum_{\mathbf{k}} \frac{V_{\mathbf{k}}}{2}(n_{\mathbf{k}}n_{-\mathbf{k}} - N) \tag{8.18}$$

with $n_{\mathbf{k}} = \sum_i e^{i \mathbf{k} \cdot \mathbf{r}_i}$, N the number of electrons, and $V_{\mathbf{k}} = 4\pi e^2/k^2$. As we chose our box to be of unit volume, $n_{\mathbf{k}}$ appears without the V^{-1} normalization.

To determine the collective excitations, we need the time evolution of $n_{\mathbf{k}}$,

$$i \dot{n}_{\mathbf{k}} = [n_{\mathbf{k}}, H]/\hbar$$

$$= \frac{1}{\hbar} \sum_i \left[\frac{\hbar^2}{m} \nabla_{\mathbf{r}_i} e^{i\mathbf{k}\cdot\mathbf{r}_i} \cdot \nabla_{\mathbf{r}_i} + \frac{\hbar^2}{2m} \nabla_{\mathbf{r}_i}^2 e^{i\mathbf{k}\cdot\mathbf{r}_i} \right]$$

$$= - \sum_i e^{i\mathbf{k}\cdot\mathbf{r}_i} \left(\frac{\mathbf{k}\cdot\mathbf{p}_i}{m} + \frac{\hbar k^2}{2m} \right). \tag{8.19}$$

The second derivative is

$$\ddot{n}_{\mathbf{k}} = \sum_i e^{i\mathbf{k}\cdot\mathbf{r}_i} \left(\frac{\mathbf{k}\cdot\mathbf{p}_i}{m} + \frac{\hbar k^2}{2m} \right)^2$$

$$+ \frac{2\pi e^2}{\hbar} \sum_i \left[e^{i\mathbf{k}\cdot\mathbf{r}_i} \frac{\mathbf{k}\cdot\mathbf{p}_i}{m}, \sum_{\mathbf{q}} \left(\frac{n_{\mathbf{q}}^\dagger n_{\mathbf{q}} - N}{q^2} \right) \right]$$

$$= \sum_i e^{i\mathbf{k}\cdot\mathbf{r}_i} \left(\frac{\mathbf{k}\cdot\mathbf{p}_i}{m} + \frac{\hbar k^2}{2m} \right)^2 - \frac{4\pi e^2}{m} \sum_{\mathbf{q}} \mathbf{k}\cdot\mathbf{q} \frac{n_{\mathbf{k}-\mathbf{q}} n_{\mathbf{q}}}{q^2}. \tag{8.20}$$

We separate the $\mathbf{k} = \mathbf{q}$ interaction term, which can be written as

$$4\pi e^2 \frac{k^2}{mk^2} n_0 n_{\mathbf{k}} = \frac{4\pi e^2 n_e}{m} n_{\mathbf{k}} = \omega_p^2 n_{\mathbf{k}}. \tag{8.21}$$

That ω_p represents the frequency of the collective oscillations of the electron gas can be seen by rewriting $\ddot{n}_{\mathbf{k}}$ in the suggestive form

$$\ddot{n}_{\mathbf{k}} + \omega_p^2 n_{\mathbf{k}} = \sum_i e^{i\mathbf{k}\cdot\mathbf{r}_i} \left(\frac{\mathbf{k}\cdot\mathbf{p}_i}{m} + \frac{\hbar k^2}{2m} \right)^2 - \frac{4\pi e^2}{m} \sum_{\mathbf{q}\neq\mathbf{k}} \mathbf{k}\cdot\mathbf{q} \frac{n_{\mathbf{k}-\mathbf{q}} n_{\mathbf{q}}}{q^2}. \tag{8.22}$$

We see, then, that the density $n_{\mathbf{k}}$ oscillates at the frequency ω_p, the plasma frequency, if the terms on the right-hand side of Eq. (8.22) are small. The first term is of order $k^2 v_F^2 n_{\mathbf{k}}$. In the next term, a product of $n_{\mathbf{q}}$'s appears. Because the density is a sum of exponential

terms with randomly varying phases, one might expect that the contribution from the second term is minimal. The approximation that
ignores this contribution is known as the *random-phase approximation* [P1963]. A well-defined plasma frequency exists at this level of
theory if

$$\omega_p^2 \gg k^2 v_F^2. \tag{8.23}$$

For a density of $n_e \sim 10^{23}$ e$^-$/cm^3, $\omega_p \sim 10^{16}$ s^{-1}, or, equivalently, the
energy in a plasma oscillation is

$$\hbar \omega_p \sim 12 \, \text{eV}. \tag{8.24}$$

Such a high-energy excitation cannot be created by thermal or phonon-
like oscillations of the ions. They also cannot be excited by a single
electron. Plasma oscillations or plasmons arise from a collective motion
of all the electrons in a solid. As such, plasmons are long-wavelength
oscillations. We estimate the maximum wave vector for which plasmons
exist by considering the ratio ω_p / κ_{TF}. Recall that $\kappa_{TF}^2 = 4me^2 p_F / \pi \hbar^3$.
As a consequence,

$$\omega_p / \kappa_{TF} = \left(\frac{4\pi e^2 n_e}{m} \frac{\pi \hbar^3}{4 p_F m e^2} \right)^{1/2}$$

$$= \frac{p_F}{\sqrt{3}m} = \frac{v_F}{\sqrt{3}}. \tag{8.25}$$

We find that $\omega_p \propto v_F \kappa_{TF}$. Comparison with Eq. (8.23) reveals that
well-defined plasma oscillations exist if $\kappa_{TF} \gg k$. For $k < \kappa_{TF}$, the
electrons act individually. Note also that $\omega_p^2 \sim 1/m$. For an interacting system, we replace m by the effective mass, m^*. In an insulator,
$m^* \to \infty$. Hence, $\omega_p \sim 0$ in an insulator. Similarly, in a conductor m^*
is finite and, as a consequence, $\omega_p \neq 0$. Consequently, the magnitude
of ω_p is a sensitive test for the insulator–metal transition in an interacting electron system [K1959].

A final observation on plasmons is that their dispersion relationship
is fundamentally tied to the dimensionality of space. If the electrons
are confined to a plane (d = 2) but the electric field lines are allowed
to live in three-dimensional space, thereby making the Coulomb interaction the standard $1/r$ potential (see Problem 8.3), there is no gap to
excite plasmon excitations. In addition (see Problem 8.1), the screening
length is independent of density in 2d, at least at the level of Thomas-

Fermi. Both of these effects illustrate how fundamentally different a 2d electron gas is from its 3d counterpart.

8.2.1 Dispersion of Light

An application in which the plasma frequency naturally appears is the propagation of transverse electromagnetic radiation in metals. Consider the Maxwell equation for the magnetic induction in the presence of a current density,

$$\nabla \times \mathbf{B} - \frac{1}{c}\frac{\partial \mathbf{E}}{\partial t} = \frac{4\pi e \mathbf{j}}{c}. \tag{8.26}$$

The time derivative of this equation is

$$\frac{1}{c}\frac{\partial}{\partial t}\nabla \times \mathbf{B} = \frac{4\pi e}{c^2}\frac{\partial \mathbf{j}}{\partial t} + \frac{1}{c^2}\frac{\partial^2 \mathbf{E}}{\partial t^2}. \tag{8.27}$$

But $\partial \mathbf{B}/\partial ct = -\nabla \times \mathbf{E}$ and $\partial \mathbf{j}/\partial t = en_e \mathbf{E}/m$. For a transverse E-field, $\nabla \cdot \mathbf{E} = 0$. As a consequence,

$$-\nabla \times (\nabla \times \mathbf{E}) = -\nabla(\nabla \cdot \mathbf{E}) + \nabla^2\mathbf{E} = \nabla^2\mathbf{E}. \tag{8.28}$$

The time derivative can now be written as

$$\left(-\frac{\partial^2}{\partial t^2} + c^2\nabla^2 - \omega_p^2\right)\mathbf{E} = 0. \tag{8.29}$$

The resultant dispersion relationship for light in a metal,

$$\omega^2 = c^2 k^2 + \omega_p^2, \tag{8.30}$$

illustrates clearly that transverse electromagnetic radiation cannot penetrate a metal for frequencies less than the plasma frequency.

8.3 LINEAR-RESPONSE THEORY

We now focus on formulating the screening problem in terms of the dielectric-response function. To facilitate this derivation, we first introduce the general formalism of treating slowly varying time-dependent perturbations, namely, linear-response theory. Ultimately, we will apply this approach to calculate the density response of a homogeneous system perturbed by the electric field of an external charge.

Consider a Hamiltonian of the form $H = H_0 + W(t)$. In the presence of the perturbation, $W(t)$, the average value of any observable, Y, will acquire a nontrivial time-dependence through the time evolution of the density matrix. The average value of any dynamical observable, Y, at any time t is determined by

$$\langle Y(t) \rangle = Tr\,[\rho(t)Y], \tag{8.31}$$

where $\rho(t)$ is the density matrix appropriately normalized, so that $Tr\rho = 1$. The time evolution of the density matrix

$$\dot{\rho} = -\frac{i}{\hbar}[H,\rho]$$

$$= -\frac{i}{\hbar}[H_0,\rho] + [W(t),\rho] \tag{8.32}$$

is governed by the Liouville equation of motion. To solve this equation, it is easiest to work in the interaction representation. For any operator O, we define $\hat{O}$ to be the interaction representation,

$$\hat{O}(t) = S^{-1}OS, \tag{8.33}$$

of O, where $S = e^{-iH_0 t/\hbar}$. To simplify notation, we have departed from the convention of using $\widehat{O}$ to denote an operator, because $\hat{O}$ now indicates an operator in the interaction representation. Rewriting the original average of Y and the Liouville equation in the interaction representation, we obtain,

$$\langle \hat{Y}(t) \rangle = Tr\,[\hat{\rho}(t)\hat{Y}(t)] \tag{8.34}$$

and

$$i\hbar\dot{\hat{\rho}} = -H_0\hat{\rho} + S^{-1}i\hbar\frac{\partial\rho}{\partial t}S + S^{-1}\rho H_0 S$$

$$= -[H_0,\hat{\rho}] + S^{-1}[H_0 + W(t),\rho]S$$

$$= [\widehat{W}(t),\hat{\rho}(t)]. \tag{8.35}$$

Let ρ_0 be the density matrix before the perturbation is turned on. For a perturbation turned on at $t = -\infty$, the solution to the Liouville equation is the time-ordered product

$$\hat{\rho}(t) = T \exp\left(\frac{1}{i\hbar}\int_{-\infty}^{t}[\widehat{W}(t'),\hat{\rho}]dt'\right)$$

$$= \rho_0 - \frac{i}{\hbar}\int_{-\infty}^{t}[\widehat{W}(t_1),\rho_0]dt_1$$

$$- \frac{1}{\hbar^2}\int_{-\infty}^{t}dt_1\int_{-\infty}^{t_1}dt_2[\widehat{W}(t_1),[\widehat{W}(t_2),\rho_0]] + \cdots. \quad (8.36)$$

Consequently, through first order in the perturbation, we find that

$$\langle\widehat{Y}(t)\rangle = \langle\widehat{Y}(t)\rangle_0 - \frac{i}{\hbar}\int_{-\infty}^{t} Tr\left(\widehat{Y}(t)[\widehat{W}(t'),\rho_0]\right)dt'. \quad (8.37)$$

Cyclically permuting under the trace leads to

$$\langle\widehat{Y}(t)\rangle = \langle\widehat{Y}(t)\rangle_0 - \frac{i}{\hbar}\int_{-\infty}^{t}\chi_{YW}(t,t')dt', \quad (8.38)$$

in which

$$\chi_{YW}(t,t') = \langle[\widehat{Y}(t),\widehat{W}(t')]\rangle_0 \quad (8.39)$$

is the two-time response function, and $\langle\cdots\rangle_0$ signifies a trace with the equilibrium or initial density matrix, $\hat{\rho}_0$. The quantum susceptibility to linear order is $\chi_{YW}(t,t')$. This quantity governs the relaxation of a quantum system. For example, the crux of quantum linear-response theory is that the fluctuation

$$\langle\delta Y(t)\rangle = \langle\widehat{Y}(t)\rangle - \langle\widehat{Y}(t)\rangle_0$$

$$= \frac{-i}{\hbar}\int_{-\infty}^{t}\chi_{YW}(t,t')dt' \quad (8.40)$$

is determined by the time integral of the average value of the commutator of the observable at time t with the perturbation at time t'. A few useful properties of $\chi_{YW}(t,t')$ are

$$\chi_{YW}(t,t') = -\chi_{WY}(t',t) = -\chi_{YW}^{*}(t,t'). \quad (8.41)$$

These relationships follow because χ_{YW} is a commutator.

8.3.1   Fluctuation-Dissipation Theorem

Consider the general fluctuation

$$S_{YW}(t,t') = \langle \delta \widehat{Y}(t) \delta \widehat{W}(t') \rangle_0$$
$$= \langle \widehat{Y}(t) \widehat{W}(t') \rangle_0 - \langle \widehat{Y}(t) \rangle_0 \langle \widehat{W}(t') \rangle_0. \qquad (8.42)$$

The fluctuation-dissipation theorem equates fundamentally the spontaneous fluctuations that occur in an equilibrium system with the relaxation of a nonequilibrium system displaced from equilibrium. The equilibrium density matrix is

$$\rho_0 = e^{-\beta H_0}. \qquad (8.43)$$

The correlation function $S_{YW}(t,t')$ is a function of the time difference $t - t'$ rather than of t and t' separately. We want to show that $S_{YW}(t,t')$ is related to χ_{YW}. To do this, we first compute

$$\langle \widehat{Y}(t) \widehat{W}(t') \rangle_0 = Tr[e^{-\beta H_0} \widehat{Y}(t) \widehat{W}(t')]$$
$$= Tr[e^{-\beta H_0} \widehat{W}(t') e^{-\beta H_0} \widehat{Y}(t) e^{\beta H_0}]$$
$$= \langle \widehat{W}(t') \widehat{Y}(t + i\beta \hbar) \rangle_0. \qquad (8.44)$$

Coupled with the identity

$$\langle \widehat{Y}(t) \rangle_0 = Tr[e^{-\beta H_0} \widehat{Y}(t)] = Tr[e^{-\beta H_0} e^{-\beta H_0} \widehat{Y}(t) e^{\beta H_0}]$$
$$= \langle \widehat{Y}(t + i\beta \hbar) \rangle_0, \qquad (8.45)$$

we arrive at the equality $S_{YW}(t,t') = S_{WY}(t', t + i\beta \hbar)$. In Fourier space, we have

$$S_{YW}(\omega) = \int_{-\infty}^{\infty} d(t - t') S_{YW}(t,t') e^{i\omega(t-t')}$$
$$= \int_{-\infty}^{\infty} d(t - t') S_{WY}(t', t + i\beta \hbar) e^{i\omega(t-t')}. \qquad (8.46)$$

Let $x = t' - t - i\beta \hbar$; $dx = d(t' - t)$. The Fourier transform of S_{YW} becomes

$$S_{YW}(\omega) = e^{\beta \hbar \omega} \int_{-\infty}^{\infty} dx S_{WY}(x) e^{-i\omega x}$$
$$= e^{\beta \hbar \omega} S_{WY}(-\omega). \qquad (8.47)$$

Combining these results to calculate $\chi_{YW}(\omega)$,

$$\chi_{YW}(\omega) = \int_{-\infty}^{\infty} d(t-t') e^{i\omega(t-t')} \langle [\widehat{Y}(t), \widehat{W}(t')] \rangle_0$$

$$= \int_{-\infty}^{\infty} d(t-t') e^{i\omega(t-t')} [S_{YW}(t,t') - S_{WY}(t',t)]$$

$$= (1 - e^{-\beta\hbar\omega}) S_{YW}(\omega). \tag{8.48}$$

Consequently, spontaneous fluctuations in equilibrium are related to the linear-response function $\chi_{YW}(\omega)$. This means that relaxation of fluctuations in a nonequilibrium system is determined by the same laws that govern the relaxation of spontaneous fluctuations in an equilibrium system. This is the fluctuation-dissipation theorem.

8.3.2   Density Response

Let us apply linear-response theory to density fluctuations. Consider a perturbation of the form

$$H'(t) = \int d\,\mathbf{r} n(\mathbf{r},t) W(\mathbf{r},t), \tag{8.49}$$

in which the electron density $n(\mathbf{r},t)$ is changed by the application of an external field, $W(\mathbf{r},t)$, which commutes with $n(\mathbf{r},t)$ and H_0. According to linear-response theory,

$$\langle \delta n(\mathbf{r},t) \rangle = \int_{-\infty}^{t} dt' d\,\mathbf{r}' \chi_{nn}(\mathbf{r}t,\mathbf{r}'t') W(\mathbf{r}',t'), \tag{8.50}$$

where

$$\chi_{nn}(\mathbf{r}t,\mathbf{r}'t') = -\frac{i}{\hbar} \langle [n(\mathbf{r},t), n(\mathbf{r}',t')] \rangle. \tag{8.51}$$

As several response functions will be introduced, we stress that χ_{nn} represents the response of the system to the external (unscreened) field. We have subsumed the $-i/\hbar$ factor into the definition of the susceptibility. In Eq. (8.50), the time evolution of the density is determined entirely by H_0.

For free electrons, we define χ_{nn}^0 to be the response function. We showed previously that

$$n(\mathbf{r}) = \frac{1}{V} \sum_{\mathbf{p},\mathbf{p}',\sigma} e^{i(\mathbf{p}-\mathbf{p}')\cdot\mathbf{r}/\hbar} a_{\mathbf{p}'\sigma}^\dagger a_{\mathbf{p}\sigma} \tag{8.52}$$

is the time-independent operator for the electron density at $\mathbf{r}$. We remind the reader that we have dropped the 'hat' on an operator because this symbol is now reserved for the interaction representation. To define $n(\mathbf{r},t)$, we need the time dependence of $a_{\mathbf{p}\sigma}$. We obtain this through the Heisenberg equations of motion

$$i\hbar \frac{\partial a_{\mathbf{p}\sigma}}{\partial t} = [a_{\mathbf{p}\sigma}, H_0]$$

$$= \sum_{\mathbf{p}',\sigma'} \epsilon_{\mathbf{p}'\sigma'} [a_{\mathbf{p}\sigma}, a_{\mathbf{p}'\sigma'}^\dagger a_{\mathbf{p}'\sigma'}]$$

$$= \epsilon_{\mathbf{p}} a_{\mathbf{p}\sigma}, \tag{8.53}$$

where H_0 is the Hamiltonian for free electrons. Integrating the above, we obtain that

$$a_{\mathbf{p}\sigma}(t) = e^{-i\epsilon_{\mathbf{p}}/\hbar} a_{\mathbf{p}\sigma}(t=0). \tag{8.54}$$

Let us now evaluate χ_{nn} for a collection of free electrons. To simplify the notation, we define

$$q_{\mathbf{p}_{12}}(\mathbf{r},t) = e^{i(\mathbf{p}_2-\mathbf{p}_1)\cdot\mathbf{r}/\hbar} e^{i(\epsilon_{\mathbf{p}_1}-\epsilon_{\mathbf{p}_2})t/\hbar}, \tag{8.55}$$

$\delta_{\mathbf{p}_{ij}} = \delta_{\mathbf{p}_i \mathbf{p}_j}$, and $f_{\mathbf{p}_1 \mathbf{p}_2} = f_{\mathbf{p}_1}(1-f_{\mathbf{p}_2})$. Combining our expression for $a_{\mathbf{p}\sigma}(t)$ together with Eq. (8.52), it follows that

$$\langle n(\mathbf{r},t)n(\mathbf{r}',t')\rangle = \frac{1}{V^2} \sum_{\substack{\mathbf{p}_1,\mathbf{p}_2,\mathbf{p}_3,\mathbf{p}_4 \\ \sigma_1,\sigma_2}} \langle a_{\mathbf{p}_1\sigma_1}^\dagger a_{\mathbf{p}_2\sigma_1} a_{\mathbf{p}_3\sigma_2}^\dagger a_{\mathbf{p}_4\sigma_2}\rangle q_{\mathbf{p}_{12}}(\mathbf{r},t) q_{\mathbf{p}_{34}}(\mathbf{r}',t')$$

$$= \langle n(\mathbf{r},t)\rangle\langle n(\mathbf{r}',t')\rangle$$

$$+ \frac{1}{V^2} \sum_{\substack{\mathbf{p}_1,\mathbf{p}_2,\mathbf{p}_3,\mathbf{p}_4 \\ \sigma_1,\sigma_2}} \delta_{\sigma_1\sigma_2}\delta_{\mathbf{p}_{14}}\delta_{\mathbf{p}_{23}} f_{\mathbf{p}_1\mathbf{p}_2} q_{\mathbf{p}_{12}}(\mathbf{r},t) q_{\mathbf{p}_{34}}(\mathbf{r}',t')$$

$$= \langle n(\mathbf{r},t)\rangle\langle n(\mathbf{r}',t')\rangle + \frac{1}{V^2} \sum_{\substack{\mathbf{p}_1,\mathbf{p}_2 \\ \sigma}} f_{\mathbf{p}_1\mathbf{p}_2} q_{\mathbf{p}_{12}}(\mathbf{r}-\mathbf{r}',t-t')$$

$$\tag{8.56}$$

Substitution of Eq. (8.55) into Eq. (8.51) illustrates immediately that the response function,

$$\chi_{nn}^0(\mathbf{r}, t, \mathbf{r}', t') = \frac{1}{i\hbar V^2} \sum_{\mathbf{p}_1, \mathbf{p}_2, \sigma} (f_{\mathbf{p}_1 \sigma} - f_{\mathbf{p}_2 \sigma}) q_{\mathbf{p}_{12}}(\mathbf{r} - \mathbf{r}', t - t') \quad (8.57)$$

and the density response function depend on the differences $\mathbf{r} - \mathbf{r}'$ and $t - t'$. Hence, this response function is independent of the particular choice of origin in space as well as in time. We will find it most useful to work with the Fourier transform of $\chi_{nn}^0(\mathbf{r}t, \mathbf{r}'t')$:

$$
\begin{aligned}
\chi_{nn}^0(\mathbf{k}, \omega) &= \int d\mathbf{x} dt e^{i\mathbf{k}\cdot\mathbf{x}} e^{-i\omega t} \chi_{nn}^0(x, t) \\
&= \frac{2}{i\hbar V} \sum_{\mathbf{p}_1, \mathbf{p}_2} \delta_{\hbar k, \mathbf{p}_2 - \mathbf{p}_1}(f_{\mathbf{p}_1} - f_{\mathbf{p}_2}) \int_{-\infty}^0 e^{-i(\hbar\omega + \epsilon_{\mathbf{p}_1} - \epsilon_{\mathbf{p}_2})t/\hbar} dt \\
&= \frac{2}{V} \sum_{\mathbf{p}_1} \frac{f_{\mathbf{p}_1} - f_{\mathbf{p}_1 + \hbar k}}{\hbar\omega + \epsilon_{\mathbf{p}_1} - \epsilon_{\mathbf{p}_1 + \hbar k}}. \quad (8.58)
\end{aligned}
$$

In Eq. (8.58), the factor of 2 comes from the spin summation, and the perturbation coupling to the density was assumed to be turned on at $t = -\infty$ and turned off at $t = 0$. It is this expression that will be used to formulate the Lindhard screening function.

Consider the zero-frequency limit of $\chi_{nn}^0(\mathbf{k}, \omega)$,

$$\chi_{nn}^0(\mathbf{k}, \omega = 0) = 2 \int \frac{d\mathbf{p}}{(2\pi\hbar)^3} \frac{\partial f_p}{\partial \epsilon_p} = -2 \int \frac{d\mathbf{p}}{(2\pi\hbar)^3} \frac{\partial f_p}{\partial \mu} = -\frac{\partial n_e}{\partial \mu}, \quad (8.59)$$

which is precisely the Thomas-Fermi approximation to the screening function. This suggests that there is a fundamental connection between the density response function and screening. To establish the connection formally, we turn to the dielectric-response function.

8.4 DIELECTRIC-RESPONSE FUNCTION

We reformulate the screening problem in terms of χ_{nn} by rewriting the perturbing field as

$$H' = \int d\mathbf{r} n(\mathbf{r}, t) U(\mathbf{r}, t), \quad (8.60)$$

where $U(\mathbf{r}, t)$ is the local electrostatic potential energy of the charge Q, which we take to be the Coulomb interaction. We must determine the net field felt by other electrons as a result of the test charge Q placed at the origin. We start by rewriting Eq. (8.9) as

$$U_{\text{eff}}(\mathbf{r}, t) = U(\mathbf{r}) + \int e^2 \frac{\langle \delta n(\mathbf{r}', t) \rangle}{|\mathbf{r} - \mathbf{r}'|} d\mathbf{r}'. \tag{8.61}$$

Fourier transforming this expression with respect to $\mathbf{r}$ and t,

$$U_{\text{eff}}(\mathbf{k}, \omega) = U(\mathbf{k}) + \frac{4\pi e^2}{k^2} \langle \delta n(\mathbf{k}, \omega) \rangle, \tag{8.62}$$

and using the linear response expression for the fluctuation (Eq. 8.51)

$$\langle \delta n(\mathbf{k}, \omega) \rangle = \chi_{nn}(\mathbf{k}, \omega) U(\mathbf{k}), \tag{8.63}$$

we obtain

$$U_{\text{eff}}(\mathbf{k}, \omega) = [1 + U(\mathbf{k}) \chi_{nn}(\mathbf{k}, \omega)] U(\mathbf{k}, \omega)$$
$$= \varepsilon^{-1}(\mathbf{k}, \omega) U(\mathbf{k}), \tag{8.64}$$

with $\varepsilon(\mathbf{k}, \omega)$ the dielectric function and $U(\mathbf{k}) = 4\pi e^2/k^2$. It is the dielectric function that contains the dynamics of the screening process described in the introduction to this chapter.

To make contact with our previous treatment of screening, we introduce a generalized screening function, χ_{sc}, through

$$\langle \delta n(\mathbf{r}, t) \rangle = \langle n(\mathbf{r}, t) \rangle - n_e = \int d\mathbf{r}' \chi_{sc}(\mathbf{r}, \mathbf{r}', t) U_{\text{eff}}(\mathbf{r}'). \tag{8.65}$$

The generalized screening function, χ_{sc}, describes the response of the system to the screened potential in contrast to χ_{nn}, which is simply the response to the bare potential. From Eq. (8.7), we see that the Thomas-Fermi screening function is simply

$$\chi_{sc}(\mathbf{r}, \mathbf{r}', t) = -\frac{\partial n_e}{\partial \mu} \delta(\mathbf{r} - \mathbf{r}') \delta(t). \tag{8.66}$$

The Fourier transform of Eq. (8.65) yields

$$\langle \delta n(\mathbf{k}, \omega) \rangle = \chi_{sc}(\mathbf{k}, \omega) U_{\text{eff}}(\mathbf{k}, \omega), \tag{8.67}$$

which, together with Eq. (8.62), implies that

$$U_{\text{eff}}(\mathbf{k}, \omega) = \left[1 - \frac{4\pi e^2}{k^2}\chi_{\text{sc}}(\mathbf{k}, \omega)\right]^{-1} U(\mathbf{k}) \tag{8.68}$$

is an equivalent expression for the total effective electrostatic potential in the presence of the test charge Q. Equating Eqs. (8.68) and (8.64), we see immediately that

$$\chi_{nn} = \varepsilon^{-1}\chi_{\text{sc}} = \frac{\chi_{\text{sc}}}{1 - \frac{4\pi e^2 \chi_{\text{sc}}}{k^2}}. \tag{8.69}$$

It is generally easier to construct a theory for χ_{sc} because it describes the response to the total field of the system. The lowest-order theory for χ_{sc} is the random-phase approximation (RPA) in which it is assumed that

$$\chi_{\text{sc}}(\mathbf{k}, \omega) = \chi_{nn}^0(\mathbf{k}, \omega). \tag{8.70}$$

Alternatively, the effective interaction is given by the geometric series

$$U_{\text{eff}}(\mathbf{k}, \omega) = U(\mathbf{k})(1 + U(\mathbf{k})\chi_{nn}^0(\mathbf{k}, \omega) + (U(\mathbf{k})\chi_{nn}^0(\mathbf{k}, \omega))^2 + \cdots)$$

$$= \frac{U(\mathbf{k})}{1 - U(\mathbf{k})\chi_{nn}^0(\mathbf{k}, \omega)}. \tag{8.71}$$

This approximation, which is shown in Fig. (8.1), leaves out exchange effects and is essentially time-dependent Hartree-Fock. As shown in the previous section, the zero-frequency limit of χ_{nn}^0 is the Thomas-Fermi approximation.

To understand the role played by the frequency dependence, we expand the denominator in Eq. (8.58) for large ω. In this limit, we find that

$$\lim_{\omega \to \infty} \chi_{nn}^0(\mathbf{k}, \omega) \to 2\int \frac{d\mathbf{p}}{(2\pi\hbar)^3}(f_{\mathbf{p}} - f_{\mathbf{p}+\hbar\mathbf{k}})\left[\frac{1}{\hbar\omega} - \frac{\epsilon_{\mathbf{p}} - \epsilon_{\mathbf{p}+\hbar\mathbf{k}}}{(\hbar\omega)^2} + \cdots\right]$$

$$= -\frac{2}{(\hbar\omega)^2}\int \frac{d\mathbf{p}}{(2\pi\hbar)^3}(f_{\mathbf{p}} - f_{\mathbf{p}+\hbar\mathbf{k}})(\epsilon_{\mathbf{p}} - \epsilon_{\mathbf{p}+\hbar\mathbf{k}})$$

$$= \frac{2k^2}{m\omega^2}\int \frac{d\mathbf{p}}{(2\pi\hbar)^3}f_{\mathbf{p}} = \frac{k^2}{m\omega^2}n_e. \tag{8.72}$$

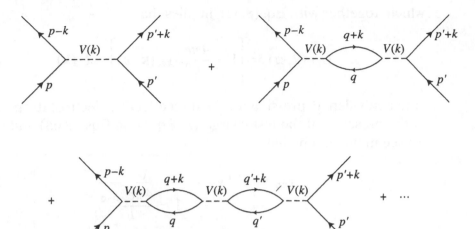

Figure 8.1 Diagramatic expansion in the RPA for the screened electron-electron interaction. Each bubble represents a particle-hole excitation. Mathematically, these excitations are described by the polarization function $\chi_{nn}^0(\mathbf{k}, \omega)$. The momentum exchanged between the particle and the hole is carried away by the Coulomb interaction at each dotted line, as indicated. As a result, the argument of the Coulomb interaction is decoupled from the momentum summation in each bubble. Consequently, all such terms can be summed exactly. The result is Eq. (8.71).

The high-frequency limit of the dielectric function,

$$\varepsilon = 1 - \frac{4\pi e^2 n_e}{mk^2}\frac{k^2}{\omega^2}$$

$$= 1 - \left(\frac{\omega_p}{\omega}\right)^2, \tag{8.73}$$

is fundamentally related to the plasma frequency. In fact, as we will see later, the plasma frequency is an exact zero of the dielectric function.

8.4.1 Structure Function

The time-dependent density response function defined in the previous section is a fundamental quantity in many-body theory. In addition to the screening function, the structure function, as well as the total energy of an interacting system, can all be written in terms of χ_{nn}. The principal reason for this is that the potential in most many-body systems is typically a sum of pair-wise interactions. In this section, we focus on calculating the structure function, as it will play a prominent role in later topics.

The structure function is defined as the autocorrelation function of the Fourier components

$$S(\mathbf{k}) = \frac{1}{N}\langle n_{\mathbf{k}} n_{-\mathbf{k}}\rangle \tag{8.74}$$

of the particle density. In equilibrium neutron-scattering experiments, the central quantity that is measured is the structure factor. As advertised, we can also write the total average energy of an interacting electron system in terms of $S(\mathbf{k}, \omega)$. This can be done trivially by computing the average value of the Hamiltonian for an interacting electron gas

$$\langle H\rangle = E_0 = \epsilon_{\text{kin}} + \sum_{\mathbf{k}} \frac{2\pi e^2}{k^2}[S(\mathbf{k}) - N] \tag{8.75}$$

in momentum space (see Eq. 8.18).

In the time domain, the structure function becomes

$$S(\mathbf{k}, t) = \frac{1}{N}\left\langle \sum_i e^{i\mathbf{k}\cdot\mathbf{r}_i(0)} \sum_j e^{-i\mathbf{k}\cdot\mathbf{r}_j(t)}\right\rangle. \tag{8.76}$$

By noting that

$$e^{i\mathbf{k}\cdot\mathbf{r}_i} = \int d\mathbf{r}\, e^{i\mathbf{k}\cdot\mathbf{r}}\delta(\mathbf{r} - \mathbf{r}_i) \tag{8.77}$$

and the density at $\mathbf{r}$ is

$$n(\mathbf{r}) = \sum_{i=1}^{N} \delta(\mathbf{r} - \mathbf{r}_i), \tag{8.78}$$

we rewrite Eq. (8.76) as

$$S(\mathbf{k}, t) = \frac{1}{N}\int d\mathbf{r}\, d\mathbf{r}'\langle n(\mathbf{r}, t = 0)n(\mathbf{r}', t)\rangle e^{i\mathbf{k}\cdot(\mathbf{r} - \mathbf{r}')}. \tag{8.79}$$

As we have seen, $\langle n(\mathbf{r}, t)n(\mathbf{r}', t')\rangle$ depends only on $\mathbf{r} - \mathbf{r}'$ and $t - t'$. Hence, our choice of the time origin at $t = 0$ does not affect our results. The dynamic structure factor is the time Fourier transform

$$S(\mathbf{k}, \omega) = \int_{-\infty}^{\infty} dt\, e^{-i\omega t} S(\mathbf{k}, t) \tag{8.80}$$

of $S(\mathbf{k}, t)$. It is $S(\mathbf{k}, \omega)$ that is measured in angle-resolved x-ray or neutron-scattering experiments. The static and dynamic structure factors are related through the simple sum rule

$$S(\mathbf{k}) = \int_{-\infty}^{\infty} \frac{d\omega}{2\pi} S(\mathbf{k}, \omega). \tag{8.81}$$

For systems with inversion symmetry, such as most solids and all fluids, $S(\mathbf{k}, \omega)$ is invariant under a change of sign of $\mathbf{k}$: $S(\mathbf{k}, \omega) = S(-\mathbf{k}, \omega)$. From Eq. (8.47), it follows that $S(\mathbf{k}, \omega) = e^{\beta \hbar \omega} S(\mathbf{k}, -\omega)$. This relationship reflects the principle of detailed balance.

From Eq. (8.51), we have

$$i \hbar \chi_{nn}(\mathbf{k}, \omega) = n_e \int_{-\infty}^{0} dt\, e^{-i\omega t} \int_{-\infty}^{\infty} \frac{d\omega'}{2\pi} e^{i\omega' t} [S(\mathbf{k}, \omega') - S(\mathbf{k}, -\omega')]$$

$$= n_e \int_{-\infty}^{\infty} \frac{d\omega'}{2\pi i} \frac{S(\mathbf{k}, \omega') - S(\mathbf{k}, -\omega')}{\omega' - \omega}$$

$$= n_e \int_{-\infty}^{\infty} \frac{d\omega'}{2\pi i} (1 - e^{-\beta \hbar \omega'}) \frac{S(\mathbf{k}, \omega')}{\omega' - \omega}. \tag{8.82}$$

We see, then, that the density response function, $\chi_{nn}(\mathbf{k}, \omega)$, can in principle be determined from experiment, once $S(\mathbf{k}, \omega)$ is known.

Parameter Differentiation of Total Energy

An expression identical to Eq. (8.75) can be derived, using parameter differentiation. We consider a variation of the ground state energy with respect to e^2:

$$\frac{\partial}{\partial e^2} E_0(e^2) = \frac{\partial}{\partial e^2} \langle \psi(e^2) | H | \psi(e^2) \rangle$$

$$= \left(\frac{\partial}{\partial e^2} \langle \psi(e^2) | \right) H | \psi(e^2) \rangle + \langle \psi(e^2) | H \left(\frac{\partial}{\partial e^2} | \psi(e^2) \rangle \right)$$

$$+ \langle \psi(e^2) | \frac{\partial H}{\partial e^2} | \psi(e^2) \rangle$$

$$= E_0(e^2) \frac{\partial}{\partial e^2} \langle \psi(e^2) | \psi(e^2) \rangle + \langle \frac{\partial H}{\partial e^2} \rangle = \langle \psi | \frac{\partial H}{\partial e^2} | \psi \rangle. \tag{8.83}$$

We have used the fact that $H|\psi(e^2)\rangle = E_0|\psi(e^2)\rangle$. Because the kinetic energy is independent of e^2 and $V \sim e^2$, we have that

$$\frac{\partial E_0}{\partial e^2} = \frac{1}{e^2}\langle\psi|V_e|\psi\rangle, \tag{8.84}$$

where V_e is the total potential for our interacting system:

$$V_e = V_{ee} + V_{ion-ion} + V_{e-ion}$$

$$= e^2\left[\frac{1}{2}\sum_{jj'}\frac{1}{|\mathbf{r}_j - \mathbf{r}_{j'}|} + \frac{1}{2}\int d\mathbf{r}d\mathbf{r}'\frac{n_e^2}{|\mathbf{r} - \mathbf{r}'|} - \sum_j\int d\mathbf{r}\frac{n_e}{|\mathbf{r} - \mathbf{r}_j|}\right]. \tag{8.85}$$

The second and third terms in the total potential represent the ion-ion and electron-ion interactions, respectively. The ions provide a homogeneous background of compensating positive charge for the electron gas. If we substitute the form for the density in Eq. (8.78), we can rewrite the total potential as

$$\frac{V_e}{e^2} = \frac{1}{2}\int d\mathbf{r}d\mathbf{r}'\frac{n(\mathbf{r})n(\mathbf{r}')}{|\mathbf{r} - \mathbf{r}'|} + \frac{n_e^2}{2}\int\frac{d\mathbf{r}d\mathbf{r}'}{|\mathbf{r} - \mathbf{r}'|} - n_e\int d\mathbf{r}d\mathbf{r}'\frac{n(\mathbf{r}')}{|\mathbf{r} - \mathbf{r}'|}$$

$$= \frac{1}{2}\int d\mathbf{r}d\mathbf{r}'\frac{(n(\mathbf{r}) - n_e)(n(\mathbf{r}') - n_e)}{|\mathbf{r} - \mathbf{r}'|}. \tag{8.86}$$

Consequently,

$$\frac{\partial E_0}{\partial e^2} = \frac{1}{2}\int d\mathbf{r}d\mathbf{r}'\frac{\langle\delta n(\mathbf{r})\delta n(\mathbf{r}')\rangle}{|\mathbf{r} - \mathbf{r}'|}$$

$$= \frac{n_e}{2}\int d\mathbf{r}d\mathbf{r}'\int\frac{d\mathbf{k}}{(2\pi)^3}\frac{e^{i\mathbf{k}\cdot(\mathbf{r}-\mathbf{r}')}}{|\mathbf{r} - \mathbf{r}'|}\int_{-\infty}^{\infty}\frac{d\omega}{2\pi}\tilde{S}(k,\omega)$$

$$= \frac{n_e V}{2}\int\frac{d\mathbf{k}}{(2\pi)^3}\frac{4\pi}{k^2}\int_{-\infty}^{\infty}\frac{d\omega}{2\pi}\tilde{S}(k,\omega), \tag{8.87}$$

where we introduced a rescaled structure factor

$$n_e\tilde{S}(\mathbf{k},\omega) = \int_{-\infty}^{\infty}dt e^{-i\omega t}e^{-i\mathbf{k}\cdot(\mathbf{r}-\mathbf{r}')}\langle\delta n(\mathbf{r})\delta n(\mathbf{r}',t)\rangle d(\mathbf{r} - \mathbf{r}'). \tag{8.88}$$

The advantage of this definition of the structure factor is that it eliminates the n_e^2-term that would normally appear in the energy. This term simply shifts the zero of the potential energy and, hence, is of no real consequence.

Using the sum rule in Eq. (8.81) for $S(\mathbf{k}, \omega)$ yields

$$E_0(e^2) = E_0(e^2 = 0) + \frac{N}{2} \int_0^{e^2} de'^2 \int \frac{d\mathbf{k}}{(2\pi)^3} \frac{4\pi}{k^2} \tilde{S}(\mathbf{k}; e'^2), \quad (8.89)$$

where we have allowed for explicit e^2-dependence in the static structure factor. This is an exact expression. If the free-particle form for $\tilde{S}(\mathbf{k})$ is used, Hartree-Fock theory results. Inclusion of the effects of screening allows us then to reduce $E_0(e^2)$ to the Gell-Mann–Brueckner perturbative expansion.

8.4.2   Evaluation of $\chi_{sc}(\mathbf{k}, \omega)$

Our goal now is to evaluate completely the effects of screening in the RPA. We start by rewriting the screening function as

$$\chi_{sc}(\mathbf{k}, \omega) = \chi_{nn}^0(\mathbf{k}, \omega) = 2 \int \frac{d\mathbf{p}}{(2\pi\hbar)^3} \frac{(f_\mathbf{p} - f_{\mathbf{p}+\hbar k})}{(\hbar\omega + \epsilon_\mathbf{p} - \epsilon_{\mathbf{p}+\hbar k})}$$

$$= 2 \lim_{\eta \to 0} \int_{-\infty}^{\infty} \frac{d\omega'}{\omega - \omega' + i\eta} \int \frac{d\mathbf{p}}{(2\pi\hbar)^3} (f_\mathbf{p} - f_{\mathbf{p}+\hbar k})$$

$$\times \delta(\hbar\omega' - \epsilon_{\mathbf{p}+\hbar k} + \epsilon_\mathbf{p}). \quad (8.90)$$

Using Eq. (8.82), we rewrite the right side in terms of the structure factor. We find that

$$4\pi\hbar \int \frac{d\mathbf{p}}{(2\pi\hbar)^3} (f_\mathbf{p} - f_{\mathbf{p}+\hbar k})\delta(\hbar\omega' - \epsilon_{\mathbf{p}+\hbar k} + \epsilon_\mathbf{p})$$

$$= n_e(1 - e^{-\beta\hbar\omega'})S_0(k, \omega'), \quad (8.91)$$

where $S_0(\mathbf{k}, \omega')$ is the structure factor for the free system. We simplify the left side of this expression by noting that $f_\mathbf{p} - f_{\mathbf{p}+\hbar k} = f_\mathbf{p}(1 - f_{\mathbf{p}+\hbar k}) - f_{\mathbf{p}+\hbar k}(1 - f_\mathbf{p})$ and $1 - f_\mathbf{p} = e^{\beta(\epsilon_\mathbf{p}-\mu)}f_\mathbf{p}$. Consequently,

$$f_\mathbf{p} - f_{\mathbf{p}+\hbar k} = f_\mathbf{p}(1 - f_{\mathbf{p}+\hbar k})(1 - e^{\beta(\epsilon_\mathbf{p}-\epsilon_{\mathbf{p}+\hbar k})}), \quad (8.92)$$

and the explicit temperature-dependent factor multiplying the free-particle structure function can be eliminated to yield

$$n_e S_0(\mathbf{k}, \omega) = 4\pi\hbar \int \frac{d\mathbf{p}}{(2\pi\hbar)^3} f_\mathbf{p}(1 - f_{\mathbf{p}+\hbar k})\delta(\hbar\omega - \epsilon_{\mathbf{p}+\hbar k} + \epsilon_\mathbf{p}).$$

$$(8.93)$$

The factor $1 - f_{\mathbf{p}+\hbar\mathbf{k}}$ is the probability that the state with momentum $\mathbf{p} + \hbar\mathbf{k}$ is empty. Hence, $S_0(\mathbf{k}, \omega)$ is determined by the number of ways a particle can exchange energy with a hole with a total energy change $\hbar\omega = \epsilon_{\mathbf{p}} - \epsilon_{\mathbf{p}+\hbar\mathbf{k}}$. In this sense, $S_0(\mathbf{k}, \omega)$ can be thought of as the effective density of states for particle-hole excitations.

To evaluate the integral in Eq. (8.93) at $T = 0$, we shift the momentum in $f_{\mathbf{p}+\hbar\mathbf{k}}$ by $-\hbar\mathbf{k}$, so that the resultant integrand

$$\int \frac{d\mathbf{p}}{(2\pi\hbar)^2}(f_{\mathbf{p}} - f_{\mathbf{p}+\hbar\mathbf{k}})\delta(\hbar\omega - \epsilon_{\mathbf{p}+\hbar\mathbf{k}} + \epsilon_{\mathbf{p}})$$

$$= \int \frac{d\mathbf{p}}{(2\pi\hbar)^2}f_{\mathbf{p}}[\delta(\hbar\omega - \epsilon_{\mathbf{p}+\hbar\mathbf{k}} + \epsilon_{\mathbf{p}}) - \delta(\hbar\omega - \epsilon_{\mathbf{p}} + \epsilon_{\mathbf{p}-\hbar\mathbf{k}})]$$

$$= \int \frac{d\mathbf{p}}{(2\pi\hbar)^2}f_{\mathbf{p}}\left[\delta\left(\hbar\omega - \frac{(\hbar k)^2}{2m} - \frac{\mathbf{p}\cdot\hbar\mathbf{k}}{m}\right) - \delta\left(\hbar\omega + \frac{(\hbar k)^2}{2m} + \frac{\mathbf{p}\cdot\hbar\mathbf{k}}{m}\right)\right]$$

$$= I_\omega - I_{-\omega}$$

$$(8.94)$$

will contain a single Fermi distribution function. We now transform to spherical coordinates and obtain

$$I_\omega = \frac{1}{\pi\hbar^2}\int_0^{PF} p^2 dp \int_{-1}^1 d\mu\, \delta\left(\hbar\omega - \frac{(\hbar k)^2}{2m} - \frac{\hbar k p \mu}{m}\right)$$

$$= \frac{m}{\pi k\hbar^3}\int_0^{PF} p\, dp\, \Theta\left(1 - \left|\frac{m}{\hbar k p}\left(\hbar\omega - \frac{(\hbar k)^2}{2m}\right)\right|\right)$$

$$= \frac{m}{\pi k\hbar^3}\int_{\frac{m}{\hbar k}\left|\hbar\omega - \frac{(\hbar k)^2}{2m}\right|}^{PF} p\, dp$$

$$= \frac{m}{2\pi k\hbar^3}\left[p_F^2 - \left(\frac{m}{\hbar k}\left(\omega - \frac{(\hbar k)^2}{2m}\right)\right)^2\right]$$

$$\times\, \Theta\left(p_F - \frac{m}{\hbar k}\left|\hbar\omega - \frac{(\hbar k)^2}{2m}\right|\right). \qquad (8.95)$$

Here $\Theta(x)$ is the Heaviside step function.

Subtracting the $\omega \to -\omega$ contribution, we find that

$$I_\omega - I_{-\omega} = \frac{m}{2\pi k \hbar^3}\left[p_F^2 - \left(\frac{m}{\hbar k}\left(\omega - \frac{(\hbar k)^2}{2m}\right)\right)^2\right]$$

$$\times \Theta\left(p_F - \frac{m}{\hbar k}\left|\hbar\omega - \frac{(\hbar k)^2}{2m}\right|\right)$$

$$- \frac{m}{2\pi k \hbar^3}\left[p_F^2 - \left(\frac{m}{\hbar k}\left(\hbar\omega + \frac{(\hbar k)^2}{2m}\right)\right)^2\right]$$

$$\times \Theta\left(p_F - \frac{m}{\hbar k}\left|\hbar\omega + \frac{(\hbar k)^2}{2m}\right|\right) \tag{8.96}$$

results. The Heaviside step function imposes the constraint

$$\frac{(\hbar k)^2}{2m} - \hbar k v_F < \omega < \frac{(\hbar k)^2}{2m} + \hbar k v_F \tag{8.97}$$

for the first term and

$$0 < \omega < \hbar k v_F - \frac{(\hbar k)^2}{2m} \tag{8.98}$$

for the second. For $\omega > 0$, the restrictions are represented graphically in Fig. (8.2) with $\omega_\pm = (\hbar k)^2/2m \pm \hbar k v_F$.

Because the range of ω for the first term in Eq. (8.96) exceeds that for the second, we consider three separate cases corresponding to

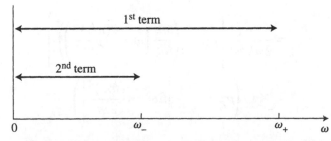

Figure 8.2 Frequency range for the zero-temperature structure function.

a) both terms contributing, b) only the first, and c) neither:

Case a) $0 < \omega < \hbar k v_F - (\hbar k)^2/2m$

$$I_\omega - I_{-\omega} = n_e S_0(\mathbf{k}, \omega) = \frac{m}{2\pi k \hbar^3}\left[p_F^2 - \left(\frac{m}{\hbar k}\left(\hbar\omega - \frac{(\hbar k)^2}{2m}\right)\right)^2 \right.$$

$$\left. - \left[p_F^2 - \left(\frac{m}{\hbar k}\left(\hbar\omega + \frac{(\hbar k)^2}{2m}\right)\right)^2 \right]\right] \qquad (8.99)$$

$$\Rightarrow n_e S_0(\mathbf{k}, \omega) = \frac{m^2 \omega}{\pi \hbar^2 k}; \qquad (8.100)$$

Case b) $\hbar k v_F - (\hbar k)^2/2m < \omega < (\hbar k)^2/2m + \hbar k v_F$

$$\Rightarrow n_e S_0(\mathbf{k}, \omega) = \frac{m}{2\pi k \hbar^3}\left[p_F^2 - \left(\frac{m}{\hbar k}\left(\hbar\omega - \frac{(\hbar k)^2}{2m}\right)\right)^2 \right]; \qquad (8.101)$$

Case c) $\hbar\omega > \dfrac{(\hbar k)^2}{2m} + \hbar k v_F$

$$\Rightarrow n_e S_0(\mathbf{k}, \omega) = 0. \qquad (8.102)$$

Figure (8.3) contains the composite graph for all three cases.

At finite temperature, an explicit expression can also be obtained for $n_e S_0(\mathbf{k}, \omega)$. We simply need to compute I_ω and then let $\omega \to -\omega$.

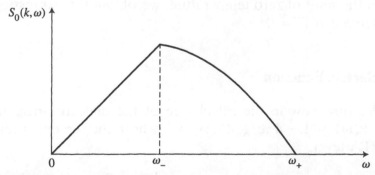

Figure 8.3 Zero-temperature function as predicted from Eqs. (8.100)–(8.102).

From Eq. (8.93), we have that

$$
\begin{aligned}
I_\omega &= \frac{m}{\pi k \hbar^3} \int_{(1/2\hbar k)|2m\hbar\omega - k^2\hbar^2|}^{P_F} \frac{p\, dp}{e^{\beta(\epsilon_p - \mu)} + 1} \\
&= \frac{m^2}{\beta \pi k \hbar^3} \int_a^{P_F} \left(\frac{\beta p}{m}\right) dp \, \frac{1}{e^{\beta(\epsilon_p - \mu)} + 1} \\
&= \frac{m^2}{\beta \pi k \hbar^3} \int_{\beta a^2/2m}^{\beta p_F^2/2m} \frac{dx}{e^{x - \beta\mu} + 1}.
\end{aligned}
\tag{8.103}
$$

With the help of the integral

$$
\int \frac{dx}{1 + b e^{cx}} = \frac{1}{\alpha c} \left[cx - \ln\left(\alpha + b e^{cx}\right) \right],
\tag{8.104}
$$

which implies that

$$
I_\omega = \frac{m^2}{\beta \pi \hbar^3 k} \left[x - \ln\left(1 + e^{x - \beta\mu}\right) \right]_{\beta a^2/2m}^{\beta p_F^2/2m},
\tag{8.105}
$$

we obtain the final expression for the temperature-dependent structure

$$
(1 - e^{-\beta\hbar\omega}) n_e S_0(\mathbf{k}, \omega) = I_\omega - I_{-\omega}
$$

$$
= \frac{m^2}{\beta \pi \hbar^3 k} \left[\beta\hbar\omega + \ln \frac{1 + \exp\left(\beta\left(\frac{1}{2m}\left(\frac{m\hbar\omega}{k} - \frac{\hbar k}{2}\right)^2 - \mu\right)\right)}{1 + \exp\left(\beta\left(\frac{1}{2m}\left(\frac{m\hbar\omega}{k} + \frac{\hbar k}{2}\right)^2 - \mu\right)\right)} \right].
$$

In the limit of zero temperature, we obtain the expression previously derived at $T = 0$.

8.4.3   Dielectric Function

We turn now to the calculation of the dielectric-response function, $\varepsilon(\mathbf{k}, \omega) = 1 - 4\pi e^2 \chi_{sc}(\mathbf{k}, \omega)/k^2$, where the screening function at the RPA level,

$$
\chi_{sc}(\mathbf{k}, \omega) = \lim_{\eta \to 0} \int_{-\infty}^{\infty} \frac{d\omega'}{2\pi\hbar} \frac{n_e S_0(\mathbf{k}, \omega')(1 - e^{-\beta\hbar\omega'})}{\omega - \omega' + i\eta},
\tag{8.106}
$$

is a convolution of the structure function. In the limit that $\eta \to 0$, the screening function will acquire real and imaginary parts through

$$\lim_{\eta \to 0} \frac{1}{\omega - \omega' + i\eta} = P \frac{1}{\omega - \omega'} - i\pi\delta(\omega' - \omega). \qquad (8.107)$$

The corresponding real $(1 + \varepsilon_R(\mathbf{k}, \omega))$ and imaginary (ε_I) parts of the dielectric function are

$$1 + \varepsilon_R(\mathbf{k}, \omega) = 1 - \frac{4\pi e^2}{k^2} P \int_{-\infty}^{\infty} \frac{d\omega'}{2\pi\hbar} \frac{n_e S_0(\mathbf{k}, \omega')(1 - e^{-\beta\hbar\omega'})}{(\omega - \omega')} \qquad (8.108)$$

and

$$\varepsilon_I = \frac{2\pi e^2}{k^2\hbar} n_e S_0(\mathbf{k}, \omega)(1 - e^{-\beta\hbar\omega}), \qquad (8.109)$$

respectively. In the limit of zero temperature, $S_0(\mathbf{k}, \omega)$ is linear in frequency and, hence, ε_I is an odd function of frequency.

The real and imaginary parts of the dielectric function are related as a result of the causal nature of the response to the test charge. From the definition of ε_R and ε_I, it follows immediately that

$$\varepsilon_R = \frac{1}{\pi} P \int_{-\infty}^{\infty} \frac{d\omega' \varepsilon_I(\mathbf{k}, \omega')}{\omega' - \omega} \qquad (8.110)$$

and

$$\varepsilon_I = \frac{-P}{\pi} \int_{-\infty}^{\infty} \frac{d\omega' \varepsilon_R(\mathbf{k}, \omega')}{\omega' - \omega}. \qquad (8.111)$$

These relationships are known as the *Kramers-Kronig relationships*. Relationships of this sort are true in general for any complex function that is analytic in either the upper- or lower-half planes. In the context of linear-response theory, they stem fundamentally from the causal nature of the response to the time-dependent perturbation.

Lindhard [L1954] has shown that at $T = 0$, ε_R is given by

$$\varepsilon_R = \frac{\kappa_{TF}^2}{k^2} \left\{ \frac{1}{2} + \frac{k_F}{4k} \left[\left[1 - \frac{(\omega - \hbar k^2/2m)^2}{k^2 v_F^2} \right] \ln \left| \frac{\omega - kv_F - \hbar k^2/2m}{\omega + kv_F - \hbar k^2/2m} \right| \right. \right.$$

$$\left. \left. + \left\{ 1 - \frac{(\omega + \hbar k^2/2m)^2}{k^2 v_F^2} \right\} \ln \left| \frac{\omega + kv_F + \hbar k^2/2m}{\omega - kv_F + \hbar k^2/2m} \right| \right] \right\}. \qquad (8.112)$$

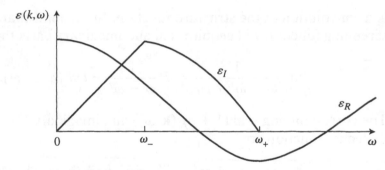

Figure 8.4 Real (ε_R) and imaginary (ε_I) parts of the dielectric function at $T = 0$.

Note first that this function is independent of the sign of ω. Hence, its parity is opposite that of ε_I. The general frequency dependence of the dielectric function is shown in Fig. (8.4). The large value of ε_R for $\omega \to 0$ indicates that the static screening is large. Another feature of the dielectric-response function is that $\varepsilon(\mathbf{k}, \omega) = 0$ at the plasma frequency. The poles of $\varepsilon^{-1}(\mathbf{k}, \omega)$ occur at the excitation frequencies of the electron gas. Recall that this is precisely the result we derived previously in the context of a small k and large ω expansion for the dielectric function. When $\varepsilon(\mathbf{k}, \omega) = 0$, fluctuations in the electron density diverge as a result of the collective nature of plasma oscillations. At this point, the whole theory we have formulated breaks down, because we assumed that the electron density was a slowly varying function of the perturbing field. Let us investigate the behavior of $\varepsilon(\mathbf{k}, \omega)$ for small ω. Setting $\omega = 0$ in our expression for ε_R results in

$$\varepsilon(x, \omega = 0) = 1 + \varepsilon_R(x, \omega = 0)$$

$$= 1 + \frac{\kappa_{TF}^2}{k^2} \left[\frac{1}{2} + \frac{1 - x^2}{4x} \ln \left| \frac{1+x}{1-x} \right| \right],$$

the dielectric function in the static limit with $x = k/2k_F$. The $x = 0$ limit of $\varepsilon(x, \omega = 0) = 1 + \kappa_{TF}^2/k^2$ is exactly the Thomas-Fermi approximation to the screening of a positive charge at the origin.

As in the Thomas-Fermi case, we can construct the spatial potential that results from this kind of screening effect. To do so, we use the equation for the effective field $U_{eff}(k, \omega) = \varepsilon^{-1}(\mathbf{k}, \omega)U(\mathbf{k}, \omega)$. For an electron gas, $U(\mathbf{k}, \omega = 0) = 4\pi e^2/k^2$. If we use our expression for $\varepsilon(\mathbf{k}, \omega = 0)$, we find that

$$\phi_{eff}(\mathbf{r}) = 4\pi e \int \frac{d\mathbf{k}}{(2\pi)^3} \frac{e^{-i\mathbf{k}\cdot\mathbf{r}}}{k^2 + \kappa_{TF}^2 Q(\mathbf{k})} \tag{8.113}$$

is the spatial dependence of the effective potential, where $Q(\mathbf{k})$ denotes the bracketed term in Eq. (8.113). When $k = 2k_F$, $Q(\mathbf{k})$ is logarithmically divergent. This divergence yields a contribution to the electrostatic potential of the form

$$\phi_{\text{eff}} \sim \frac{\cos 2k_F r}{r^3}, \tag{8.114}$$

as $r \to \infty$. This oscillatory behavior of the electrostatic potential is a consequence solely of screening and is known as a *Friedel oscillation* [F1954]. At long distances, then, we find that the charge is not sufficiently screened to give rise to the $e^{-\kappa_{\text{TF}}r}/r$ of Thomas-Fermi theory. Algebraic decay of the electrostatic potential signifies that a localized external charge affects the charge density everywhere in the electron gas. Kohn [K1959] was first to argue that this slow decay of the screened electrostatic potential arises from the sharpness of the Fermi surface. This effect shows up in the phonon spectrum of a metal for excitations with net momentum transfer $k > 2k_F$. He also pointed out with Luttinger [KL1965] that the negative contribution from ϕ_{eff} gives rise to a superconducting instability in an electron gas at $T = 0$. This observation is significant because it illustrates that if left alone, an electron gas with bare repulsive interactions can become superconducting without the assistance of phonons.

8.5 STOPPING POWER OF A PLASMA

When an electron is injected into a plasma with some incoming energy, it is expected to be slowed as a result of the Coulomb interactions with the electrons in the plasma. On these grounds, Bethe argued that the rate of energy loss of the injected electron should be proportional to $|V_{\text{int}}(\mathbf{p})|^2$, where V_{int} is the Coulomb interaction between the plasma and the electron. For a Coulomb potential, $V_{\text{int}} \sim p^{-2}$, as shown previously. Summing over all incoming momentum values, we find that the energy loss is given by

$$\frac{dE}{dt} \propto \int d\mathbf{p} \, p |V_{\text{int}}(\mathbf{p})|^2 \propto \int_0^{p_F} \frac{dp}{p}. \tag{8.115}$$

This integral is logarithmically divergent at the lower limit. As a result, this simple account produces a divergent energy loss, which is clearly incorrect. We see immediately that, for a screened interaction, the divergence at the lower limit would vanish, thereby making dE/dt finite. This is the primary failure of the Bethe approach. We will now formulate this problem in a rigorous way that gets around the Bethe divergence by including the effects of screening.

The physical problem at hand is that of a metal in some initial state $|i\rangle$ and an electron with initial momentum $\mathbf{p}$ impinging on a metal. Upon interacting with the metal, the electron will have a new momentum, $\mathbf{p} - \hbar\mathbf{k}$, and the metal will be in some new state $|f\rangle$. We assume the states of the metal form a complete orthonormal set $\langle n|m\rangle = \delta_{nm}$. At the level of Fermi's golden rule, the transition rate between the initial and final states is

$$W_{\mathbf{p},\mathbf{p}-\hbar\mathbf{k}} = \frac{2\pi}{\hbar} \sum_f |\langle f, \mathbf{p} - \hbar\mathbf{k}|V_{\text{int}}|i, \mathbf{p}\rangle|^2 \delta\left(\frac{\mathbf{p}^2}{2m} + E_i - \frac{(\mathbf{p} - \hbar\mathbf{k})^2}{2m} - E_f\right),$$

(8.116)

where E_i and E_f are the total energies of the states $|i\rangle$ and $|f\rangle$, respectively. The interaction energy

$$V_{\text{int}}(\mathbf{r}) = \int d\mathbf{r}'[n(\mathbf{r}') - n_e]\frac{e^2}{|\mathbf{r}' - \mathbf{r}|}$$

(8.117)

includes the ion as well as the electron Coulomb energy. The initial state of the electron is a plane wave of the form $|\mathbf{p}\rangle = e^{i\mathbf{p}\cdot\mathbf{r}/\hbar}/\sqrt{V}$ and the final electron state is $|\mathbf{p} - \hbar\mathbf{k}\rangle = e^{i(\mathbf{p}-\hbar\mathbf{k})\cdot\mathbf{r}/\hbar}/\sqrt{V}$. With these states, we rewrite the matrix element in Eq. (8.116) as

$$\langle f, \mathbf{p} - \hbar\mathbf{k}|V_{\text{int}}|i, \mathbf{p}\rangle = \int d\mathbf{r}\,d\mathbf{r}'\langle f|n(\mathbf{r}') - n_e|i\rangle\frac{e^{i\mathbf{k}\cdot\mathbf{r}}}{V}\frac{e^2}{|\mathbf{r} - \mathbf{r}'|}$$

$$= \frac{4\pi e^2}{k^2}\int \frac{d\mathbf{r}}{V}e^{i\mathbf{k}\cdot\mathbf{r}}\langle f|n(\mathbf{r})|i\rangle.$$

(8.118)

Using the integral representation of the δ-function,

$$2\pi\hbar\delta(\hbar\omega + E_i - E_f) = \int_{-\infty}^{\infty} e^{i(\hbar\omega + E_i - E_f)t/\hbar}dt,$$

we recast the transition rate as

$$W_{\mathbf{p},\mathbf{p}-\hbar\mathbf{k}} = \left(\frac{4\pi e^2}{V\hbar k^2}\right)^2 \int_{-\infty}^{\infty} e^{i(\hbar\omega + E_i - E_f)t/\hbar}dt \sum_{f\neq i} |\langle f|\int d\mathbf{r}\,n(\mathbf{r})e^{i\mathbf{k}\cdot\mathbf{r}}|i\rangle|^2$$

$$= \left(\frac{4\pi e^2}{V\hbar k^2}\right)^2 \int_{-\infty}^{\infty} dt\,e^{i\omega t}\sum_{f\neq i}\langle f|n(\mathbf{k})|i\rangle\langle i|e^{iE_i\frac{t}{\hbar}}n(\mathbf{k})e^{-iE_f\frac{t}{\hbar}}|f\rangle,$$

(8.119)

with $\hbar\omega = \epsilon_{\mathbf{p}} - \epsilon_{\mathbf{p}-\hbar\mathbf{k}}$. In the interaction representation,

$$\hat{n}(\mathbf{k}, t) = e^{iH_0\frac{t}{\hbar}} n(\mathbf{k}) e^{-iH_0\frac{t}{\hbar}}. \tag{8.120}$$

Consequently,

$$W_{\mathbf{p},\mathbf{p}-\hbar\mathbf{k}} = \left(\frac{4\pi e^2}{V\hbar k^2}\right)^2 \int_{-\infty}^{\infty} e^{i\omega t}\, dt \sum_{f\neq i} \langle i|\hat{n}(\mathbf{k}, t)|f\rangle\langle f|n(\mathbf{k})|i\rangle, \tag{8.121}$$

which can be simplified to

$$W_{\mathbf{p},\mathbf{p}-\hbar\mathbf{k}} = \left(\frac{4\pi e^2}{\hbar k^2}\right)^2 \frac{n_e}{V}\left[S(\mathbf{k}, \omega) - \frac{1}{V}2\pi\delta(\omega)n_e\right] \tag{8.122}$$

using the definition of the structure function and the completeness relation for the metal states, $\sum |f\rangle\langle f| = 1$. As expected, it is the dynamic structure factor that determines the response of our system to the incident electron. Screening effects are implicitly included in $S(\mathbf{k}, \omega)$. The Bethe result arises from the zero-frequency part of the transition rate.

We are primarily interested in the rate of energy loss to the plasma. This is determined by summing over all of the energy differences, $\epsilon_{\mathbf{p}} - \epsilon_{\mathbf{p}-\hbar\mathbf{k}}$, weighted by the transition rate, W:

$$\frac{dE}{dt} = -\sum_{\mathbf{k}} (\epsilon_{\mathbf{p}} - \epsilon_{\mathbf{p}-\hbar\mathbf{k}}) W_{\mathbf{p},\mathbf{p}-\hbar\mathbf{k}}$$

$$= -\frac{\hbar n_e}{V} \sum_{\mathbf{k}} \omega \left(\frac{4\pi e^2}{\hbar k^2}\right)^2 S(\mathbf{k}, \omega)$$

$$= -n_e \int \frac{d\mathbf{k}}{(2\pi)^3} \left(\frac{4\pi e^2}{\hbar k^2}\right)^2 \int_{-\infty}^{\infty} \omega d\omega S(\mathbf{k}, \omega) \times \delta(\hbar\omega - \epsilon_{\mathbf{p}} + \epsilon_{\mathbf{p}-\hbar\mathbf{k}}). \tag{8.123}$$

To evaluate this quantity, we switch to polar coordinates and perform first the θ integral for the angle between $\mathbf{p}$ and $\mathbf{p} - \hbar\mathbf{k}$:

$$2\pi \int_0^{\pi} d\theta \sin\theta \delta(\hbar\omega - \epsilon_{\mathbf{p}} + \epsilon_{\mathbf{p}-\hbar\mathbf{k}}) = 2\pi \int_{-1}^{1} dx \delta\left(\hbar\omega + \frac{\hbar^2 k^2}{2m} - \frac{p\hbar kx}{m}\right)$$

$$= \frac{2\pi m}{p\hbar k}\Theta(kv_p - |\omega + \hbar k^2/2m|). \tag{8.124}$$

With this result, the rate of energy loss to the plasma simplifies to

$$\frac{1}{v_p}\frac{dE}{dt} = \frac{-4e^4n_e}{\hbar^2 v_p^2}\int_0^\infty \frac{dk}{k^3}\int_{-kv_p-\frac{\hbar k^2}{2m}}^{kv_p-\frac{\hbar k^2}{2m}}\omega d\omega S(\mathbf{k},\omega), \qquad (8.125)$$

where v_p is the incoming velocity of the incident electron.

Complete stoppage of the electron by the plasma is most likely to occur if the electron gas acts collectively, that is, if plasma oscillations dominate. Thus, the electron gas obtains maximum stopping power if $|\omega| = \omega_p$, the plasma frequency. We seek, then, an expression for the structure function in the limit of high frequency. From the definition of the dielectric function (see Eqs. 8.64 and 8.82), we express the imaginary part of $\varepsilon(\mathbf{k},\omega)^{-1}$,

$$\mathrm{Im}\varepsilon^{-1} = -\frac{2\pi e^2 n_e}{\hbar k^2}\left(1 - e^{-\beta\hbar\omega}\right)S(\mathbf{k},\omega), \qquad (8.126)$$

in terms of the structure function and use the high-frequency expansion for the dielectric function

$$\varepsilon(\omega) \sim \lim_{\eta\to 0}\left(1 - \frac{\omega_p^2}{(\omega + i\eta)^2}\right). \qquad (8.127)$$

In this limit, the imaginary part of ε^{-1},

$$\mathrm{Im}\varepsilon^{-1}(\omega) = \lim_{\eta\to 0}\mathrm{Im}\left[\frac{\omega^2}{(\omega + i\eta)^2 - \omega_p^2}\right]$$

$$= -\frac{\pi\omega_p}{2}\left[\delta(\omega - \omega_p) - \delta(\omega + \omega_p)\right], \qquad (8.128)$$

is a sum of two δ functions at $\pm\omega_p$. With the aid of Eq. (8.126), we see immediately that in the $k \to 0$ limit,

$$S(\mathbf{k},\omega) = \frac{\hbar\pi k^2}{m}\frac{\left[\delta(\omega - \omega_p) - \delta(\omega + \omega_p)\right]}{1 - e^{-\beta\hbar\omega}}. \qquad (8.129)$$

The ω-integral in Eq. (8.125) is now straightforward:

$$\int_{\omega_\ell}^{\omega_u}\omega d\omega S(\mathbf{k},\omega) = \frac{\hbar\pi k^2}{m}(1 + g_p)\Theta(\omega_\ell < \omega_p < \omega_u)$$

$$- \frac{\pi k^2}{m}g_p\Theta(\omega_\ell < -\omega_p < \omega_u), \qquad (8.130)$$

where $\omega_\ell = -kv_p - \hbar k^2/2m$ and $\omega_u = kv_p - \hbar k^2/2m$. In evaluating this integral, we introduced $g_p = (e^{\beta\hbar\omega_p} - 1)^{-1}$, which determines the number of plasmons thermally excited at a temperature T. The energy loss is transformed to

$$\frac{1}{v_p}\frac{dE}{dt} = -\left(\frac{\omega_p e}{\hbar v_p}\right)^2 \int \frac{dk}{k}(1 + g_p)\Theta\left(\hbar k v_p - \frac{(\hbar k)^2}{2m} - \hbar\omega_p > 0\right)$$

$$- g_p \Theta\left(\frac{(\hbar k)^2}{2m} - \hbar k v_p < \hbar\omega_p < \frac{(\hbar k)^2}{2m} + \hbar k v_p\right). \quad (8.131)$$

The first term represents the energy loss upon the creation of a plasmon and the latter the energy transferred to the electron by a plasmon thermally excited in the medium. As $T \to 0$, the probability that a plasmon will be thermally excited vanishes. As a result, plasmons can be excited only by an impinging electron. In this limit, the energy loss takes on the simple form:

$$\frac{1}{v_p}\frac{dE}{dt} = -\left(\frac{\omega_p e}{\hbar v_p}\right)^2 \int \frac{dk}{k}\Theta\left(\hbar k v_p - \frac{\hbar^2 k^2}{2m} - \hbar\omega_p > 0\right)$$

$$= -\left(\frac{\omega_p e}{\hbar v_p}\right)^2 \ln\frac{k_+}{k_-}, \quad (8.132)$$

where $k_\pm$ are the solutions to

$$\hbar^2 k^2 - 2m\hbar k v_p + 2m\hbar\omega_p = 0, \quad (8.133)$$

or, equivalently,

$$\hbar k_\pm = p \pm \sqrt{p^2 - 2m\hbar\omega_p}. \quad (8.134)$$

For the incident electron to excite a plasmon, $\frac{p^2}{2m} > \omega_p$. If we expand $k_\pm$ in this limit, we find that

$$k_\pm = p \pm (p - m\hbar\omega_p/p) = \left\{\begin{array}{c} 2p - m\hbar\omega_p/p \\ m\hbar\omega_p/p \end{array}\right\}. \quad (8.135)$$

We expect an absence of collective plasmon oscillations if $k \lesssim a^{-1}$, where a is the interparticle spacing. We should then cut off k_+ at $2p_F$.

As a consequence,

$$\frac{dE}{dt} = -\frac{\omega_p^2 e^2}{v_p^2} \ln \frac{2pp_F}{m\hbar\omega_p}, \tag{8.136}$$

which is completely well behaved and finite, unlike the Bethe result.

8.6 SUMMARY

Electron gases exhibit a myriad of collective long-wavelength phenomena, such as plasma oscillations and screening, which alter significantly the independent electron picture of an electron gas. In the context of screening, an electron gas acts collectively to decrease the bare interaction from $V_{\mathbf{q}}$ to

$$V_{\text{eff}} = \frac{V_{\mathbf{q}}}{\varepsilon(q,\omega)}, \tag{8.137}$$

where $\varepsilon(q,\omega)$ is the dielectric function. At metallic densities, the RPA is sufficient to describe the screening effects. In this approximation, uncorrelated particle-hole scattering to all orders of perturbation theory are summed to obtain the polarization function, or polarization bubble. Exchange effects are not included in the RPA. In sufficiently dilute electron systems, $r_s > 3$, where the Coulomb energy is comparable to or greater than the electron kinetic energy, a theory beyond the RPA is necessary. Methods, such as the local-field approach [STLS1968], which attempt to model the correlation hole around each electron, have proven quite successful in this context (see Problem 5).

PROBLEMS

1. Redo the Thomas-Fermi screening problem for a $+Q$ charge located at the origin in dimensions d = 1 and d = 2. Assume the unscreened potential is Coulombic. Calculate explicitly the Thomas-Fermi screening length in d = 1 and d = 2. Show explicitly that in d = 2, the Thomas-Fermi screening length is independent of density and given by $1/\kappa_{\text{TF}} = a_0/2$.

2. In the Thomas-Fermi treatment of screening, evaluate explicitly the term proportional to U_{eff}^2 in Eq. (8.7). Under what condition can this term be ignored?

3. Use the method developed in the plasmon section to show that plasmons in $d = 2$ are described by a dispersion relationship that scales as $\omega_p \propto \sqrt{k}$.

4. Redo Bethe's argument in $d = 1$ and $d = 2$ for the stopping power of a plasma. Why is the result finite?

5. In the limit of small wave vectors, the dielectric function takes on the form [N1964],

$$\lim_{q \to 0} \varepsilon(q,0) = 1 + \frac{q_{\mathrm{TF}}^2}{q^2} \frac{\kappa}{\kappa_F}, \tag{8.138}$$

where κ is the compressibility of the electron gas and κ_F is the compressibility of the free system. For the free system, $\kappa_F = 3/(n_e \epsilon_F)$. Show that $\kappa/\kappa_F = 1$ in the RPA. Hubbard-[H1957] showed that exchange hole effects can be included by introducing the factor $G(q) = q^2/2(q^2 + k_F^2)$ into the dielectric function as

$$\varepsilon(q,\omega) = 1 - \frac{v_{\mathbf{q}} \chi_{nn}^0}{1 + v_{\mathbf{q}} \chi_{nn}^0 G(q)}. \tag{8.139}$$

Using Eqs. (8.138) and (8.139), derive an expression for the compressibility in the Hubbard approximation. Plot κ as a function of r_s. Determine the value of r_s at which κ changes sign. Physically, what does a negative compressibility, $\kappa < 0$, mean?

REFERENCES

[F1928] E. Fermi, *Z. Phys.* **48**, 73 (1928).

[F1954] H. Friedel, *Adv. Phys.* **3**, 446 (1954).

[H1957] J. Hubbard, *Proc. R. Soc. London Ser.* A **243**, 336 (1957).

[K1959] W. Kohn, *Phys. Rev. Lett.* **2**, 393 (1959).

[KL1965] W. Kohn, J. H. Luttinger, *Phys. Rev. Lett.* **15**, 524 (1965).

[L1954] J. Lindhard, *Kgl. Danske Videnskab. Selskab, Mat-fys. Medd.* **28**, 8 (1954).

[N1964] P. Noziéres, *Interacting Fermi Systems* (Benjamin, Reading, Mass., 1964), p. 287.

[P1963] D. Pines, *Elementary Excitations in Solids* (Benjamin, Reading, Mass., 1963).

[STLS1968] K. S. Singwi, M. P. Tosi, R. H. Rand, A. Sjölander, *Phys. Rev. Lett.* **176**, 589–599 (1968).

[T1927] L. H. Thomas, *Proc. Camb. Phil. Soc.* **23**, 542 (1927).

– 9 –

Bosonization

Let us look closely at the effective equation of motion for the Fourier components of the electron density in the limit where plasma oscillations occur. We showed in the previous chapter that in the limit $\omega_p^2 \gg k^2 v_F^2$, the equation of motion

$$\ddot{\rho}_k + \omega_p^2 \rho_k = 0 \tag{9.1}$$

resembles that of a harmonic oscillator. However, harmonic excitations obey Bose statistics. This suggests that plasma oscillations in some way represent bosonic excitations of the interacting electron gas. Equivalently, plasma oscillations can be thought of as harmonic excitations of the electrons in the gas. That this state of affairs holds profound consequences for the electron gas was pointed out initially by Tomonaga [T1950], who gave an explicit prescription for constructing the sound wave spectrum in a dense Fermi system. Three years later, Bohm and Pines [BP1953] showed that there is a natural connection between the sound wave (Bose) spectrum and the random-phase approximation. Since then, the equivalence between long-wavelength excitations in a dense Fermi system and a collection of bosons has proven to be of fundamental importance in solid state physics as well as in relativistic field theories. In this chapter, we focus on how a collection of interacting electrons in 1d obeying standard anticommutation relations can be described by an equivalent set of boson modes. The problem of constructing such a boson field theory for a collection of fermions is known as *bosonization*. In this context, we will first bosonize the 1d Hubbard [H1964] model and, in so doing, obtain the Luttinger liquid [L1960]. Luttinger liquids form the general basis for analyzing the properties of interacting electrons in 1d, in so far as such systems are dominated by short-range Coulomb interactions.

9.1 LUTTINGER LIQUID

Before we introduce the Luttinger model [L1960], let us review the form of the excitation spectra for noninteracting electrons. Electrons on a 1d lattice are described by the symmetric energy band, $\epsilon_k = -2t\cos ka - \epsilon_F$, where $-t$ is the nearest-neighbor-hopping integral among the lattice sites, with a the lattice constant. Electrons in free space obey the parabolic band $\epsilon_k = \hbar^2 k^2/2m - \epsilon_F$. Both of these bands are plotted in Fig. (9.1). At any filling of the single-particle

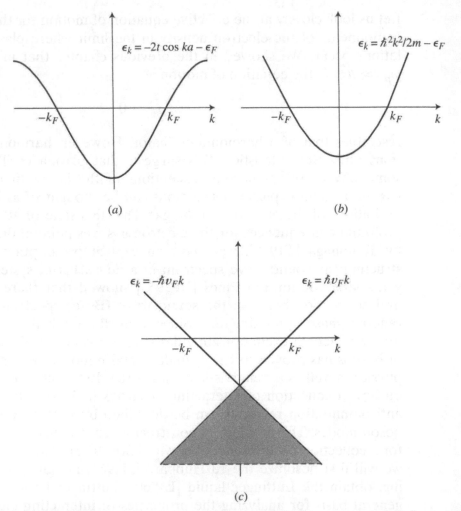

Figure 9.1 (a) Energy bands in the lattice, and (b) free-space models for noninteracting electrons in 1d. (c) Linearized energy band in the vicinity of the Fermi level. The shaded region represents the filling of the negative energy states, the Dirac sea.

states, the dispersion relation in the vicinity of the Fermi level can be linearized, resulting in the spectrum shown in Fig. (9.1c). In both the Tomonaga and the **Luttinger** models, the dispersion relations are linearized around the Fermi level. However, in the Tomonaga model, the negative energy states are neglected; that is, the dispersion relation is taken to be positive semi-definite.

Both of these models can be derived from the 1d Hubbard model,

$$H_L = -\frac{t}{2}\sum_{n\sigma}[\Psi_{n\sigma}^\dagger\Psi_{n+1\sigma} + h.c.] + \mu\sum_n\Psi_{n\sigma}^\dagger\Psi_{n\sigma} + U\sum_{n\sigma}\rho_{n\sigma}\rho_{n-\sigma}$$
$$= H_F + H_U, \tag{9.2}$$

in which the cost of doubly occupying any lattice site with electrons of opposite spin is the on-site Coulomb energy, U. In Chapter 6, on the Anderson impurity, we defined the on-site repulsion in terms of the on-site atomic orbitals. In Eq. (9.2), t is the hopping matrix element, $\Psi_{n\sigma}$ annihilates an electron on site n with spin σ, μ is the chemical potential for the ordered system, and $\rho_{n\sigma} = \Psi_{n\sigma}^\dagger\Psi_{n\sigma}$ is the electron density at the nth lattice site. Because of the factor of $-1/2$ in the definition of the hopping part of the Hamiltonian, the energy of the ordered band is $\mu - t\cos ka$, with a the lattice constant. In the ground state of the free system, all states with momentum $|k| \le k_F$ are doubly occupied. Hence, the chemical potential is defined as $\mu = t\cos k_F a$, so that $\epsilon_F = \mu$. Our goal is to recast H_L in an equivalent bosonic description in the linear approximation for the single-particle spectrum in the vicinity of the Fermi points, $\pm k_F$.

We now demonstrate how the dispersion relationship can be linearized for a 1d Fermi surface. We can accomplish this formally by writing the site amplitudes for each spin in the vicinity of $\pm k_F$ as a linear combination

$$\Psi_{n\sigma} = e^{ik_F na}\Psi_{n\sigma+} + e^{-ik_F na}\Psi_{n\sigma-} = R_\sigma(n) + L_\sigma(n) \tag{9.3}$$

of left-moving, $\Psi_{n\sigma-}$, and right-moving, $\Psi_{n\sigma+}$, fermion fields. The left-moving and right-moving fields are assumed to be slowly varying on the scale of the lattice constant. Further, they have mean momentum centered narrowly around $\pm k_F$. Consequently, we expand the fermion field for site $n + 1$ in terms of the amplitude for site n,

$$\Psi_{n+1\sigma\pm} = \Psi_{n\sigma\pm} + a\partial_x\Psi_{n\sigma\pm} + \cdots, \tag{9.4}$$

and retain only the linear term. Substituting this expression along with Eq. (9.3) into the kinetic energy part of the Hamiltonian will result in

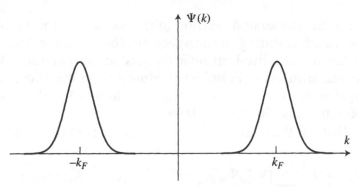

Figure 9.2 Fourier components of the fermion fields. In the bosonization procedure, we retain only the momentum components of the fermion fields at $\pm k_F$.

cross terms involving oscillating Fermi factors of the form $e^{\pm 2ik_F na}$. The contribution of such terms is determined by the overlap of the left- and right-moving fields. However, this overlap is zero because we are assuming the momentum spread of the $\pm$ fields is narrow, as depicted in Fig. (9.2). Hence, the only terms that survive in the linearized Hamiltonian,

$$H_F = -\hbar v_F \sum_\sigma \int_{-L/2}^{L/2} [\Psi_{\sigma+}^\dagger(x)(i\partial_x)\Psi_{\sigma+}(x)dx +$$

$$\Psi_{\sigma-}^\dagger(x)(-i\partial_x)\Psi_{\sigma-}(x)dx], \qquad (9.5)$$

are those involving products of fields with the same parity. The Fermi velocity is $\hbar v_F = at \sin k_F a$. In deriving Eq. (9.5), we transformed to the continuum representation. This limit is somewhat tricky. The continuum and lattice fields are related as follows: $\Psi_{n\sigma\pm} = \sqrt{a}\Psi_{\sigma\pm}(x)$. Hence, $\Psi_{\sigma\pm}(x)$ has units of $1/\sqrt{L}$. The second term in Eq. (9.5) is obtained by integrating by parts the Hermitian conjugate of the first term. To see that Eq. (9.5) is equivalent to a linearized fermion theory, substitute the momentum-space representation of $\Psi_{\sigma\pm}(x)$,

$$\Psi_{\sigma\pm}(x) = \int_{-\infty}^{\infty} \frac{dk}{2\pi} e^{ikx} \Psi_{\sigma\pm}(k), \qquad (9.6)$$

into Eq. (9.5). The resultant dispersion relationship is explicitly of the form $E_\pm(k) = \pm\hbar v_F k$, where the energy of left and right movers is distinct. Inclusion of negative-energy states is a key difference between the Tomonaga and Luttinger models.

The negative-energy states introduce, however, a subtle complexity into the Luttinger model. In the ground state, the negative-momentum branch of the right-moving states and the positive-momentum branch of the left-moving states are occupied. This is illustrated in Fig. (9.1b) and can be thought of as the filling of the negative-energy Dirac sea. The analogy here is between the positron states in Dirac theory and the negative-energy single-particle states in the Luttinger model. However, the Hamiltonian, Eq. (9.5), is expressed in terms of creation operators for right- and left-moving fields, $\Psi^\dagger_\pm$, that act on the space of zero particle states. In this space, the interaction term between the electrons vanishes. Put another way, our Hamiltonian lacks a ground state energy. Luttinger [L1960] alleviated this problem by performing a canonical transformation of the form

$$\Psi_{\sigma+}(k) \rightarrow \begin{cases} b_{k\sigma} & k \geq 0 \\ c^\dagger_{k\sigma} & k < 0 \end{cases} \qquad (9.7)$$

$$\Psi_{\sigma-}(k) \rightarrow \begin{cases} b_{k\sigma} & k < 0 \\ c^\dagger_{k\sigma} & k \geq 0 \end{cases} \qquad (9.8)$$

to the left- and right-moving fields, which effectively filled the negative-energy Dirac sea. The $b_{k\sigma}$'s and $c_{k\sigma}$'s obey the usual anticommutation relations. A consequence of this transformation is that the elementary excitations in momentum state k are pair excitations consisting of a particle and a hole. Further, it is linear combinations of these excitations that now describe the collective or plasma excitations of the electron gas. Luttinger, however, was not aware that this canonical transformation affected also the value of commutators involving the electron density. This is the profound change that the canonical transformation introduces into the Luttinger model. Mattis and Lieb [ML1965] were the first to point this out and correctly solve the Luttinger model.

To illustrate the changes this switch of basis introduces, we rewrite the Hamiltonian in terms of the new particle-hole operators. The transformed free Hamiltonian

$$H_F = \hbar v_F \int_{-\infty}^{\infty} |p| dp [b_p^\dagger b_p + c_p^\dagger c_p] + W \qquad (9.9)$$

contains the constant

$$W = -\hbar v_F \int_0^\infty p\, dp + \hbar v_F \int_{-\infty}^0 p\, dp, \qquad (9.10)$$

which represents the infinite energy of the filled Dirac sea. This result could have been obtained by normal ordering the operators in the original Hamiltonian. Consider an arbitrary reference state, $|\Omega\rangle$, and an operator, $\hat{A}$, which can be written as a product of creation and annihilation operators. Normal ordering the creation and annihilation operators in $\hat{A}$ means moving all the creation operators to the left and all the annihilation operators to the right. As a consequence, infinities arising from commutators are removed. We denote normal ordering by the symbol $: \hat{A} :$. Consequently, $: \hat{A} : |\Omega\rangle = 0$ and any average of $: \hat{A} :$ with respect to the ground state, $|\Omega\rangle$, identically vanishes, $\langle\Omega| : \hat{A} : |\Omega\rangle = 0$. Hence, the transformation in Eq. (9.7) is equivalent to normal ordering the left-right moving-field operators with respect to the filled Dirac sea. In so doing, one can extract the infinite energy of the filled negative-energy states.

Let us look also at the momentum-space representation of the density of right and left movers:

$$n_{\sigma\pm}(x) = \Psi^{\dagger}_{\sigma\pm}(x)\Psi_{\sigma\pm}(x)$$

$$= \int_{-\infty}^{\infty} \frac{dp}{2\pi}\frac{dq}{2\pi} \Psi^{\dagger}_{\sigma\pm}(p)\Psi_{\sigma\pm}(q)e^{i(q-p)x} \qquad (9.11)$$

with

$$n_{\sigma\pm}(q) = \int_{-\infty}^{\infty} dx e^{-iqx} n_{\sigma\pm}(x). \qquad (9.12)$$

Substituting the transformed form for the left and right movers into Eq. (12.15) and taking the Fourier transform (according to Eq. 9.12), we find that the form of $n_{\sigma\pm}(k)$,

$$n_{\sigma+}(k > 0) = \int_{0}^{\infty} dq \, [b^{\dagger}_q b_{k+q} + c_{-k-q}c^{\dagger}_{-q} + c_{-q}b_{k-q}\,\theta(k-q)]$$

$$n_{\sigma+}(k < 0) = \int_{0}^{\infty} dq \, [b^{\dagger}_{k-q}b_q + c_{-q}c^{\dagger}_{k-q} + b^{\dagger}_q c^{\dagger}_{k+q}\,\theta(-k-q)]$$

$$n_{\sigma-}(k > 0) = \int_{0}^{\infty} dq \, [b^{\dagger}_{-k-q}b_{-q} + c_q c^{\dagger}_{k+q} + b^{\dagger}_{-q}c^{\dagger}_{k-q}\,\theta(k-q)]$$

$$n_{\sigma-}(k < 0) = \int_{0}^{\infty} dq \, [b^{\dagger}_{-q}b_{k-q} + c_{q-k}c^{\dagger}_q + c_q b_{k+q}\,\theta(-k-q)],$$

depends on the sign of the momentum. The form derived here for the momentum components of the density is equivalent to the form given by Mattis and Lieb [ML1965]. The difference arises in the definition of the Fourier transform of the density defined in Eq. (9.12). The dependence of the density on the sign of the momentum arises from the pair nature of the fundamental excitations.

It is also from this complexity that differences arise when certain commutators are calculated in the original left-right and particle-hole bases. Consider the current operators for the total electron density

$$j_0^\sigma = \Psi_{\sigma+}^\dagger(x)\Psi_{\sigma+}(x) + \Psi_{\sigma-}^\dagger(x)\Psi_{\sigma-}(x) \qquad (9.13)$$

and the difference

$$j_1^\sigma = \hbar v_F(\Psi_{\sigma+}^\dagger(x)\Psi_{\sigma+}(x) - \Psi_{\sigma-}^\dagger(x)\Psi_{\sigma-}(x)) \qquad (9.14)$$

of the electron densities. In the original left-right basis, commutators involving j_0^σ and j_1^σ identically vanish. However, if the new particle-hole basis is used, Eq. (9.7), we find that the equal-time commutation relations between the components of the current,

$$[j_0^\sigma(x),j_1^\sigma(y)] = -\frac{i\hbar v_F}{\pi}\partial_x\,\delta(x-y) \qquad (9.15)$$

$$[j_0^\sigma(x),j_0^\sigma(y)] = [j_1^\sigma(x),j_1^\sigma(y)] = 0, \qquad (9.16)$$

obey Bose statistics. The right-hand side of the first commutator is a c-number. This discrepancy is a bit surprising but not totally unexpected if we write the density in first-quantized form,

$$n_{\sigma\pm}(k) = \int_{-L/2}^{L/2} dx\, e^{-ikx}. \qquad (9.17)$$

These components commute trivially. However, this is not the physically relevant form for the density operators when the Dirac sea is filled. Schwinger [S1959] anticipated this apparent paradox that arises when the very existence of a ground state forces certain commutators to be nonzero, which would vanish trivially otherwise in the first-quantized language. The full consequences for the Luttinger model were worked out by Mattis and Lieb [ML1965].

We turn now to the construction of an equivalent boson theory of the free part of the Luttinger model. To this end, we focus on the Heisenberg equations of motion,

$$-i\partial_t j_0^\sigma(x) = [H_F, j_0^\sigma(x)]. \qquad (9.18)$$

It turns out that the value of the commutator in Eq. (9.18) is independent of the filling in the Dirac sea. That is, this commutator, unlike those involving the components of the density, is invariant to normal ordering. The commutators of the densities for left and right movers with the free Hamiltonian,

$$[H_F, n_{\sigma+}(\pm k)] = \mp k n_{\sigma+}(\pm k)$$

$$[H_F, n_{\sigma-}(\pm k)] = \pm k n_{\sigma-}(\pm k), \tag{9.19}$$

depend on the sign of the momentum. Note the sign change relative to that of Mattis and Lieb. Here again, this difference arises from our definition of the Fourier transform, Eq. (9.12). Consequently,

$$\left[H_F, \int_{-\infty}^{\infty} dq\, e^{iqx} j_0^{\sigma}(q)\right] = \hbar v_F \int_0^{\infty} dq\, e^{iqx} q\, (n_{\sigma-}(q) - n_{\sigma+}(q))$$

$$+ \hbar v_F \int_0^{\infty} dq\, e^{-iqx} q\, (n_{\sigma+}(-q) - n_{\sigma-}(-q))$$

$$= \hbar v_F \int_{-\infty}^{\infty} dq\, e^{iqx} q\, (n_{\sigma-}(q) - n_{\sigma+}(q))$$

$$= i\partial_x j_1^{\sigma}, \tag{9.20}$$

or, equivalently,

$$\partial_t j_0^{\sigma} = -\partial_x j_1^{\sigma}. \tag{9.21}$$

If we introduce the notation, $t \to x_0$ and $x \to x_1$ with $x \equiv (x_0, x_1)$, we can rewrite the Heisenberg equation of motion as a conservation equation:

$$\partial_\mu j_\mu^{\sigma} = 0. \tag{9.22}$$

Hence, the total fermion current, normal ordered with respect to the full Dirac sea, is conserved.

Conservation of the fermion current suggests that if we define a boson field $\Phi_\sigma(x)$ with conjugate momentum $\Pi_\sigma(x)$ such that

$$j_0^{\sigma}(x) = \frac{1}{\sqrt{\pi}} \partial_x \Phi_\sigma(x) \tag{9.23}$$

and

$$j_1^{\sigma}(x) = -\frac{1}{\sqrt{\pi}} \partial_t \Phi_\sigma(x) = -\frac{1}{\sqrt{\pi}} \Pi_\sigma(x), \tag{9.24}$$

the Heisenberg equation of motion,

$$\partial_t[\partial_x\Phi_\sigma(x)] = \partial_x\Pi_\sigma(x), \qquad (9.25)$$

will resemble an equation of motion for two conjugate fields. To satisfy the Bose statistics of the currents,

$$[\Phi_\sigma(x), \Pi_{\sigma'}(y)] = i\delta_{\sigma\sigma'}\delta(x - y). \qquad (9.26)$$

In the Bose basis, the free Hamiltonian density,

$$H_F = \frac{\hbar v_F}{2}\sum_\sigma\int dx\,[\Pi_\sigma^2(x) + (\partial_x\Phi_\sigma(x))^2], \qquad (9.27)$$

yields an equation of motion that is equivalent to the Heisenberg evolution equation for the densities in the fermion basis. Hence, we have reformulated our fermion theory in terms of an equivalent boson theory. We emphasize that the boson theory has been constructed by an analogy based on the equations of motion. Hence, Eq. (9.27) is not a unique choice for the equivalent boson Hamiltonian density. By the construction of an equivalent boson theory, we imply no more than an equivalence between the equations of motion in the two accounts.

We can take the bosonization procedure a step further and construct explicitly the mapping between the fermion fields $\Psi_{\sigma\pm}(x)$ and the new boson field, $\Phi_\sigma(x)$. Such a mapping is problematic because the $\Psi_{\sigma\pm}(x)$'s are not the operators associated with the physical states of the electron gas. Recall that the $\psi_{\sigma\pm}$'s describe particle-hole excitations. Hence, processes in which the electron number is changed are forbidden. For the moment, we set this problem aside and develop a mapping involving the boson fields that preserves the anticommutation relations of the fermion fields. The key idea in bosonization is to associate (not equate) the left- and right-moving fermion fields with boson fields of the form,

$$\Psi_{\sigma\pm}(x) = \frac{1}{\sqrt{2\pi a}}e^{-i\sqrt{\pi}[\int_{-\infty}^{x}\Pi_\sigma(x')dx'\mp\Phi_\sigma(x)]}$$

$$= \frac{1}{\sqrt{2\pi a}}e^{\pm i\sqrt{\pi}\Phi_{\sigma\pm}(x)} \qquad (9.28)$$

where

$$\Phi_{\sigma\pm}(x) = \Phi_\sigma(x) \mp \int_{-\infty}^{x}\Pi_\sigma(x')dx'. \qquad (9.29)$$

It is crucial that the operator equivalence, Eq. (9.28), be understood strictly in the normal-ordered sense. For example, if Eq. (9.28) were naively substituted into the continuum version of the Hamiltonian, Eq. (9.5), the resultant boson Hamiltonian would correspond to twice Eq. (9.27). Such an equation would yield the incorrect equations of motion. This should underscore our disclaimer that the fermion fields should be associated, not equated, with the boson field, Eq. (9.28). True equalities arise in the two theories when average values of physical observables are computed. The physical motivation for Eq. (9.28) is as follows. The amplitude that a particle is at x is $\Psi_{\sigma\pm}(x)$. Classically, the evolution of a particle from x to $x + a$ is brought about by the displacement operator, $e^{a\partial_x}$. Quantum mechanically, this quantity becomes the exponential of the momentum operator. Hence, $\Psi_{\sigma\pm}(x)$ must be proportional to the exponential of the momentum operator. We call this quantity $\Pi_\sigma(x)$. However, if this is the only dependence, then $\Psi_{\sigma\pm}(x)$ would commute with itself. The simplest form that ensures the fermion anticommutation relations is Eq. (9.28).

To solve the problem that the bosonized form of the fermion fields cannot be used, in their current form, to connect electronic states that differ in particle number, we introduce ladder operators [KS1996, H1981], which change the particle number by integer values. Such ladder operators (traditionally known as *Klein factors*) must lie outside the space of the bosonic operators because bosons are necessarily composites of even numbers of fermions. No combination of bosonic operators can ever create a single electron. Let N_σ represent the deviation of the electron occupation number from the ground state value. Following the notation of Kotliar and Si [KS1996], we introduce the operator F_σ ($F_\sigma^\dagger$), which lowers (raises) N_σ by one. The F_σ's commute with all the boson fields in $\psi_{\sigma\pm}$ for $q \neq 0$. Hence, F_σ and N_σ represent the zero-momentum modes of the electron gas. The $q \neq 0$ modes are particle-hole excitations. The commutation relations of the F_σ's are

$$F_\sigma^\dagger F_\sigma = F_\sigma F_\sigma^\dagger = 1$$
$$F_\sigma^\dagger F_{\sigma'} = -F_{\sigma'} F_\sigma^\dagger$$
$$F_\sigma F_{\sigma'} = -F_{\sigma'} F_\sigma. \tag{9.30}$$

The physical states of the electron are now captured by

$$\tilde{\psi}_{\sigma\pm}^\dagger(x) = F_\sigma^\dagger e^{\frac{2\pi i x N_\sigma}{L}} \psi_{\sigma\pm}^\dagger. \tag{9.31}$$

For processes that do not conserve particle number, Klein factors must be included [KS1996, H1981, VDS1998]. However, as the problems on which we focus conserve particle number, we will omit the Klein factors, as they cannot change the physics. Nonetheless, they are part of the complete story of bosonization.

We now turn to the bosonization of the interaction terms. In the continuum limit, the Hubbard interaction can be written as

$$H_U = aU \int_{-L/2}^{L/2} dx \, (R_\uparrow^\dagger(x) + L_\uparrow^\dagger(x))(R_\uparrow(x) + L_\uparrow(x))$$

$$\times (R_\downarrow^\dagger(x) + L_\downarrow^\dagger(x))(R_\downarrow(x) + L_\downarrow(x)), \tag{9.32}$$

where $R_\sigma(x)(L_\sigma(x)) = \Psi_{n\sigma\pm}e^{\pm ik_F x}/\sqrt{a}$. To implement the bosonization scheme, we must normal order the operators in the Hubbard interaction. As in the free system, normal ordering of the operators in the Hubbard interaction,

$$H_U = : H_U : + Q, \tag{9.33}$$

will generate a constant term (Q) that is again infinite. In this case, the infinite term is associated with the infinite charge in the negative energy states. The important point here is that the infinite term is constant. Hence, it can be ignored.

If we now expand Eq. (9.32), keeping only the nonoscillatory terms, we find that the Hubbard interaction reduces to

$$H_U \to H_{\text{int}} = aU \int_{-L/2}^{L/2} dx \, [(: \Psi_{\uparrow+}^\dagger \Psi_{\uparrow+} : + : \Psi_{\uparrow-}^\dagger \Psi_{\uparrow-} :)$$

$$\times (: \Psi_{\downarrow+}^\dagger \Psi_{\downarrow+} : + : \Psi_{\downarrow-}^\dagger \Psi_{\downarrow-} :)$$

$$+ aU \, [: \Psi_{\uparrow+}^\dagger \Psi_{\uparrow-} :: \Psi_{\downarrow-}^\dagger \Psi_{\downarrow+} : + h.c.]]. \tag{9.34}$$

Diagrams illustrating the two types of scattering processes retained in H_{int} are shown in Fig. (9.3). In Fig. 9.3a, the net momentum transfer across each vertex is zero. Hence, the first term in Eq. (9.34) corresponds to forward scattering of the electrons. In the second term (see Fig. 9.3b), particles are converted from left (right) into right (left) movers. Hence, this term corresponds to a backscattering process. We have dropped the terms which have a prefactor of $e^{\pm i4k_F}$. At half-filling, $k_F = \pi/2$, and the exponential prefactor reduces to unity. The $4k_F$ term is referred to as an *Umklapp process* in which two right (left) movers are destroyed and two left (right) movers are created. We will come back to these terms later.

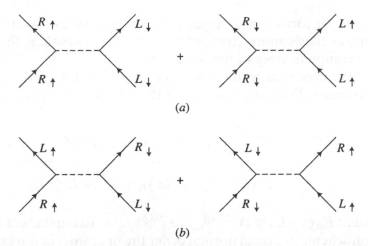

(a)

(b)

Figure 9.3 (a) Forward-scattering process corresponding to the first term in Eq. (9.34). $R(L)$ represent right- and left-moving fields. At each vertex of a forward-scattering term, the momentum change is zero. The wavy line indicates the Coulomb interaction aU. (b) Backscattering corresponds to the second term in Eq. (9.34). The magnitude of the momentum change at each vertex is $|2k_F|$.

To complete the bosonization scheme, we need a rule for writing products of fermion operators in Eq. (9.34) in terms of the Bose fields. To proceed, we use the Baker-Hausdorff identity,

$$e^A e^B = e^{A+B} e^{\frac{1}{2}[A,B]}. \tag{9.35}$$

Using the commutation relation, Eq. (9.26), we reduce the operator product,

$$\Psi^\dagger_{\sigma+} \Psi_{\sigma-} = \frac{1}{2\pi a} e^{-i\sqrt{4\pi}\Phi_\sigma(x)} e^{-\pi i/2}$$

$$= -i\frac{1}{2\pi a} e^{-i\sqrt{4\pi}\Phi_\sigma(x)}, \tag{9.36}$$

in the backscattering terms to a simple exponential of the Bose field. The factor of $-i$ arises from the commutator,

$$[\Phi_{\sigma+}(x), \Phi_{\sigma-}(y)] = \left[\Phi_\sigma(x), \int_{-\infty}^{y} \Pi(x')dx'\right] + \left[\Phi_\sigma(y), \int_{-\infty}^{x} \Pi_\sigma(x')dx'\right]$$

$$= i\int_{-\infty}^{y} dx'\delta(x-x') + i\int_{-\infty}^{x} dx'\delta(y-x')$$

$$= i. \tag{9.37}$$

Further simplification requires additional identities. The fermion field $\psi_{\sigma\pm}$ is an exponential of two fields that obey Bose statistics. In analogy with the harmonic oscillator, we partition the exponential into a sum of creation and annihilation operator parts. Consequently, the $\psi_{\sigma\pm}$ fields are of the form

$$\psi_{\sigma\pm} = e^A = e^{A^+ + A^-}, \tag{9.38}$$

where $A^+(A^-)$ represents the creation (annihilation) operator part of A. Using the Baker-Hausdorff operator identity, Eq. (9.35), we find that

$$e^A = e^{A^+} e^{A^-} e^{-\frac{1}{2}[A^+, A^-]}. \tag{9.39}$$

Because the normal ordering places all the creation operators to the left and the annihilation operators to the right,

$$: e^A := e^{A^+} e^{A^-} = e^{\frac{1}{2}[A^+, A^-]} e^A \tag{9.40}$$

is the corresponding normal-ordered form of the $\psi_{\sigma\pm}$ fields. This expression is of the utmost utility in the bosonization procedure. For example, it is common in the literature [M1975] to write the left-moving and right-moving fermion fields in the normal-ordered form. The normal-ordered form differs from Eq. (9.28) by the phase factor in Eq. (9.40). In the context of the back-scattering terms, we can use Eq. (9.40) to evaluate a product of two normal-ordered operators. Upon using Eq. (9.40) twice, we find that

$$: e^A :: e^B := e^{[A^-, B^+]} : e^{A+B} :, \tag{9.41}$$

where $[A^\pm, B^\pm] = 0$. In general $[A^\pm, B^\mp]$ is a c-number. For the case that $\Phi_{\sigma+} = e^A$ and $\Phi_{\sigma-} = e^B$, the commutator in the exponential factor of Eq. (9.41) has been evaluated by Mattis [M1974] and Mandelstam [M1975]. Its value is unity. As a consequence,

$$: \Psi_{\uparrow+}^\dagger \Psi_{\uparrow-} :: \Psi_{\downarrow-}^\dagger \Psi_{\downarrow+} : + h.c. = \frac{1}{2\pi^2 a^2} : \cos\left(\sqrt{4\pi}(\Phi_\uparrow(x) - \Phi_\downarrow(x))\right) : . \tag{9.42}$$

The forward-scattering terms are slightly more difficult because operator products of the form $\Psi_{\uparrow+}^\dagger(x + \epsilon)\Psi_{\uparrow+}(x - \epsilon)$ are not well defined in the limit that $\epsilon \to 0$. Naively, one might expect this quantity to equal unity. This is not so. We can evaluate this product by considering the fermion correlator,

$$
\langle \Psi_{\sigma+}^{\dagger}(x)\Psi_{\sigma+}(0)\rangle = \int_{-\infty}^{\infty} \frac{dk}{2\pi} \int_{-\infty}^{\infty} \frac{dq}{2\pi} e^{-iqx} \langle \Psi_{\sigma+}^{\dagger}(q)\Psi_{\sigma+}(k)\rangle
$$

$$
= \int_{0}^{\infty} \frac{dk}{2\pi} e^{ikx}. \tag{9.43}
$$

The double integrals contract to $2\pi\delta(k-q)$, with $k < 0$. The latter constraint arises because right movers exist only in the negative-momentum states in the ground state. Eq. (9.43) is not well behaved for large momentum, but it can be regularized by introducing the exponential factor, e^{-ak}, where the lattice constant acts as a short-distance cutoff. We find then that

$$
\langle \Psi_{\sigma+}^{\dagger}(x)\Psi_{\sigma+}(0)\rangle = \frac{1}{2\pi}\frac{1}{a-ix}. \tag{9.44}
$$

For left movers, the corresponding correlator is the complex conjugate of Eq. (9.44).

If we use the bosonized form of these field operators to evaluate the density correlator, we must obtain the same result. To proceed, we simplify a product of the form $e^A e^B$ using Eqs. (9.35) and (9.41),

$$
e^A e^B = : e^{A+B} : e^{[A^-,B^+]+\frac{[A^-,A^+]+[B^-,B^+]}{2}}. \tag{9.45}
$$

In general, the commutators appearing in Eq. (9.45) are c-numbers. Further, because $\langle (A^{\pm})^2\rangle = \langle A^+ A^-\rangle = 0$, it follows immediately that $\langle [A^-,A^+]\rangle = \langle A^2\rangle$. Consequently,

$$
\langle e^A e^B\rangle = : e^{A+B} : e^{\langle AB + \frac{A^2+B^2}{2}\rangle}. \tag{9.46}
$$

The average value of the exponential of any normal-ordered operator is unity, $\langle : e^A :\rangle = 1$. Hence, we obtain the all-important rule for evaluating a correlation function,

$$
\langle e^A e^B\rangle = e^{\langle AB + \frac{A^2+B^2}{2}\rangle}, \tag{9.47}
$$

from which it follows that

$$
\langle \Psi_{\sigma+}^{\dagger}(x)\Psi_{\sigma+}(0)\rangle = \frac{1}{2\pi a} e^{\pi\langle \Phi_{\sigma+}(x)\Phi_{\sigma+}(0)-\Phi_{\sigma+}^2(0)\rangle}. \tag{9.48}
$$

By analogy with Eq. (9.44), we find that

$$\langle \Psi_{\sigma+}^{\dagger}(x)\Psi_{\sigma+}(0)\rangle = \frac{1}{2\pi a}e^{\pi G_{+}(x)}$$

$$= \frac{1}{2\pi a}\frac{a}{a-ix} \qquad (9.49)$$

or, equivalently,

$$\langle e^{-i\eta\Psi_{\sigma\pm}}e^{i\eta\Psi_{\sigma\pm}}\rangle = \left(\frac{a}{a\mp ix}\right)^{\eta^2/\pi} \equiv e^{\eta^2 G_{\pm}}. \qquad (9.50)$$

We now have the tools to evaluate the original fermion product:

$$: \Psi_{\sigma\pm}^{\dagger}(x)\Psi_{\sigma\pm}(x) : = \lim_{\epsilon\to 0}\frac{1}{2\pi a} : e^{\mp i\sqrt{\pi}\Phi_{\sigma\pm}(x+\epsilon)\pm\Phi_{\sigma\pm}(x-\epsilon)} : e^{\pi G_{\pm}}$$

$$= \lim_{\epsilon\to 0}\frac{1}{2\pi a} : 1\mp\sqrt{\pi}\partial_x\Phi_{\sigma\pm}\epsilon + \cdots : \left(\pm\frac{ia}{\epsilon\pm ia}\right)$$

$$= \lim_{\epsilon\to 0}\frac{\pm i}{2\pi(\epsilon+ia)} + : \frac{1}{\sqrt{\pi}}\partial_x\Phi_{\sigma\pm} + O(\epsilon) :. \qquad (9.51)$$

For arbitrarily small a, the first term in this expression is infinite, a reflection of the infinite number of right and left movers in the negative-energy states. Normal ordering the operators in the charge density results in the removal of the divergent charge density in the ground state. As a consequence, the normal-ordered charge density,

$$: \Psi_{\sigma\pm}^{\dagger}(x)\Psi_{\sigma\pm}(x) : = \frac{1}{\sqrt{\pi}}\partial_x\Phi_{\sigma\pm}(x), \qquad (9.52)$$

has a well-defined interpretation in the boson basis. Noting that

$$: \Psi_{\sigma+}^{\dagger}(x)\Psi_{\sigma+}(x) : + : \Psi_{\sigma-}^{\dagger}(x)\Psi_{\sigma-}(x) : = \frac{1}{\sqrt{\pi}} : \partial_x\Phi_{\sigma}(x) : \qquad (9.53)$$

and using Eq. (9.42), we simplify the interaction terms to

$$H_{\text{int}} = aU\int dx \sum_{\sigma}\left[\frac{:\partial_x\Phi_{\uparrow}\partial_x\Phi_{\downarrow}:}{\pi}\right.$$

$$\left. + \frac{1}{2\pi^2 a^2} : \cos\left(\sqrt{4\pi}(\Phi_{\uparrow}(x) - \Phi_{\downarrow}(x))\right) :\right]. \qquad (9.54)$$

The utility of this expression and the bosonization procedure is made clear by introducing the charge, Φ_c, and spin, Φ_s, boson fields,

$$\Phi_c = \frac{\Phi_\uparrow + \Phi_\downarrow}{\sqrt{2}}, \qquad \Phi_s = \frac{\Phi_\uparrow - \Phi_\downarrow}{\sqrt{2}}. \tag{9.55}$$

The new charge and spin fields obey the usual Bose commutation relations, as they are simply sums and differences of Bose fields. If we substitute these expressions into the bosonized pieces of the Hamiltonian, we obtain a Hamiltonian, $H_B = H_c + H_s$, in which the charge,

$$H_c = \frac{\hbar v_F}{2} \int dx \, [\Pi_c^2 + g_c^2 (\partial_x \Phi_c)^2], \tag{9.56}$$

and spin degrees of freedom,

$$H_s = \frac{\hbar v_F}{2} \int dx \, [\Pi_s^2 + g_s^2 (\partial_x \Phi_s)^2] + \frac{U}{2\pi^2 a} \int dx : \cos \sqrt{8\pi} \Phi_s :, \tag{9.57}$$

are completely decoupled. The coupling constants

$$g_c^2 = 1 + \frac{aU}{2\pi\hbar v_F} \tag{9.58}$$

and

$$g_s^2 = 1 - \frac{aU}{2\pi\hbar v_F}, \tag{9.59}$$

which are now a function of the interactions, can be used to define new velocities for the spin and charge degrees of freedom. Let $v_F^c = v_F g_c$ and $v_F^s = v_F g_s$. To see that the spin and charge sectors are now moving with different velocities, we consider the transformation $\Phi_\nu \sqrt{g_\nu} \to \tilde{\Phi}_\nu$ and $\Pi_\nu 1/\sqrt{g_\nu} \to \tilde{\Pi}_\nu$, where $\nu = c, s$. This transformation leaves intact the Bose commutation relations, Eq. (9.37). In terms of the rescaled fields, the charge

$$H_c = \frac{\hbar v_F^c}{2} \int dx \, [\tilde{\Pi}_c^2 + (\partial_x \tilde{\Phi}_c)^2] \tag{9.60}$$

degrees of freedom resemble a collection of noninteracting fermions but with a new Fermi velocity, v_F^c, that increases as the strength of the on-site repulsions increases. In the spin sector,

$$H_s = \frac{\hbar v_F^s}{2} \int dx \, [\tilde{\Pi}_s^2 + (\partial_x \tilde{\Phi}_s)^2] + \frac{U}{2\pi^2 a} \int dx : \cos\sqrt{8\pi/g_s}\,\tilde{\Phi}_s :$$

(9.61)

and the new velocity, v_F^s, of the spins lags behind that of the charge degrees of freedom for repulsive interactions between the electrons. Hence, the spin and charge degrees of freedom move with completely different velocities. Another difference between the spin and charge sectors is the $:\cos\tilde{\Phi}_s:$ term. In the bosonized language, this factor is equivalent to a mass term in the original fermion Hamiltonian. Physically, the mass term represents an energy gap. However, only in the case of attractive Coulomb interactions [S1990], in which case electrons form bound antiferromagnetic pairs, does a spin gap arise, thereby heightening the difference between the spin and charge sectors in 1d. As a result, for 1d interacting electron liquids, the electron falls apart into distinct spin and charge quasi particles. We can think of an electron then as a composite particle made out of two distinct entities. We will refer to the entity that carries the charge as the *holon* (or *eon*) and the spin part as the *spinon*. Holons (or eons) and spinons obey Bose statistics. The emergence of holon and spinon excitations in 1d represents a marked departure from Fermi-liquid behavior. In Fermi-liquid theory, there is a one-to-one correspondence between the excited states of the interacting and noninteracting systems. Spin and charge separation in 1d fundamentally destroy this correspondence because now the electron gives rise to two excitations rather than the single excitation indicative of Fermi-liquid theory.

9.2 PAIR BINDING: CAN ELECTRONS DO IT ALONE?

It is well known from the early argument of Kohn and Luttinger (see Chapter 8) that any Fermi system has a superconducting instability (albeit at an ultra-low temperature on the order of 10^{-5}K) when electronic vertex corrections are considered. Since their work, there have been numerous arguments presented for purely electronic mechanisms of superconductivity. Rather than present a specific mechanism, we focus instead on a general way of understanding pair binding from purely electronic considerations by appealing to the holon-spinon construct. Let Φ_n be the energy of an electronic system containing n extra

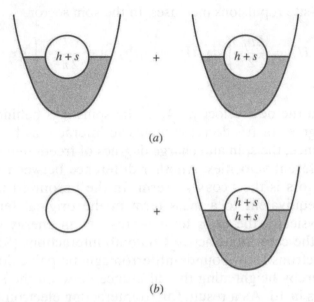

(a)

(b)

Figure 9.4 The two electronic systems that are used to compute the pair-binding energy for $n = 1$. Pair binding occurs if the energy of (b) is lower than the energy of (a). Whether or not this state of affairs obtains for purely electronic reasons depends on the relative magnitudes of the holon-holon, spinon-spinon, and holon-spinon interaction strengths.

electrons. The pair-binding question is as follows: do two isolated systems, each containing $n + 1$ and $n - 1$ extra electrons, have lower total energy than two separate systems with n extra electrons? The energy difference

$$E_{\text{pair}}^{(n)} = 2\Phi_n - \Phi_{n+1} - \Phi_{n-1} \qquad (9.62)$$

is the pair-binding energy. If $E_{\text{pair}}^{(n)} > 0$, pairing is expected, as a net attraction favors disproportionation into $n + 1$ and $n - 1$ states. To illustrate that pair binding can result from repulsive interactions alone, we adopt the spin-charge dichotomy of the Luttinger liquid. For simplicity, we consider the case in which $n = 1$, as illustrated in Fig. (9.4). To compute the pair-binding energy, we must evaluate $2\Phi_1 - \Phi_2 - \Phi_0$. We set $\Phi_0 = 0$ and let E_h and E_s be the energy of the holon and spinon, respectively. Also, we define V_{hh} and V_{hs} to be the holon-holon and holon-spinon interaction. The relevant physical system is illustrated in Fig. (9.4). Within an additive constant, the energy of $2\Phi_1$,

$$2\Phi_1 = 2(E_h + E_s + V_{hs}), \qquad (9.63)$$

is simply the sum of the energy of two independent electronic systems with an extra electron added and that of the doubly occupied state is

$$\Phi_2 = 2E_h + V_{hh} + 4V_{hs} + V_{ss} + 2E_s . \tag{9.64}$$

The factor of 4 arises from the two holon-spinon interactions for each electron and the cross interaction between the two electrons. We have assumed that the spinon energy in the singlet is zero. The energy difference between these states is

$$E_{pair}^{(1)} = -(2V_{hs} + V_{hh} + V_{ss}). \tag{9.65}$$

Now for an electron to be stable, $V_{hs} < 0$. Also Coulomb's law dictates that $V_{hh} > 0$. The spinon-spinon interaction energy then is the crucial quantity that determines whether or not $E_{pair}^{(1)} \gtrless 0$. In some situations, V_{ss} is of the right magnitude, such that $E_{pair}^{(1)} > 0$. In this case, pair binding can occur from purely repulsive electron interactions.

9.3   EXCITATION SPECTRUM

If we compare Eq. (9.60) with the standard Hamiltonian for a harmonic oscillator, $H = P^2 + Q^2$, where P and Q are the canonical momentum and position, we find an equivalence between the bosonized charge sector and a collection of harmonic oscillators. As advertised, charge excitations in the Luttinger model are the usual bosonic modes, known more commonly as *plasma excitations*. What is the appropriate density of states for these excitations? It might be suspected that they are governed by the standard Bose-Einstein distribution. However, this is not the case. Charge excitations in a Luttinger liquid [L1960] have a vanishing density of states at the Fermi level of the form, $|k - k_F|^\alpha$, which is due entirely to the correlations among the electrons. The power law is given by $\alpha = (g_c + g_c^{-1})/2 - 1$. Hence, rather than the smooth density of states indicative of Fermi liquids, Luttinger liquids develop a soft gap at the Fermi level for any $U \neq 0$. The source of this effect can be traced to the renormalization of v_F^c by g_c in the bosonized form of the electron gas.

We establish this result by calculating the total charge correlator

$$G(x) = \left\langle \sum_\sigma (\Psi_{\sigma+}(x)\Psi_{\sigma+}^\dagger(0) + \Psi_{\sigma-}(x)\Psi_{\sigma-}^\dagger(0)) \right\rangle. \tag{9.66}$$

To compact the notation, we define

$$\Theta_\sigma(x) = \int_{-\infty}^{x} \Pi_\sigma(x')dx'. \tag{9.67}$$

For each spin component, we make the substitution $\Phi_{\uparrow(\downarrow)} = (\Phi_c \pm \Phi_s)/\sqrt{2}$ and $\Theta_{\uparrow(\downarrow)} = (\Theta_c \pm \Theta_s)/\sqrt{2}$, where the upper sign applies to up spins and the lower sign to down spins. In the absence of a magnetic field, $n_\uparrow(x) = n_\downarrow(x)$. As a result, all averages linear in either Φ_s or Π_s vanish identically. We simplify the average in Eq. (9.66) even further by using the transformed fields $\tilde\Phi_\nu(x)$ and $\tilde\Pi_\nu(x)$. The average in Eq. (9.66) will now be over the ground state of the noninteracting bosonized system characterized by the renormalized Fermi velocities, v_F^c and v_F^s. Consequently, we can use the standard rules for simplifying the average of a product of exponentials of boson operators. Using the third form for the Baker-Hausdorff identity (Eq. 9.45), we find that the correlator of the total electron density simplifies to

$$G(x) = \frac{1}{\pi a} e^{i\,\text{phase}} e^{\frac{\pi}{2}\sum_{\nu=c,s}(g_\nu^{-1}a_\nu + g_\nu b_\nu)}, \tag{9.68}$$

where

$$a_\nu = \langle \tilde\Phi_\nu(x)\tilde\Phi_\nu(0) - \tilde\Phi_\nu^2(0)\rangle \tag{9.69}$$

and

$$b_\nu = \langle \tilde\Theta_\nu(x)\tilde\Theta_\nu(0) - \tilde\Theta_\nu^2(0)\rangle. \tag{9.70}$$

The phase is determined by the cross terms. For both charge and spin fields, the cross terms are of the form $\langle \tilde\Phi_\nu(0)\tilde\Theta_\nu(0) + \tilde\Theta_\nu(0)\tilde\Phi_\nu(0)\rangle - \langle \tilde\Phi_\nu(x)\tilde\Theta_\nu(0) + \tilde\Theta_\nu(x)\tilde\Phi_\nu(0)\rangle$. This term is identically zero (see Problem 4).

The simplest way to evaluate the averages in $G(x)$ is to introduce the Fourier expansion for free boson fields:

$$\tilde\Phi_\nu(x) = \int_{-\infty}^{\infty} \frac{dp}{2\pi\sqrt{2|p|}} e^{-a|p|/2}[\phi_\nu(x)e^{ipx} + \phi_\nu^\dagger(p)e^{-ipx}]$$

$$\tilde\Pi_\nu(x) = \int_{-\infty}^{\infty} \frac{dp\,|p|}{2\pi\sqrt{2|p|}} e^{-a|p|/2}[-i\phi_\nu(x)e^{ipx} + i\phi_\nu^\dagger(p)e^{-ipx}], \tag{9.71}$$

where

$$[\phi_\nu(p), \phi_\nu^\dagger(q)] = 2\pi\delta(q-p) \tag{9.72}$$

and $\nu = c, s$. From these expressions, it follows that

$$\langle \tilde{\Phi}_\nu(x)\tilde{\Phi}_\nu(0) - \tilde{\Phi}_\nu^2(0)\rangle = \frac{1}{4\pi}\ln\left(\frac{a^2}{a^2+x^2}\right). \tag{9.73}$$

The derivation of the analogous result for the conjugate field is a bit more complicated, and, hence, we perform the calculation explicitly. From the definition of $\tilde{\Theta}(x)$, we reduce the correlator of $\tilde{\Theta}_\nu(x)$ to

$$\langle \tilde{\Theta}_\nu(x)\tilde{\Theta}_\nu(0) - \tilde{\Theta}_\nu^2(0)\rangle = \int_0^x dx' \int_{-\infty}^0 dx'' \langle \tilde{\Pi}_\nu(x')\tilde{\Pi}_\nu(x'')\rangle. \tag{9.74}$$

To simplify the average over the conjugate momentum fields, we note that $\langle \phi_\nu(p)\phi_\nu^\dagger(q)\rangle = 2\pi\delta(p-q)$. Consequently,

$$\langle \tilde{\Pi}_\nu(x')\tilde{\Pi}_\nu(x'')\rangle = \int_{-\infty}^\infty \frac{dp\,|p|}{4\pi}e^{ip(x'-x'')}e^{-a|p|}$$

$$= \frac{-1}{2\pi}\left[\frac{(x'-x'')^2 - a^2}{(a^2+(x'-x'')^2)^2}\right]. \tag{9.75}$$

Substituting this expression into Eq. (9.74), we find that

$$\langle \tilde{\Theta}_\nu(x)\tilde{\Theta}_\nu(0) - \tilde{\Theta}_\nu^2(0)\rangle = \frac{1}{2\pi}\int_0^x dx' \int_\infty^{x'} dy\, \frac{y^2 - a^2}{(a^2+y^2)^2}$$

$$= \frac{-1}{2\pi}\int_0^x dx'\, \frac{y}{a^2+y^2} = \frac{1}{4\pi}\ln\frac{a^2}{a^2+x^2} \tag{9.76}$$

and, hence, the correlator (Eq. 9.68) of the total electron density

$$G(x) \propto |x|^{-\frac{g_c + g_c^{-1} + g_s + g_s^{-1}}{4}} \tag{9.77}$$

decays algebraically as a function of distance. Algebraic decay of the electron correlator is the key defining feature of 1d correlated-electron systems. The momentum-distribution function about k_F is defined through the Fourier transform

$$n(k) = \int_{-\frac{L}{2}}^{\frac{L}{2}} dx\, e^{i(k-k_F)x}G(x)$$

$$\propto \int_{-\frac{L}{2}}^{\frac{L}{2}} dx\, e^{i(k-k_F)x}|x|^{-\frac{g_c + g_c^{-1} + g_s + g_s^{-1}}{4}}. \tag{9.78}$$

Changing variables to $y = (k - k_F)x$, we find that the momentum distribution function for electronic excitations,

$$n(k) \propto |k - k_F|^{\frac{g_c + g_c^{-1} + g_s + g_s^{-1}}{4} - 1},\qquad(9.79)$$

vanishes algebraically in the vicinity of the Fermi level, as depicted in Fig. (9.5). For spinless electrons, the factor of $1/\sqrt{2}$ that arises from the transformation to charge and spin fields does not appear. As a result, the factor of $\pi/2$ in the exponent of Eq. (9.68) is replaced by π. As a consequence, the corresponding exponent for spinless electrons is $(g_c + g_c^{-1})/2 - 1$. In both cases, however, we recover the Fermi-liquid condition that the density of states is a nonzero constant at the Fermi level by setting $g_\nu = 1$. Algebraic vanishing of the distribution function at the Fermi level signifies that there are no well-defined quasi particles in a Luttinger liquid, unlike the Fermi-liquid case.

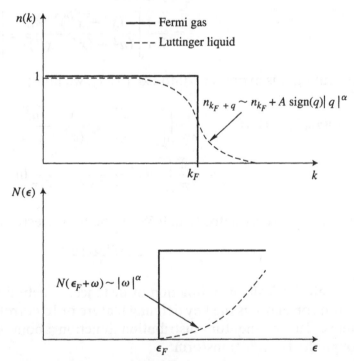

Figure 9.5 Shown here are the momentum distribution (upper figure) and tunneling density of states (lower figure) for a Luttinger liquid (dashed line) and a Fermi gas (solid line). The characteristic algebraic dependence of the momentum-distribution function in the vicinity of the Fermi energy is shown. In the Fermi gas limit, $\alpha = 0$ and the constant $A = -1/2$. The vanishing of the tunneling density of states at the Fermi level is responsible for the algebraic temperature dependence of the I-V characteristics shown in Fig. (9.6).

The characteristic value of $k - k_F$ at low temperature is given by $|k^2/2m - k_F^2/2m| \sim |k - k_F| v_F \sim T$. Hence, algebraic vanishing of the momentum-distribution function at the Fermi level translates into an algebraic temperature dependence of the form

$$n(T) \propto T^{\frac{g_c + g_c^{-1}}{2} - 1} \qquad (9.80)$$

for the excitation spectrum. Such algebraic scaling is expected to have profound experimental consequences. Consider a tunneling experiment in which electrons traverse a barrier into a Luttinger liquid. Transport through any system is determined by the distribution of electronic states available to the charge carriers. Hence, one would expect a vanishing of the conductance (transmittance) as $T \to 0$ across a barrier that separates two Luttinger liquids. As a consequence, rather than increase exponentially with a characteristic activation energy away from $T = 0$, the conductance should increase as a power law [KF1992]. This power law should appear in measurements of the current-voltage (I-V) characteristics for transport in Luttinger liquids. Experiments illustrating the power law dependence of the I-V characteristics in a system exhibiting Luttinger-liquid behavior were performed by Chang and co-workers [CPW1996]. Their data are shown in Fig. (9.6). In these particular experiments, the Luttinger liquid was the edge of a fractional quantum Hall state. For these samples, Kane and Fisher [KF1992] showed that the current scales as

$$I \propto T^{\alpha}[x + x^{\alpha}], \qquad (9.81)$$

with $\alpha = 3$ and $x = eV/2\pi k_F T$. Excellent agreement with theory was obtained for $\alpha = 2.7$ and the $1/2\pi$ scale factor between the voltage and the temperature. Similar results have also been obtained by Webb and co-workers [MUW1996] in quantum Hall systems and by Bockrath and colleagues on carbon nanotubes [B1999]. As a result, that electron correlations are at the heart of the power-law dependence of the current-voltage characteristics for 1d electronic systems is on firm experimental footing.

To close this chapter, we comment briefly on what happens when the Umklapp terms are retained at half-filling. In this case, a cosine term appears in the charge sector, which implies that there is a gap to excitations for repulsive interactions in the charge sector at half-filling. This signifies that the holons are not free to conduct without a huge energy cost. This situation arises because, at half-filling, there is exactly one electron per site. Hence, the energy cost to move an electron from site to site is U, the magnitude of the charge gap. If U were negative,

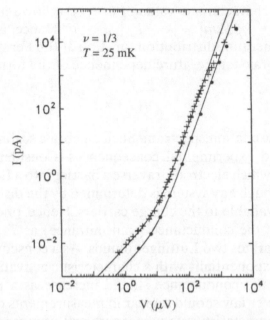

Figure 9.6 Current-voltage (I-V) characteristics for tunneling from bulk-doped n + GaAs into the edge of the fractional quantum Hall state with a filling factor in the lowest Landau level of $\nu = 1/3$. The solid circles and crosses correspond to magnetic fields of $B = 10.8T$ and $B = 13.4T$, respectively. The edge states in the fractional quantum Hall state are Luttinger liquids. Algebraic dependence of I versus V is expected to occur for tunneling into a Luttinger liquid as a result of the algebraic dependence of the tunneling density of states (see Eq. 9.79). The solid curves represent fits to the theoretical universal form, Eq. (9.81).

then the ground state at half-filling would correspond to two electrons tightly bound (with a binding energy of order U) per lattice site. Such an entity costs no energy to move. Hence, at half-filling, a charge gap exists only for repulsive interactions.

9.4 SUMMARY

In 1d, it is possible to develop a bosonized view of the collective excitations of an electron gas in the presence of short-range interactions. This level of description enables a clear demonstration that spin and charge move with fundamentally different velocities in 1d. It is yet unknown how spin and charge separation can be formally established for $d > 1$. In addition, Coulomb interactions in 1d give rise to algebraic decay of the density of states at the Fermi level. Such characteristic decay of the density of states is the signature of Luttinger-liquid behavior and is experimentally observable in tunneling experiments.

PROBLEMS

1. Show that if Eq. (9.28) were substituted into the continuum version of the Hamiltonian (Eq. 9.5), the resultant boson Hamiltonian would correspond to twice Eq. (9.27).

2. Use Eq. (9.40) and the Baker-Hausdorff identity (Eq. 9.35) to prove that

$$: e^A :: e^B := : e^{A+B} : e^{\frac{1}{2}[A^+, B^-]}. \tag{9.82}$$

3. Prove that the equal-time commutation relations between the components of the current are given by

$$[j_0^\sigma(x), j_1^\sigma(y)] = -\frac{i\hbar v_F}{\pi} \partial_x \delta(x - y) \tag{9.83}$$

$$[j_0^\sigma(x), j_0^\sigma(y)] = [j_1^\sigma(x), j_1^\sigma(y)] = 0. \tag{9.84}$$

4. Show that the phase factor in Eq. (9.68) identically vanishes. You might find the following plan of attack helpful: note first that Θ_ν and Φ_ν scale inversely with respect to g_ν. Hence, the cross terms are independent of the rescaling factors. From the continuity equation $\partial_\tau \Phi = \partial_x \theta$, you can establish that

$$\theta(x, \tau) = \int_{-\infty}^{x} dx' \partial_\tau \phi(x', \tau) \tag{9.85}$$

and, hence, that

$$\langle \tilde{\Phi}_\nu(x, \tau) \tilde{\Theta}_\nu(0) - \tilde{\Theta}_\nu(0) \tilde{\Phi}_\nu(0) \rangle$$

$$= -\int_{\infty}^{x} dx' \left\{ \partial_\tau G(x, x', \tau, 0)|_{\tau=0} - \partial_\tau G(0, x', \tau)|_{x'=0, \tau=0} \right\}, \tag{9.86}$$

where $G(x, x', \tau, \tau') = \langle \Phi_\nu(x, 0) \Phi_\nu(x', \tau) \rangle$. You should then be able to show that the remaining contribution enters with the opposite sign, and, hence, the sum vanishes.

REFERENCES

[B1999] M. Bockrath, D. H. Cobden, L. Jia, A. G. Rinzler, R. E. Smalley, L. Balents, P. L. McEuen, *Nature* (London) **397**, 598 (1999).
[BP1953] D. Bohm, D. Pines, *Phys. Rev.* **92**, 609 (1953).

[CPW1996] A. Chang, L. Pfeiffer, K. West, *Phys. Rev. Lett.* **77**, 2538 (1996).

[H1981] F. D. M. Haldane, *J. Phys. C: Sol. State Phys.* **14**, 2585 (1981).

[H1964] J. Hubbard, *Proc. Roy. Soc.* A **277**, 237, and 281 (1964).

[KF1992] C. L. Kane, M. P. A. Fisher, *Phys. Rev. Lett.* **68**, 1220 (1992).

[KS1996] G. Kotliar, Q. Si, *Phys. Rev.* B **53**, (1996).

[L1960] J. M. Luttinger, *Phys. Rev.* **119**, 1153 (1960).

[M1975] S. Mandelstam, *Phys. Rev.* D **11**, 3026 (1975).

[M1974] D. Mattis, *J. Math. Phys.* (N.Y.) **15**, 609 (1974).

[ML1965] D. Mattis, E. Lieb, *J. Math. Phys.* **6**, 304 (1965).

[MUW1996] F. P. Milliken, C. P. Umbach, R. A. Webb, *Solid State Commun.* **97**, 309 (1996).

[S1990] H. J. Schulz, *Phys. Rev. Lett.* **64**, 2831 (1990).

[S1959] J. Schwinger, *Phys. Rev. Lett.* **3**, 296 (1959).

[T1950] S. Tomonaga, *Prog. Theor. Phys.* (Kyoto) **5**, 544 (1950).

[VDS1998] J. von Delft, H. Schoeller, *Annalen Phys.* **7**, 225 (1998).

– 10 –

Electron-Lattice
Interactions

So far, we have focused on the electron problem, treating the ions as fixed in place at their equilibrium positions $\mathbf{R}_i^0$. In the context of the electron gas, we adopted an even simpler view of the ions, namely, that they provide a uniform background of compensating positive charge. To be able to describe the range of physics observed in a solid, we must invoke some realism into our treatment of ion motion. The coupling of electronic degrees of freedom with the motion of the ions is the electron-phonon problem. Phonons in a solid arise from collective motion of the ions. Such motion is quantized and fundamentally responsible for 1) polaron formation, 2) the electron attraction in superconductivity, and 3) the temperature dependence of the resistivity in metals. In this chapter, we focus on the general formulation of the electron-phonon problem and its subsequent application to the low-temperature resistivity in metals.

10.1 HARMONIC CHAIN

We begin with a brief review [M1981] of a 1d chain of N atoms joined by harmonic springs. Let x_i denote the deviation of each oscillator from its equilibrium position, ω the frequency of oscillation of each spring, and M the mass of each atom. The total Hamiltonian for this harmonic chain is

$$H = \sum_i \frac{P_i^2}{2M} + \frac{M\omega^2}{2} \sum_i (x_i - x_{i+1})^2. \qquad (10.1)$$

We diagonalize this Hamiltonian by Fourier transforming the momentum

$$P_n = \frac{1}{\sqrt{N}} \sum_k e^{ikna} P_k \tag{10.2}$$

and the displacement operators

$$x_n = \frac{1}{\sqrt{N}} \sum_k e^{ikna} x_k, \tag{10.3}$$

where a is the lattice constant. By noting that

$$\sum_n P_n^2 = \sum_k P_k P_{-k} \tag{10.4}$$

and

$$\sum_n x_n x_{n+m} = \sum_k x_k x_{-k} e^{-ikma}, \tag{10.5}$$

we rewrite the Hamiltonian in k-space:

$$H = \frac{1}{2M} \sum_k P_k P_{-k} + \frac{M}{2} \sum_k \omega_k^2 x_k x_{-k}, \tag{10.6}$$

where $\omega_k^2 = 2\omega^2(1 - \cos ka) = 4\omega^2 \sin^2 ka/2$. As a consequence, the $k = 0$ mode costs no energy to excite. This is a defining feature of acoustic phonons. The $k = 0$ mode corresponds to a uniform translation of the ions. By translational invariance, such a transformation cannot change the energy. Such long-wavelength bosonic excitations which cost no energy are called Goldstone modes [G1963]. In magnetic systems, such as ferromagnets and antiferromagnets, analogous long-wavelength excitations exist, which at $k = 0$ cost no energy. Such excitations, known as spin waves or magnons, constitute the low-energy excitations in magnetic systems and hence determine the magnetic contribution to the specific heat, for example.

Let us define new operators

$$\tilde{Q}_k = x_k \left(\frac{M\omega_k}{2\hbar} \right)^{1/2} \tag{10.7}$$

and

$$\tilde{P}_k = \frac{P_k}{(2M\omega_k \hbar)^{1/2}}, \tag{10.8}$$

which allow us to recast H in the suggestive form

$$H = \sum_k \hbar\omega_k [\tilde{P}_k \tilde{P}_{-k} + \tilde{Q}_k \tilde{Q}_{-k}]. \tag{10.9}$$

We can factorize H once we define the creation

$$b_k^\dagger = (\tilde{Q}_{-k} - i\tilde{P}_k) \tag{10.10}$$

and annihilation

$$b_k = (\tilde{Q}_k + i\tilde{P}_{-k}) \tag{10.11}$$

operators. The commutation relations obeyed by b_k and $b_k^\dagger$,

$$[b_k^\dagger, b_{k'}] = -\delta_{kk'}, \tag{10.12}$$

follow from the canonical commutator

$$[\tilde{P}_k, \tilde{Q}_k] = -\frac{i}{2}. \tag{10.13}$$

The factorized Hamiltonian takes on the familiar oscillator form

$$H = \sum_k \hbar\omega_k \left(b_k^\dagger b_k + \frac{1}{2}\right), \tag{10.14}$$

which is indicative of a collection of bosons. The time dependence of the b_k's,

$$b_k(t) = b_k(t = 0)e^{-i\omega_k t}, \tag{10.15}$$

is obtained by solving the Heisenberg equations of motion,

$$-i\hbar\dot{b}_k = [H, b_k] = -\hbar\omega_k b_k. \tag{10.16}$$

The operators $b_k^\dagger(t)$ create a collective lattice distortion with frequency ω_k at time t. The spatial resolution of this distortion is given by solving Eqs. (10.10) and (10.11) for x_k

$$x_k(t) = \frac{1}{2}\left(\frac{2\hbar}{M\omega_k}\right)^{1/2} (b_k(t) + b_{-k}^\dagger(t))$$

$$= \left(\frac{\hbar}{2M\omega_k}\right)^{1/2} (b_k e^{-i\omega_k t} + b_{-k}^\dagger e^{i\omega_k t}) \tag{10.17}$$

and then Fourier transforming

$$x_\ell(t) = \sum_k \left(\frac{\hbar}{2MN\omega_k}\right)^{1/2} (b_k e^{-i\omega_k t} + b^\dagger_{-k} e^{i\omega_k t}) e^{ik\ell a}. \quad (10.18)$$

This expression for $x_\ell(t)$ tells us the amplitude of the lattice vibration on site ℓ at time t. The sum over k is restricted to the first Brillouin zone.

10.2 ACOUSTIC PHONONS

To make more concrete contact with a solid, we consider a general pair-wise potential of interaction between ions:

$$V_{ion} = \sum_{i<j} V(\mathbf{R}_i - \mathbf{R}_j). \quad (10.19)$$

The equilibrium positions, $\{\mathbf{R}_i^0\}$, of the ions are determined by the condition that the net force on each ion vanishes:

$$\mathbf{F}_j = \sum_i \nabla V(\mathbf{R}_i^0 - \mathbf{R}_j^0) = 0. \quad (10.20)$$

Consequently, we represent the actual position of each ion

$$\mathbf{R}_i = \mathbf{R}_i^0 + \mathbf{Q}_i \quad (10.21)$$

by an expansion about the equilibrium positions. Here $\{\mathbf{Q}_i\}$ plays the role of $\{x_i\}$ in the linear chain, as they represent the displacement of each ion from its home position. The vanishing of the forces on each ion at the home positions guarantees that the Taylor expansion for the ion potential

$$V_{ion} = \sum_{i<j} V(\mathbf{R}_i^0 - \mathbf{R}_j^0) + \frac{1}{2}\sum_{i<j}(\mathbf{Q}_i - \mathbf{Q}_j)_\mu (\mathbf{Q}_i - \mathbf{Q}_j)_\nu F_{\mu\nu}, \quad (10.22)$$

with

$$F_{\mu\nu} = \frac{\partial^2}{\partial R_\mu \partial R_\nu} V(\mathbf{R}_i^0 - \mathbf{R}_j^0), \quad (10.23)$$

is harmonic to lowest order in the fluctuation about the minimum, where repeated indices are summed over.

As in the 1d chain, we diagonalize this interaction by defining the collective coordinate

$$\mathbf{Q}_i(t) = \sum_{\mathbf{k},\lambda} \left(\frac{\hbar}{2MN\omega_{\mathbf{k},\lambda}}\right)^{1/2} (b_{\mathbf{k},\lambda}\lambda_{\mathbf{k}}e^{-i\omega_{\mathbf{k},\lambda}t} + b^{\dagger}_{-\mathbf{k},\lambda}\lambda^{*}_{-\mathbf{k}}e^{i\omega_{\mathbf{k},\lambda}t})e^{i\mathbf{k}\cdot\mathbf{R}^0_i},$$

$$(10.24)$$

where $\lambda_{\mathbf{k}}$ is a polarization vector of unit length. For a longitudinal phonon, $\lambda_{\mathbf{k}}$ is parallel to $\mathbf{k}$, while $\lambda_{\mathbf{k}}$ is perpendicular to $\mathbf{k}$ for a transverse phonon. Because $\mathbf{Q}_i$ is Hermitian, we choose the polarization vectors to be purely real. With this definition of $\mathbf{Q}_i(t)$, we rewrite the ion-potential

$$V_{\text{ion}} = \sum_{i<j} V(\mathbf{R}^0_i - \mathbf{R}^0_j) + \frac{M}{2}\sum_{\mathbf{k},\lambda}\omega^2_{\mathbf{k},\lambda}\mathbf{Q}_{\mathbf{k},\lambda}\mathbf{Q}_{-\mathbf{k},\lambda} \qquad (10.25)$$

in terms of the phonon or harmonic modes, $\mathbf{Q}_{\mathbf{k},\lambda}$. For acoustic phonons, $\mathbf{Q}_{\mathbf{k},\lambda}$ describes a distortion in which neighboring ions move in the same direction. These correspond to long-wavelength modes of the crystal. We will focus only on distortions of this sort. The conjugate momentum for a phonon mode is defined by computing $\mathbf{P}_i = M\dot{\mathbf{Q}}_i$, for example, in the site representation. $\mathbf{P}_i$ and $\mathbf{Q}_i$ obey the canonical commutation relations in Eq. (10.13).

10.3   ELECTRON-PHONON INTERACTION

In our general many-body Hamiltonian, the interaction between each ion and the electrons is of the form

$$V_{ei} = \sum_j V_{ei}(\mathbf{r}_j)$$

$$= \sum_{i,j} V_{ei}(\mathbf{r}_j - \mathbf{R}_i). \qquad (10.26)$$

To make contact with the phonon expansion introduced in the previous section, we write the ion coordinate in terms of a deviation from the home position; $\mathbf{R}_i = \mathbf{R}^0_i + \mathbf{Q}_i$ and Taylor series expand the electron-ion potential around $\mathbf{R}^0_i$. To first order, we have that

$$V_{ei} = \sum_{i,j} V_{ei}(\mathbf{r}_j - \mathbf{R}^0_i) - \sum_{i,j} \mathbf{Q}_i \cdot \nabla_j V_{ei}(\mathbf{r}_j - \mathbf{R}^0_i) + O(Q^2) + \cdots.$$

$$(10.27)$$

The first term defines the periodic potential seen by a conduction electron and, hence, contains no new information regarding the coupling of
the electrons to the lattice distortion. Such information is contained in
the second term. To simplify this term, we introduce the Fourier transform of the electron-ion potential,

$$V_{ei}(\mathbf{r}) = \frac{1}{N} \sum_{\mathbf{k}} V_{ei}(\mathbf{k}) e^{i\mathbf{k}\cdot\mathbf{r}}. \tag{10.28}$$

With this definition in hand, we write the electron-ion potential as

$$V_{ei} = V_0 - \frac{i}{N} \sum_{\mathbf{k},i,j} V_{ei}(\mathbf{k}) e^{i\mathbf{k}\cdot\mathbf{r}_j}$$

$$\times \mathbf{k}\cdot \sum_{\mathbf{q},\lambda} \left(\frac{\hbar}{2MN\omega_{\mathbf{q},\lambda}}\right)^{\frac{1}{2}} \lambda_{\mathbf{q}}(b_{\mathbf{q},\lambda} + b^{\dagger}_{-\mathbf{q},\lambda}) e^{i(\mathbf{q}-\mathbf{k})\cdot\mathbf{R}_i^0}, \tag{10.29}$$

where we have set the first term in Eq. (10.27) equal to V_0. We restrict
the sums over $\mathbf{q}$ and $\mathbf{k}$ to the first Brillouin zone, such that

$$\frac{1}{N} \sum_{i} e^{i(\mathbf{q}-\mathbf{k})\cdot\mathbf{R}_i^0} = \sum_{\mathbf{L}} \delta_{\mathbf{k},\mathbf{q}+\mathbf{L}}, \tag{10.30}$$

where the sum over $\mathbf{L}$ is over all reciprocal lattice vectors. We note also
that the kth component of the electron density is

$$\rho_{\mathbf{k}} = \sum_{j} e^{i\mathbf{k}\cdot\mathbf{r}_j}. \tag{10.31}$$

We introduce the electron-phonon coupling constant

$$M_{\mathbf{q},\mathbf{L},\lambda} = -i \left(\frac{\hbar}{2MN\omega_{\mathbf{q},\lambda}}\right)^{1/2} (\mathbf{q}+\mathbf{L})\cdot\lambda_{\mathbf{q}} V_{ei}(\mathbf{q}+\mathbf{L}) \tag{10.32}$$

and recast the electron-phonon term as

$$V_{ei} = V_0 + \sum_{\mathbf{q},\mathbf{L},\lambda} M_{\mathbf{q},\mathbf{L},\lambda}(b_{\mathbf{q},\lambda} + b^{\dagger}_{-\mathbf{q},\lambda})\rho_{\mathbf{q}+\mathbf{L}} = V_0 + H_{e-ph}. \tag{10.33}$$

The electron-phonon coupling constant contains the product $(\mathbf{q}+\mathbf{L})\cdot$
$\lambda_{\mathbf{q}}$. As a consequence, when $\lambda_{\mathbf{q}}$ is perpendicular to $\mathbf{q}+\mathbf{L}$, $M_{\mathbf{q}+\mathbf{L},\lambda} = 0$. That is, only the longitudinal acoustic phonon modes couple to the
electrons. As a result, we can drop the λ subscript, as there is only one

longitudinal acoustic mode. This is an important result. However, there are certainly longitudinal optical phonons that couple to the electron motion. Such processes arise from a Coulombic rather than an elastic deformation coupling to the electron motion. In the optical phonon case, the linear q dependence of the coupling constant, M_q, is replaced by a q^{-2} dependence. The inverse $1/q^2$ term arises from the Fourier transform of the Coulomb interaction. In polar crystals, optical phonons dominate over the acoustic modes. Because we are primarily interested in superconductivity, we limit our discussion solely to the acoustic case.

Let us now compute matrix elements of $H_{\mathrm{e-ph}}$. To do this, we consider the composite electron-phonon state

$$|\psi_{\mathrm{e-ph}}\rangle = |\{n_{\mathbf{k}}\}; \{N_{\mathbf{q},\lambda}\}\rangle. \tag{10.34}$$

Here, the electronic state $|\{n_{\mathbf{k}}\}\rangle$ represents a many-body state in which $n_{\mathbf{k}}$ electrons are in the single-particle Bloch state $\langle \mathbf{r}|\mathbf{k}\rangle \equiv e^{i\mathbf{k}\cdot\mathbf{r}}U_{\mathbf{k}}(\mathbf{r})$, and $|\{|N_{\mathbf{q},\lambda}\}\rangle$ denotes a many-body phonon state in which $N_{\mathbf{q},\lambda}$ phonons are in the qth lattice mode of polarization $\lambda_{\mathbf{q}}$. The function $U_{\mathbf{q}}(\mathbf{r})$ has the same periodicity of the lattice as does the Fourier coefficient, $e^{i\mathbf{q}\cdot\mathbf{r}}$, namely, $U_{\mathbf{k}}(\mathbf{r}) = U_{\mathbf{k}}(\mathbf{r} + \mathbf{R}_i^0)$. To evaluate matrix elements of the electron-phonon interaction, it is helpful to express $H_{\mathrm{e-ph}}$ in second-quantized form. The only electron operator in $H_{\mathrm{e-ph}}$ is the electron density, $\hat{\rho}_{\mathbf{k}}$. In second-quantized form, $\hat{\rho}_{\mathbf{k}}$ becomes

$$\hat{\rho}_{\mathbf{k}} = \sum_{\mathbf{k}_1,\mathbf{k}_2} \langle \mathbf{k}_1|e^{i\mathbf{k}\cdot\mathbf{r}}|\mathbf{k}_2\rangle a_{\mathbf{k}_1}^{\dagger} a_{\mathbf{k}_2}, \tag{10.35}$$

where the operator $a_{\mathbf{k}}^{\dagger}$ creates an electron in the momentum state $\mathbf{k}$. The electron-phonon interaction can now be written as

$$H_{\mathrm{e-ph}} = \sum_{\mathbf{k},\mathbf{L},\mathbf{k}_1,\mathbf{k}_2,\mathbf{q}} M_{\mathbf{q},\mathbf{L}} \langle \mathbf{k}_1|e^{i(\mathbf{q}+\mathbf{L})\cdot\mathbf{r}}|\mathbf{k}_2\rangle a_{\mathbf{k}_1}^{\dagger} a_{\mathbf{k}_2}(b_{\mathbf{q}} + b_{-\mathbf{q}}^{\dagger}). \tag{10.36}$$

Assuming the electron wave functions are simply plane waves, we find that the matrix element of the density operator is given exactly by

$$\langle \mathbf{k}_1|e^{i\mathbf{k}\cdot\mathbf{r}}|\mathbf{k}_2\rangle = \int \frac{d\mathbf{r}}{V} e^{i\mathbf{r}\cdot(\mathbf{k}+\mathbf{k}_2-\mathbf{k}_1)} = \delta_{\mathbf{k}_1,\mathbf{k}+\mathbf{k}_2}. \tag{10.37}$$

In general, the electron wave functions need not be plane waves. We define

$$\alpha_{\mathbf{q}_1,\mathbf{q}_2} = \langle \mathbf{q}_1|\mathbf{q}_2\rangle \tag{10.38}$$

to be the general overlap between two electronic states. Because the $U_{\mathbf{k}}$'s have the periodicity of the lattice, the condition in Eq. (10.37) still holds, even when the electronic wave functions are more complicated than plane waves. Consequently, the full electron-phonon Hamiltonian reduces to

$$H_{\text{e}-\text{ph}} = \sum_{\mathbf{q},\mathbf{k},\mathbf{L}} M_{\mathbf{q},\mathbf{L}} \alpha_{\mathbf{k}+\mathbf{q}+\mathbf{L},\mathbf{k}} a^{\dagger}_{\mathbf{q}+\mathbf{L}+\mathbf{k}} a_{\mathbf{k}} (b_{\mathbf{q}} + b^{\dagger}_{-\mathbf{q}}), \quad (10.39)$$

when Eqs. (10.36–10.38) are combined.

As is evident, this Hamiltonian contains a myriad of electron-phonon processes, some of which involve the electron's moving from one Brillouin zone to another, $\mathbf{L} \neq 0$. All such processes in which the electron wave vector is changed by $\mathbf{q} + \mathbf{L} + \mathbf{k}$ are called *Umklapp processes*. In German, *umklappen* means "to flip over." Normal processes refer to those in which momentum transfer does not result in an electron's changing Brillouin zones. In such cases, a phonon of wave vector $\mathbf{q}$ scatters an electron with momentum $\mathbf{k}$ and yields an electron state with wave vector $\mathbf{q} + \mathbf{k}$. Diagrams illustrating the various kinds of scattering processes are shown in Fig. (10.1).

We are interested primarily in the amplitude for emission and absorption. In emission, a phonon is created. Hence, only the $b^{\dagger}_{-\mathbf{q}}$ term

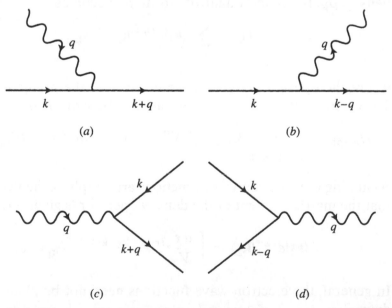

Figure 10.1 Electron-phonon scattering. A wavy line represents a phonon, and incoming and outgoing arrows represent electrons and holes, respectively. (*a*) and (*c*) represent absorption, whereas (*b*) and (*d*) correspond to emission of a phonon.

contributes. Likewise in absorption, a phonon is annihilated. For an emission process, the initial and final states must be of the form

$$|\text{init}\rangle = |n_{\mathbf{k+q}}, n_{\mathbf{k}}; N_{-\mathbf{q}}, N_{\mathbf{q}}\rangle \tag{10.40}$$

$$|\text{efin}\rangle = |n_{\mathbf{k+q}} + 1, n_{\mathbf{k}} - 1; N_{-\mathbf{q}} + 1, N_{\mathbf{q}}\rangle. \tag{10.41}$$

The amplitude for emission involves the matrix elements

$$\langle\text{efin}|H_{e-\text{ph}}|\text{init}\rangle = \sum_{\mathbf{q,k}} M_{\mathbf{q}}\alpha_{\mathbf{k+q,k}}\langle\text{efin}|a_{\mathbf{q+k}}^{\dagger}a_{\mathbf{k}}b_{-\mathbf{q}}^{\dagger}|\text{init}\rangle. \tag{10.42}$$

Because $n_{\mathbf{k+q}}$ and $n_{\mathbf{k}}$ are restricted to be 1 or 0, $a_k|\text{init}\rangle$ is nonzero only if $n_{\mathbf{k}} = 1$. Likewise, $a_{\mathbf{q+k}}^{\dagger}|\text{init}\rangle$ will yield a nonzero result only if $1 - n_{\mathbf{q+k}} = 1$. Consequently,

$$\langle\text{efin}|H_{e-\text{ph}}|\text{init}\rangle = \sum_{\mathbf{q,k}} M_{\mathbf{q}}\alpha_{\mathbf{k+q,k}}\sqrt{(1 - n_{\mathbf{q+k}})n_{\mathbf{k}}(N_{-\mathbf{q}} + 1)}. \tag{10.43}$$

In the event that $n_{\mathbf{k}} = 1$ and $n_{\mathbf{q+k}} = 0$, the energy difference between the initial and final states is

$$\Delta E_{\text{emis}} = E_{\text{fin}} - E_{\text{init}} = E(\mathbf{q} + \mathbf{k}) - E(\mathbf{k}) + \hbar\omega_{\mathbf{q}}, \tag{10.44}$$

where $E(\mathbf{k})$ is the energy of an electron state with momentum $\mathbf{k}$.

In the absorption process, the final state in this case is

$$|\text{afin}\rangle = |n_{\mathbf{k+q}} + 1, n_{\mathbf{k}} - 1; N_{-\mathbf{q}}, N_{\mathbf{q}} - 1\rangle. \tag{10.45}$$

As a consequence,

$$\langle\text{afin}|H_{e-\text{ph}}|\text{init}\rangle = \sum_{\mathbf{q,k}} M_{\mathbf{q}}\alpha_{\mathbf{k+q,k}}\sqrt{(1 - n_{\mathbf{q+k}})n_{\mathbf{k}}N_{\mathbf{q}}}. \tag{10.46}$$

The energy difference here is

$$\Delta E_{\text{abs}} = E(\mathbf{k} + \mathbf{q}) - E(\mathbf{k}) - \hbar\omega_{\mathbf{q}}. \tag{10.47}$$

Applying Fermi's golden rule to the emission and absorption amplitudes yields

$$W_{\mathbf{k}\to\mathbf{k+q}}^{\text{abs}} = \frac{2\pi}{\hbar}|M_{\mathbf{q}}\alpha_{\mathbf{k+q,k}}|^2 n_{\mathbf{k}}N_{\mathbf{q}}(1 - n_{\mathbf{q+k}})\delta(E(\mathbf{k} + \mathbf{q}) - E(\mathbf{k}) - \hbar\omega_{\mathbf{q}})$$

$$\tag{10.48}$$

and

$$W_{k \to k+q}^{emis} = \frac{2\pi}{\hbar} |M_q \alpha_{k+q,k}|^2 n_k (1 - n_{q+k})$$
$$\times (N_{-q} + 1) \delta(E(k+q) - E(k) + \hbar\omega_q) \quad (10.49)$$

for the emission and absorption rates, respectively. The δ-functions ensure that energy is conserved. For the sake of generality, we have included explicitly the electron occupation numbers, although $n_k = 1$ and $n_{q+k} = 0$. It is customary at this stage of our calculation to replace the electron and phonon occupation numbers by their equilibrium Fermi-Dirac and Bose forms. This simplification is valid only if the electron-phonon system is in equilibrium before the transition occurs. We will find this simplification useful when we treat superconductivity.

10.4 ULTRASONIC ATTENUATION

Imagine that we send a beam of phonons into a metal. The total rate at which the phonons are absorbed is determined by the direct absorption into the metal and emission back into the beam. Consequently, the rate at which a beam loses $N_{q,\lambda}$ phonons per unit time is given by a kinetic gain-loss equation:

$$\frac{dN_{q,\lambda}}{dt} = -\sum_p (W_{p \to p+q}^{abs} - W_{p \to p-q}^{emis}). \quad (10.50)$$

The first term represents the absorption of phonons from the beam and the second term the reemission of phonons into the beam. Inclusion of the reemission term is essential to describe the correct physics.

We can simplify our kinetic equation by recalling that the structure function for free electrons is given by

$$n_e S_0(p, \omega) = \frac{2}{V} \sum_{p'} f_{p'} (1 - f_{p+p'}) 2\pi\hbar \delta(\hbar\omega - \epsilon_{p+p'} + \epsilon_{p'}). \quad (10.51)$$

Physically, $S_0(p, \omega)$ is the density of electron-hole excitations separated by an energy $\hbar\omega$. Inspection of the expressions for the phonon absorption and emission rates reveals that they are directly proportional to the right-hand side of the equation for the structure function. Let us assume that the electronic states are perfect plane waves with free-particle energies $\epsilon_p = p^2/2m$. As a result, the matrix element in

Eq. (10.38) is equal to unity: $\alpha_{k+q,q} = 1$. Consequently, we rewrite the net absorption and emission rates as

$$\sum_p W_{p\to p+q}^{abs} = \frac{n_e V}{2\hbar^2}|M_q|^2 N_{q,\lambda} S_0(q, \hbar\omega_q) \qquad (10.52)$$

and

$$\sum_p W_{p\to p-q}^{emis} = \frac{n_e V}{2\hbar^2}|M_q|^2 (N_{q,\lambda} + 1) S_0(-q, -\hbar\omega_q). \qquad (10.53)$$

In the context of the fluctuation-dissipation theorem, we showed that

$$S_0(p, \hbar\omega) = e^{\beta\hbar\omega} S_0(-p, -\hbar\omega). \qquad (10.54)$$

Substitution of this result into the equation of motion for $N_{q,\lambda}(t)$ yields

$$\dot{N}_{q,\lambda} = -\frac{n_e V}{2\hbar^2}|M_q|^2 S_0(q, \omega_q\hbar)[N_{q,\lambda} - e^{-\beta\hbar\omega_q}(N_{q,\lambda} + 1)]. \qquad (10.55)$$

We define the net rate of phonons absorbed to be

$$\frac{1}{\tau_{ph}} = \frac{n_e V}{2\hbar^2}|M_q|^2 (1 - e^{-\beta\hbar\omega_q}) S_0(q, \hbar\omega_q) \qquad (10.56)$$

and the equilibrium phonon distribution, $N_{q,\lambda}^{eq}$, to be the standard Bose-Einstein distribution function,

$$N_{q,\lambda}^{eq} = \frac{1}{e^{\beta\hbar\omega_q} - 1}. \qquad (10.57)$$

Consequently, our equations of motion become

$$\dot{N}_{q,\lambda} = \frac{-1}{\tau_{ph}}[N_{q,\lambda} - N_{q,\lambda}^{eq}], \qquad (10.58)$$

and the solution to this linear differential equation has the characteristic

$$N_{q,\lambda}(t) = N_{q,\lambda}^{eq} + e^{-t/\tau_{ph}}(N_{q,\lambda}(t = 0) - N_{q,\lambda}^{eq}) \qquad (10.59)$$

exponential form. We find, then, that the number of phonons absorbed relaxes to an equilibrium value at long times with a rate $1/\tau_{ph}$. This effect is known as *ultrasonic attenuation*, the loss of phonons to a medium as a result of interactions with electrons. Because electrons in a superconductor are bound together in pairs, with a binding energy proportional to the gap, they can absorb phonons only if the phonon frequency exceeds a critical value. As a consequence, ultrasound attenuation is used as a tool for measuring the gap in a superconductor. We will investigate this further in the next chapter.

A final observation is that the calculation we have performed here is valid only if electron interactions are negligible. That is, if τ_{ee} is the effective time scale for electron scattering, then our calculation is valid if $\omega_q \tau_{ee} \gg 1$. If this condition does not hold and $\tau_{ee}\omega_q < 1$, then sound waves are attenuated via electron scattering rather than by phonon-induced electron-hole pairs. For completeness, let us now evaluate τ_{ph}. At $T = 0$, we showed that at intermediate frequencies, $n_e S_0(\mathbf{k}, \omega) = m^2\omega/\pi k \hbar^2$. Because $|M_\mathbf{q}|^2 \sim q^2 V_{ei}(\mathbf{q})/\omega_q$, $1/\tau_{ph} \sim q|V_{ei}(\mathbf{q})|^2$. Focusing only on the zero frequency part of the structure function, we find that $\epsilon_\mathbf{p} = \epsilon_{\mathbf{p+k}}$. For $p = p_F$, the transferred momentum is $k = 2p_F$. Hence, at low frequencies, τ_{ph} is determined by particle scattering across the Fermi surface.

10.5   ELECTRICAL CONDUCTION

The conductivity in a metal is measured by applying a voltage or an electric field to the material. The resultant current density

$$\mathbf{j} = \sigma\mathbf{E} \tag{10.60}$$

is directly proportional to the electric field through the conductivity, σ. The current density is in the direction of the carrier velocity, $\mathbf{v}$. The constant of proportionality is the net charge density, $-en_e$. Consequently,

$$\mathbf{j} = -n_e e\mathbf{v}. \tag{10.61}$$

Let us now express $\mathbf{v}$ in terms of $\mathbf{E}$. In the absence of an electric field, $\langle\mathbf{v}\rangle = 0$ because $\mathbf{v}$ is randomized. Let τ represent the collision time of the electron. That is, over a time period τ, an electron is moving with constant velocity that we estimate as follows. The force exerted on an electron by the electric field is simply $-e\mathbf{E}$. The acceleration of the electron is, then, $-e\mathbf{E}/m$. If the acceleration is constant over a time τ, then the average velocity is $\mathbf{v}_{avg} = -e\mathbf{E}\tau/m$ and the current density is

$$\mathbf{j} = \frac{e^2 n_e}{m}\mathbf{E}\tau = \sigma\mathbf{E}, \tag{10.62}$$

or, equivalently,

$$\sigma = \frac{n_e e^2 \tau}{m}. \tag{10.63}$$

This is the Drude formula for the conductivity.

We have expressed the conductivity, then, in terms of a single unknown quantity, τ, the relaxation or collision time. Experimentally, once the conductivity is measured, τ can be extracted. Listed below are a few relaxation times for the alkali earth metals [AM1976].

Element	77 K	273 K
Li	$7.3 \times 10^{-14} s$	$8.8 \times 10^{-15} s$
Na	$1.7 \times 10^{-13} s$	$3.2 \times 10^{-14} s$
K	$1.8 \times 10^{-13} s$	$4.1 \times 10^{-14} s$
Rb	$1.4 \times 10^{-13} s$	$2.8 \times 10^{-14} s$
Cs	$8.6 \times 10^{-14} s$	$2.1 \times 10^{-14} s$

We now want to develop a general theory that can account for the relaxation time and, hence, the conductivity. In a pure metal, the primary source of resistance is via interactions with lattice phonons. Any successful account of σ in a metal must explain the following: 1) σ is independent of E for moderate values of E, 2) The wide variation of σ from metal to metal, 3) The Wiedemann-Franz law that the ratio of $\kappa/\sigma = T$, where κ is the thermal conductivity, and 4) $\sigma \sim 1/T$ in most metals with a transition to $\sigma \sim T^{-5}$ at $T \to 0$. In this chapter, we focus entirely on the crossover from $1/T$ to T^{-5} behavior at low temperatures.

10.5.1 Boltzmann Equation

To proceed, we develop the Boltzmann transport theory. We introduce the distribution function $f(\mathbf{r}, \mathbf{k}, t)$, which defines the probability that a quantum "state" is occupied with momentum $\mathbf{k}$ and position $\mathbf{r}$ at time t. Although we are interested in only one band, f can be generalized to include all bands in a solid. The distribution function f specifies both the position and the momentum of an electron in a quantum state. Adopting such a distribution function is valid strictly at long wavelengths, that is, $\lambda \gg \hbar v_F / \kappa_B T$. Otherwise, the uncertainty principle is violated.

Consider the volume element $d\mathbf{k} d\mathbf{r} / (2\pi\hbar)^3 = d\Omega$. The product of this differential volume element with f, $f d\Omega$, defines the number of electrons in $d\,\Omega$. In the problem at hand, interactions with phonons alter the occupation in phase space. Let us refer to such processes as *lattice collisions*. Clearly, df/dt would be zero if no such collisions occurred. In fact, for a solid in equilibrium, f is simply the Fermi-Dirac distribution. For the nonequilibrium case, f must be determined from the general equations of motion. The total time derivative of f,

$$\frac{df}{dt} = \frac{\partial f}{\partial t} + \dot{\mathbf{k}} \cdot \nabla_{\mathbf{k}}f + \dot{\mathbf{r}} \cdot \nabla_{\mathbf{r}}f = \left.\frac{\partial f}{\partial t}\right|_{\text{coll}}, \qquad (10.64)$$

is determined by all the terms that either implicitly or explicitly depend on time. This is the Boltzmann equation. Because each volume element should be equivalent, the average number of electrons entering and leaving a volume element should be a constant. As a result, $\partial f/\partial t = 0$. In addition, our system is homogeneous, even in the presence of an electric field. As a consequence, the spatial derivative of f vanishes and

$$\frac{\partial \mathbf{k}}{\partial t} \cdot \nabla_{\mathbf{k}}f = \left.\frac{\partial f}{\partial t}\right|_{\text{coll}}, \qquad (10.65)$$

the steady-state Boltzmann equation, results. Physically, $\partial \mathbf{k}/\partial t$ is the force on the electrons in the Fermi sea. In an electric field,

$$\frac{\partial \mathbf{k}}{\partial t} = -e\,\mathbf{E}, \qquad (10.66)$$

and, consequently, all the electrons are accelerated equally by the field.

To apply this equation to an electron-lattice problem, we must include an analogous Boltzmann equation for the phonons. The appropriate Boltzmann equation is one in which the momentum term is absent, because in an unstrained crystal, there is no force on phonons. Let g be the phonon distribution function. It follows that

$$\left.\frac{\partial g}{\partial t}\right|_{\text{coll}} = \dot{\mathbf{r}} \cdot \nabla_{\mathbf{r}}g. \qquad (10.67)$$

In all of our calculations to follow, we will replace g by its equilibrium value. Hence, we will not spend much time discussing g, though our treatment of f can be paralleled to solve for g. Because f is determined by electron-phonon exchanges, the most general expression we can write for the collisions is one in which all possible electron-phonon processes are summed over. Let W^{eq} represent an emission term and W^{aq} a phonon absorption process with wave vector $\mathbf{q}$. The general collision terms that enter the Boltzmann equation are shown in Fig. (10.2) and can be written as a gain-loss master equation,

$$\left.\frac{\partial f}{\partial t}\right|_{\text{coll}} = \sum_{\mathbf{q}} (W_{\mathbf{k+q}\to\mathbf{k}}^{e\,\mathbf{q}} + W_{\mathbf{k+q}\to\mathbf{k}}^{a\,-\mathbf{q}} - W_{\mathbf{k}\to\mathbf{k+q}}^{a\,\mathbf{q}} - W_{\mathbf{k}\to\mathbf{k+q}}^{e\,-\mathbf{q}})$$

$$= \sum_{\mathbf{q}} [\text{gain}(\mathbf{k}) - \text{loss}(\mathbf{k})], \tag{10.68}$$

for electron states with momentum $\mathbf{k}$ and $\mathbf{k+q}$. From the exact expressions for W in Eqs. (10.48) and (10.49), it is convenient to define

$$W_{\mathbf{q}}^0 = \frac{2\pi}{\hbar}|M_{\mathbf{q}}|^2. \tag{10.69}$$

We have assumed that the electron states are plane waves. Hence, $\alpha_{\mathbf{k},\mathbf{q}} = 1$. The collision terms in Eq. (10.69) are easily computed if they are grouped as emission-absorption pairs:

$$\left.\frac{\partial f}{\partial t}\right|_{\text{coll}} = \sum_{\mathbf{q}} (W_{\mathbf{k+q}\to\mathbf{k}}^{e\,\mathbf{q}} - W_{\mathbf{k}\to\mathbf{k+q}}^{a\,\mathbf{q}}) - (W_{\mathbf{k}\to\mathbf{k+q}}^{e\,-\mathbf{q}} - W_{\mathbf{k+q}\to\mathbf{k}}^{a\,-\mathbf{q}})$$

$$= \sum_{\mathbf{q}} W_{\mathbf{q}}^o [(1 - f(\mathbf{k}))(f(\mathbf{k+q}))(g(\mathbf{q}) + 1)\delta(\epsilon_{\mathbf{k}} - \epsilon_{\mathbf{k+q}} + \hbar\omega_{\mathbf{q}})$$

$$- (1 - f(\mathbf{k+q}))f(\mathbf{k})g(\mathbf{q})\delta(\epsilon_{\mathbf{k+q}} - \epsilon_{\mathbf{k}} - \hbar\omega_{\mathbf{q}})]$$

$$+ W_{\mathbf{q}}^o [(1 - f(\mathbf{k}))f(\mathbf{k+q})g(-\mathbf{q})\delta(\epsilon_{\mathbf{k}} - \epsilon_{\mathbf{k+q}} - \hbar\omega_{\mathbf{q}})]$$

$$- (1 - f(\mathbf{k+q})f(\mathbf{k})(g(-\mathbf{q}) + 1)\delta(\epsilon_{\mathbf{k+q}} - \epsilon_{\mathbf{k}} + \hbar\omega_{\mathbf{q}})]. \tag{10.70}$$

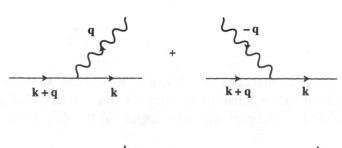

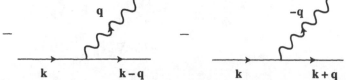

Figure 10.2 Electron-phonon collision terms that enter the Boltzmann equation.

There are three common simplifications that are used to solve the Boltzmann equation for $f(\mathbf{k})$. The first is to assume that $g(\mathbf{q}) = g_{equil}$, which is known as the *Bloch assumption*. The Bose-Einstein distribution is just g_{equil}. Let $N_{\mathbf{q}} = g_{equil} = N_{-\mathbf{q}}$. We also define

$$W_{\mathbf{q}}(\mathbf{k}, \mathbf{k}') = W_{\mathbf{q}}^0[\delta(\epsilon_{\mathbf{k}'} - \epsilon_{\mathbf{k}} + \hbar\omega_{\mathbf{q}})(N_{\mathbf{q}} + 1) + N_{\mathbf{q}}\delta(\epsilon_{\mathbf{k}'} - \epsilon_{\mathbf{k}} - \hbar\omega_{\mathbf{q}})].$$

$$(10.71)$$

The right-hand side of the Boltzmann equation now simplifies to

$$\frac{\partial f}{\partial t}\Big|_{coll} = \sum_{\mathbf{q}}[W_{\mathbf{q}}(\mathbf{k} + \mathbf{q}, \mathbf{k})f(\mathbf{k} + \mathbf{q})(1 - f(\mathbf{k}))$$
$$- W_{\mathbf{q}}(\mathbf{k}, \mathbf{k} + \mathbf{q})f(\mathbf{k})(1 - f(\mathbf{k} + \mathbf{q}))]. \qquad (10.72)$$

10.5.2   Relaxation-Time Approximation

In the next step, we assume that on average, $f(\mathbf{k})$ is slowly varying when the field is applied. Collisions with phonons return the system to the equilibrium Fermi-Dirac distribution function, $f_0 = n_{\mathbf{k}}$. We write $f(\mathbf{k})$ as

$$f(\mathbf{k}) \simeq n_{\mathbf{k}} + \delta f_{\mathbf{k}}, \qquad (10.73)$$

with $\delta f_{\mathbf{k}}$ the variation of $f(\mathbf{k})$ induced by the electric field. We suspect that $\delta f_{\mathbf{k}}$ is proportional to the acceleration, $\partial \mathbf{k}/\partial t$. To see how this comes about, we make the ansatz that collision-induced changes of $f(\mathbf{k})$ relax the system back to $n_{\mathbf{k}}$ with a mean relaxation time $\tau(\mathbf{k})$, such that

$$\frac{\partial f}{\partial t}\Big|_{coll} = -\frac{\delta f_{\mathbf{k}}}{\tau(\mathbf{k})}. \qquad (10.74)$$

Note that introduction of a relaxation time at this stage can be done only at the expense of making τ $\mathbf{k}$-dependent. Using Eqs. (10.65) and (10.66) and linearizing with respect to the fluctuation $\delta f_{\mathbf{k}}$, we find that

$$\frac{\partial f}{\partial t}\Big|_{coll} = -\frac{\delta f_{\mathbf{k}}}{\tau} = -e\mathbf{E} \cdot \nabla_{\mathbf{k}}f(\mathbf{k})$$
$$= -e\mathbf{E} \cdot \nabla_k \epsilon_{\mathbf{k}} \frac{\partial n_{\mathbf{k}}}{\partial \epsilon_{\mathbf{k}}}$$
$$= e\frac{\mathbf{E} \cdot \mathbf{k}}{m}n_{\mathbf{k}}(1 - n_{\mathbf{k}})\beta, \qquad (10.75)$$

where we have used the free-particle dispersion relation, $\epsilon_{\mathbf{k}} = \mathbf{k}^2/2m$. Because $\delta f_{\mathbf{k}} = f(\mathbf{k}) - n_{\mathbf{k}}$, we obtain

$$f(\mathbf{k}) = n_{\mathbf{k}} + \beta\delta\Phi_{\mathbf{k}}n_{\mathbf{k}}(1 - n_{\mathbf{k}}), \qquad (10.76)$$

with

$$\delta\Phi_{\mathbf{k}} = -e\frac{\mathbf{E}\cdot\mathbf{k}}{m}\tau(\mathbf{k}). \qquad (10.77)$$

We see, then, that once $\tau(\mathbf{k})$ is determined, we can find the distribution function $f(\mathbf{k})$ immediately and that the conductivity can be obtained through the Drude formula.

10.5.3 Low-Temperature Resistivity

To simplify the Boltzmann equation, we observe that $W_{\mathbf{q}}(\mathbf{k},\mathbf{k}')$ obeys the symmetry relationship

$$W_{\mathbf{q}}(\mathbf{k},\mathbf{k}')e^{\beta\epsilon_{\mathbf{k}'}} = W_{\mathbf{q}}(\mathbf{k}',\mathbf{k})e^{\beta\epsilon_{\mathbf{k}}}. \qquad (10.78)$$

This statement is simply one of detailed balance. An equivalent, more useful way of writing Eq. (10.78) is

$$W_{\mathbf{q}}(\mathbf{k},\mathbf{k}') = e^{\beta(\epsilon_{\mathbf{k}}-\epsilon_{\mathbf{k}'})}W_{\mathbf{q}}(\mathbf{k}',\mathbf{k}) \qquad (10.79)$$

$$= \frac{n_{\mathbf{k}'}(1 - n_{\mathbf{k}})}{n_{\mathbf{k}}(1 - n_{\mathbf{k}'})}W_{\mathbf{q}}(\mathbf{k}',\mathbf{k}). \qquad (10.80)$$

This identity implies that the quantity

$$Z_{\mathbf{q}}(\mathbf{k},\mathbf{k}') = W_{\mathbf{q}}(\mathbf{k},\mathbf{k}')n_{\mathbf{k}}(1 - n_{\mathbf{k}'}) \qquad (10.81)$$

is symmetric with respect to interchange of its arguments. An immediate consequence of this identity is that the collision terms in the Boltzmann equation vanish identically, when $f(\mathbf{k}) = n_{\mathbf{k}}$:

$$W_{\mathbf{q}}(\mathbf{k}',\mathbf{k})(1 - n_{\mathbf{k}})n_{\mathbf{k}'} - W\mathbf{q}(\mathbf{k},\mathbf{k}')(1 - n_{\mathbf{k}'})n_{\mathbf{k}} = Z_{\mathbf{q}}(\mathbf{k}',\mathbf{k}) - Z_{\mathbf{q}}(\mathbf{k},\mathbf{k}')$$

$$= 0. \qquad (10.82)$$

The consequences of this identity are immediate. Recall that we have approximated $f(\mathbf{k})$ as $f(\mathbf{k}) \simeq n_{\mathbf{k}} + \beta n_{\mathbf{k}}(1 - n_{\mathbf{k}})\delta\Phi_{\mathbf{k}}$. Hence, only the terms with at least a linear variation of δf survive in the Boltzmann equation:

$$\frac{\partial f}{\partial t}\bigg|_{coll} = \sum_{\mathbf{q}} W_{\mathbf{q}}(\mathbf{k}', \mathbf{k})(\delta f_{\mathbf{k}'}(1 - n_{\mathbf{k}}) - n_{\mathbf{k}'} \delta f_{\mathbf{k}}) - \mathbf{k} \Leftrightarrow \mathbf{k}'$$

$$= \beta \sum_{\mathbf{q}} Z_{\mathbf{q}}(\mathbf{k}', \mathbf{k})[\delta\Phi_{\mathbf{k}'}(1 - n_{\mathbf{k}'}) - \delta\Phi_{\mathbf{k}} n_{\mathbf{k}} - \mathbf{k} \Leftrightarrow \mathbf{k}']$$

$$= \beta \sum_{\mathbf{q}} Z_{\mathbf{q}}(\mathbf{k}', \mathbf{k})[\delta\Phi_{\mathbf{k}'} - \delta\Phi_{\mathbf{k}}]. \qquad (10.83)$$

In deriving Eq. (10.83), we dropped the $O(\delta\Phi^2)$ terms, thus obtaining the linearized Boltzmann equation.

As a result of the variation $\delta\Phi_{\mathbf{k}} \sim \tau(\mathbf{k})$, the Boltzmann equation is in general an integral equation that must be solved self-consistently by some ansatz. As in all integral equations, a variational principle applies, and we are guaranteed that a trial solution for $\delta\Phi$ will result in a distribution function, f, that produces a higher energy than the true ground-state energy. In the relaxation-time approximation, $\delta\Phi_{\mathbf{k}} = -e\,\mathbf{E} \cdot \mathbf{k}\tau(\mathbf{k})/m$. In an electric field, the drift velocity is $\mathbf{v}_d = -e\,\mathbf{E}\tau/m$. As a consequence, $\delta\Phi_k \equiv \mathbf{v_d} \cdot \mathbf{k}$ is known as the *drift-velocity ansatz*. Physically, this ansatz signifies that the electrons are in equilibrium with a drifting distribution. The drifting distribution is equivalent to the equilibrium Fermi-Dirac distribution with $\mathbf{q} \to \mathbf{q} - m\mathbf{v}_d$. As depicted in Fig. (10.3), the drift-velocity ansatz amounts to an overall translation of the Fermi surface by an amount

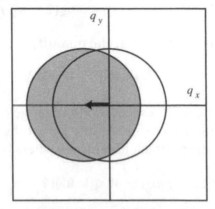

Figure 10.3 Equilibrium Fermi surface and the Fermi surface in the presence of a uniform electric field. The drift-velocity ansatz states that once the electric field is turned on, the new Fermi-Dirac distribution function is equivalent to the original one by simply translating $\mathbf{q} \to \mathbf{q} - m\mathbf{v}_d$.

proportional (in linear order) to $m\mathbf{v}_d$. We close the Boltzmann equation by averaging the collision terms over the electron momentum:

$$
\begin{aligned}
\frac{\partial}{\partial t}\langle \mathbf{k}\rangle|_{\text{coll}} = \langle \dot{\mathbf{k}}\rangle &= 2\int \mathbf{k}\, \frac{\partial f}{\partial t}\bigg|_{\text{coll}} \frac{d\mathbf{k}}{(2\pi\hbar)^3} \\
&= -2\int \mathbf{k}\frac{d\mathbf{k}}{(2\pi\hbar)^3}\frac{f - f_0}{\tau} \\
&= -2\int \mathbf{k}\frac{d\mathbf{k}}{(2\pi\hbar)^3}\frac{f_0(\mathbf{k} - m\mathbf{v}_d)}{\tau} \\
&= -\frac{m}{\tau}\mathbf{v}_d n_e.
\end{aligned}
\tag{10.84}
$$

In deriving Eq. (10.84), we used the fact that $\langle \mathbf{k}\rangle$ in equilibrium vanishes.

Noting that the collision terms are antisymmetric with respect to interchange of $\mathbf{k}$ and $\mathbf{k}'$, we find upon substituting Eq. (10.84) into the Boltzmann equation that

$$
\begin{aligned}
\frac{-m\mathbf{v}_d n_e}{\tau} &= \frac{\beta}{V}\sum_{\mathbf{k}}\mathbf{k}\sum_{\mathbf{k}'} W_{\mathbf{q}}(\mathbf{k},\mathbf{k}')n_{\mathbf{k}}(1 - n_{\mathbf{k}'})[\mathbf{v}_d\cdot\mathbf{k}' - \mathbf{v}_d\cdot\mathbf{k}] \\
&= \frac{\beta}{2V}\sum_{\mathbf{k}}\sum_{\mathbf{k}'}(\mathbf{k} - \mathbf{k}')W_{\mathbf{q}}(\mathbf{k},\mathbf{k}')n_{\mathbf{k}}(1 - n_{\mathbf{k}'})[\mathbf{v}_d\cdot\mathbf{k}' - \mathbf{v}_d\cdot\mathbf{k}] \\
&= -\frac{\beta}{V}\sum_{\mathbf{k}}\sum_{\mathbf{q}}\mathbf{q}(\mathbf{v}_d\cdot\mathbf{q})W_{\mathbf{q}}^0 n_{\mathbf{k}}(1 - n_{\mathbf{k}+\mathbf{q}})[\delta(\epsilon_{\mathbf{k}+\mathbf{q}} - \epsilon_{\mathbf{k}} + \hbar\omega_{\mathbf{q}}) \\
&\quad\times (N_{\mathbf{q}} + 1) + N_{\mathbf{q}}\delta(\epsilon_{\mathbf{k}+\mathbf{q}} - \epsilon_{\mathbf{k}} - \hbar\omega_{\mathbf{q}})],
\end{aligned}
\tag{10.85}
$$

with $\mathbf{k}' = \mathbf{k} + \mathbf{q}$. Recalling the definition of the free-particle structure function (see Eq. 10.51), we reduce the linearized Boltzmann equation to

$$
\begin{aligned}
\frac{m\mathbf{v}_d n_e}{\tau} &= \frac{n_e\beta}{8\pi\hbar}\sum_{q}\mathbf{q}(\mathbf{v}_d\cdot\mathbf{q})W_{\mathbf{q}}^0[S_0(\mathbf{q},\hbar\omega_{\mathbf{q}})N_{\mathbf{q}} + (N_{\mathbf{q}} + 1)S_0(\mathbf{q},-\hbar\omega_{\mathbf{q}})] \\
&= \frac{n_e\beta}{4\pi\hbar}\sum_{q}\mathbf{q}(\mathbf{v}_d\cdot\mathbf{q})W_{\mathbf{q}}^0 N_{\mathbf{q}}S_0(\mathbf{q},\hbar\omega_{\mathbf{q}}).
\end{aligned}
\tag{10.86}
$$

To simplify Eq. (10.86) further, we replace $\mathbf{q}(\mathbf{v}_d\cdot\mathbf{q})$ with its angular average

$$
\langle \mathbf{q}(\mathbf{q}\cdot\mathbf{v}_d)\rangle = \frac{1}{3}q^2\mathbf{v}_d.
\tag{10.87}
$$

Substitution of Eq. (10.87) into Eq. (10.86) results in the general expression

$$\frac{1}{\tau} = \frac{\beta}{24\pi^2 M \hbar^3 m n_e n_{\text{ion}}} \int_0^{q_D} \frac{q^6 dq}{\hbar \omega_{\mathbf{q}}} |V_{ei}(q)|^2 \frac{S_0(q,\hbar\omega_{\mathbf{q}})}{e^{\beta\hbar\omega_{\mathbf{q}}} - 1} \quad (10.88)$$

for the relaxation time. In Eq. (10.88), q_D is the momentum cut-off on the phonon spectrum.

We need an expression for S_0 that captures the essential physics at low temperatures. We showed in the previous chapter that at $T = 0$, $S_0 \propto 1/q$. Asymptotically, this expression vanishes as $q \to \infty$ and, hence, is expected to be valid, as long as the phonon momentum is cut off. The Debye cut-off in the relaxation time justifies our use of the $T = 0$ limit. Away from $T = 0$, the explicit temperature dependence can be introduced by including the factor of $(1 - \exp(-\beta\hbar\omega_{\mathbf{q}}))$, which appears in the original definition of $S_0(k,\omega)$. Consequently, we write

$$S_0(q,\hbar\omega_{\mathbf{q}}) = \frac{m^2 \hbar\omega_{\mathbf{q}}}{\pi n_e q \hbar^2 (1 - e^{-\beta\hbar\omega_{\mathbf{q}}})}. \quad (10.89)$$

We also need an expression for $V_{ei}(q)$. In the Thomas-Fermi treatment of screening, we showed that

$$V_{ei}(q) = -\frac{4\pi Z e^2}{V(q^2 + \kappa_{TF}^2)}. \quad (10.90)$$

At low temperatures, we focus on the limit ($q \to 0$) of V_{ei}. Hence, we approximate $V_{ei}(q)$ with

$$V_{ei}(q) = -\frac{4\pi Z e^2}{V\kappa_{TF}^2}$$

$$= -\frac{\pi^2 Z \hbar^3}{V m p_F} = -\frac{Z}{N(\epsilon_F)}. \quad (10.91)$$

The relaxation time can be written as

$$\frac{1}{\tau} = \frac{Z^2 \beta m}{24\pi^3 n_e^2 \hbar^3 N^2(\epsilon_F) M n_{\text{ion}}} \int_0^{q_D} \frac{q^5 dq}{(1 - e^{-\beta\hbar\omega_{\mathbf{q}}})(e^{\beta\hbar\omega_{\mathbf{q}}} - 1)}. \quad (10.92)$$

For the phonon spectrum, we use the linear-dispersion relationship, $\omega_q = sq/\hbar$, where s is a constant. Let $x = \beta\omega_q$. With the observation that

$$\frac{1}{(e^x - 1)(1 - e^{-x})} = -\frac{\partial}{\partial x}\frac{1}{e^x - 1}, \tag{10.93}$$

we rewrite the relaxation time as

$$\frac{1}{\tau} = -\frac{Z^2\beta m}{24\pi^3\hbar^7 n_e^2 M N^2(\epsilon_F)(\beta s)^6}\int_0^{q_D s\hbar/k_B T} x^5\frac{\partial}{\partial x}\frac{1}{(e^x - 1)}$$
$$- \alpha_0 T^5 J_5(T_D/T), \tag{10.94}$$

with

$$J_5(y) = -\int_0^y x^5 dx\frac{\partial}{\partial x}\frac{1}{e^x - 1}, \tag{10.95}$$

$T_D = q_D s\hbar/k_B$ is the Debye temperature, and

$$\alpha_0 = \frac{m Z^2 k_B^5}{\hbar^7 24\pi^3 n_e^2 n_{\text{ion}} M N^2(\epsilon_F)s^6}. \tag{10.96}$$

There are two cases of interest. At low temperatures, $T_D/T \gg 1$, implying that the integral can be extended to infinity, leaving

$$J_5(\infty) = -\int_0^\infty x^5 dx\frac{\partial}{\partial x}\frac{1}{e^x - 1}$$
$$= 5!\zeta(5), \tag{10.97}$$

where $\zeta(n)$ is the Riemann-Zeta function. As a consequence,

$$\frac{1}{\tau} = 5!\zeta(5)\alpha_0 T^5 \tag{10.98}$$

for $T \ll T_D$. From the Drude formula, we have that the resistivity $\rho \sim 1/\tau$. We see, then, that for $T \ll T_D$, the resistivity scales as $\rho \sim T^5$. The origin of the T^5 contribution is as follows. A factor of T^3 arises from the number of phonons present at $T = 0$. The remaining factors of T arise from momentum transfer and the fraction of electrons in the vicinity of T_F that can scatter. Each of these processes scales as T. Note the factor α_0 correctly represents the scaling of the resistivity in terms of the ion mass M, the electron density n_e, and the density of states $N(\epsilon_F)$.

Consider now the high-temperature limit. In this case, $T \gg T_D$, and the upper limit in Eq. (10.95) is $y \ll 1$. We can then series expand the integrand to obtain

$$J_5(y) = -\int_0^y x^5 dx \, \frac{\partial}{\partial x}\left(\frac{1}{x} + \cdots\right) = \int_0^y x^3 dx = \frac{y^4}{4}. \quad (10.99)$$

Because $y = T_D/T$, we find that at high temperatures,

$$\rho \sim \frac{1}{\tau} \propto T^5 \left(\frac{T_D}{T}\right)^4 \sim T. \quad (10.100)$$

Linear behavior sets in for $T/T_D > 0.2$. Of current interest is the linear-T resistivity in the normal state of the high-temperature copper oxide materials [GF1987]. As this behavior persists until the onset of superconductivity, the linear-T resistivity is of fundamentally different (and currently unknown) origin than the high-temperature linear-T resistivity that results from phonon scattering.

Of course, normal nonmagnetic impurities also contribute to the resistivity. When the concentration (n_{imp}) of nonmagnetic impurities is small, the first Born approximation can be used. In this limit, impurity scattering contributes a constant term to the relaxation rate, proportional to ϵ_F at $T = 0$. This is a reflection that scattering at normal impurities is governed by the electrons at the Fermi energy. For normal impurities of charge Z, the basic result for the impurity relaxation rate is

$$\frac{1}{\tau_{imp}} \propto \frac{2n_{imp}Z^2 \epsilon_F}{\hbar n_e}. \quad (10.101)$$

Consequently, disorder in a metal is expected to lead to a nonzero resistance at $T = 0$, which is commonly referred to as the *residual resistance* in a metal. Of course, the situation changes dramatically in the strong-disorder regime. In this limit, perturbation theory breaks down. A transition to an Anderson localized state occurs when the strength of the disorder exceeds a critical value for $d > 2$ (see Chapter 12). In the localized regime, the electronic states decay exponentially with distance. For $d \leq 2$, the transition is particularly striking, as any amount of disorder leads to complete localization of all the electronic eigenstates. Localization of the eigenstates results in a vanishing of the dc-conductivity and the onset of an insulating state. In three dimensions, the disorder must exceed a critical value before insulating behavior obtains. In an Anderson localized system, charge carriers must be thermally excited if they are to transport at all. Consequently, activated transport typically obtains in insulators above $T = 0$.

10.6   HYDRODYNAMIC LIMIT: PHONON DRAG

We have been considering the Boltzmann equation in the context of the phonon contribution to the conductivity at low temperatures. In the course of the derivation, we assumed that the effective particle distribution in the presence of the electric field could be described by a drifting distribution, in which the electron momentum is replaced by $\mathbf{p} \rightarrow \mathbf{p} - m_e \mathbf{v}_d$. We treat here the situation in which long-wavelength and low-frequency variations dominate. This is the hydrodynamic limit:

$$\omega \tau_{ee} \ll 1 \qquad (10.102a)$$

$$k\ell \ll 1, \qquad (10.102b)$$

where ω and k are the frequency and wave vectors of the electron and τ_{ee} and ℓ are the collision times and mean-free paths, respectively.

We define the momentum density to be

$$\mathbf{g} = 2 \int \frac{d\mathbf{p}}{(2\pi\hbar)^3} \mathbf{p} f_{\mathbf{p}}(\mathbf{r}, t). \qquad (10.103)$$

We are interested in the rate of change of the momentum density. Specifically, our focus is on the rate at which the collision processes limit momentum exchange. Denoting the explicit time dependence of the momentum density in the time derivative as $\partial \mathbf{g}/\partial t$, we write the collision terms as

$$\frac{\partial \mathbf{g}}{\partial t}\bigg|_{coll} \equiv \frac{\partial \mathbf{g}}{\partial t} + 2 \int \frac{d\mathbf{p}}{(2\pi\hbar)^3} \mathbf{p} (\dot{\mathbf{p}} \cdot \nabla_{\mathbf{p}} f_{\mathbf{p}} + \dot{\mathbf{r}} \cdot \nabla_{\mathbf{r}} f_{\mathbf{p}})$$

$$= \frac{\partial \mathbf{g}}{\partial t} + 2 \int \frac{d\mathbf{p}}{(2\pi\hbar)^3} \mathbf{p} \frac{\partial f_{\mathbf{p}}}{\partial t}\bigg|_{coll}. \qquad (10.104)$$

We will find it expedient to rewrite the collision terms, using the definitions

$$\dot{\mathbf{p}} = -\nabla_{\mathbf{r}} \epsilon_{\mathbf{p}} \qquad (10.105a)$$

$$\dot{\mathbf{r}} = \nabla_{\mathbf{p}} \epsilon_{\mathbf{p}} = \mathbf{v}_{\mathbf{p}} \qquad (10.105b)$$

in steady state. Note that in steady state, $\dot{f} = 0$. In the presence of an electric field, we replace the single-particle energy levels with $\epsilon_{\mathbf{p}} \rightarrow \epsilon_{\mathbf{p}} + e\Phi(\mathbf{r})$, where $\Phi(\mathbf{r})$ is the electrical potential. As a consequence,

$\nabla_{\mathbf{r}}\epsilon_{\mathbf{p}} = -e\,\mathbf{E}$. If we substitute these definitions into Eq. (10.104), we obtain

$$\frac{\partial \mathbf{g}}{\partial t}\bigg|_{\text{coll}} = \frac{\partial \mathbf{g}}{\partial t} + 2\int \frac{d\mathbf{p}}{(2\pi\hbar)^3}\mathbf{p}(\nabla_{\mathbf{p}}\epsilon_{\mathbf{p}}\cdot\nabla_{\mathbf{r}}f_{\mathbf{p}} - \nabla_{\mathbf{r}}\epsilon_{\mathbf{p}}\cdot\nabla_{\mathbf{p}}f_{\mathbf{p}}) \qquad (10.106)$$

$$= \frac{\partial \mathbf{g}}{\partial t} + 2\int \frac{d\mathbf{p}}{(2\pi\hbar)^3}(\nabla_{\mathbf{r}}\epsilon_{\mathbf{p}})f_{\mathbf{p}} + 2\int \frac{d\mathbf{p}}{(2\pi\hbar)^3}\mathbf{p}\nabla_{\mathbf{r}}\cdot(\mathbf{v}_{\mathbf{p}}f_{\mathbf{p}}) \qquad (10.107)$$

$$= \frac{\partial \mathbf{g}}{\partial t} - e\,\mathbf{E}\tilde{n}_e + \nabla_{\mathbf{r}}\cdot T, \qquad (10.108)$$

where T is a stress tensor with components

$$T_{ij} = 2\int \frac{d\mathbf{p}}{(2\pi\hbar)^3}p_i\,v_{\mathbf{p}_j}f_{\mathbf{p}}, \qquad (10.109)$$

with $i,j = x,y,z$, and $\tilde{n}_e$ is the electron density in the presence of the electric field. In deriving Eq. (10.107) from Eq. (10.106), we integrated the last term in Eq. (10.106) by parts.

We compute T_{ij} assuming the electrons are in equilibrium with the drifting distribution created by an electric field. The distribution function for a drifting velocity distribution is simply $f_{\mathbf{p}} \to f_{\mathbf{p}-m\mathbf{v}_d}$. In evaluating $\partial\mathbf{g}/\partial t$, we translate $\mathbf{p} \to \mathbf{p} + m\mathbf{v}_d$; as a consequence, the distribution function will remain unchanged from the static Fermi-Dirac distribution. The stress tensor then becomes

$$T_{ij} = 2\int \frac{d\mathbf{p}}{(2\pi\hbar)^3}(p_i + mv_{d_i})(v_{\mathbf{p}_j} + v_{d_j})f_{\mathbf{p}} \qquad (10.110)$$

$$= 2\int \frac{d\mathbf{p}}{(2\pi\hbar)^3}(p_i\,v_{\mathbf{p}_j} + mv_{d_i}\,v_{d_j})f_{\mathbf{p}}, \qquad (10.111)$$

because $\langle\mathbf{p}\rangle = 0$. For an isotropic system, $\langle\mathbf{p}\mathbf{v}_{\mathbf{p}}\rangle = \frac{1}{3}\langle\mathbf{p}\cdot v_{\mathbf{p}}\rangle$. In the context of the noninteracting electron gas, we showed that the pressure exerted by the electron gas is

$$P_e = \frac{2}{3}\int \frac{d\mathbf{p}}{(2\pi\hbar)^3}\mathbf{p}\cdot v f_{\mathbf{p}}. \qquad (10.112)$$

As a consequence, the stress tensor reduces to

$$T_{ij} = \delta_{ij}P_e + mv_{d_i}\,v_{d_j}n_e, \qquad (10.113)$$

and the equation of motion for the momentum density

$$\left.\frac{\partial \mathbf{g}}{\partial t}\right|_{\text{coll}} = \frac{\partial \mathbf{g}}{\partial t} + \nabla P_e + \nabla \cdot (n_e m \mathbf{v}_d \mathbf{v}_d) - e \mathbf{E} \tilde{n}_e \qquad (10.114)$$

contains the gradient of the pressure.

In the absence of collisions of any sort, momentum is conserved and the right-hand side of Eq. (10.114) vanishes, except for the electric field term. Consequently,

$$\left.\frac{\partial \mathbf{g}}{\partial t}\right|_{\text{coll}} + e \mathbf{E} \tilde{n}_e = 0 \qquad (10.115)$$

is the basic equation underlying electrical transport.

When ions are present, we can replace the left-hand side of Eq. (10.114) with the relaxation-time approximation

$$\left.\frac{\partial \mathbf{g}}{\partial t}\right|_{\text{coll}} = -\frac{\mathbf{g}}{\tau} = -\frac{\tilde{n}_e m \mathbf{v}_d}{\tau}. \qquad (10.116)$$

We expand the range of applicability of Eq. (10.116) even further by incorporating the ion velocity. Because the ions are moving, collisions with ions will cause equilibration to the ion velocity rather than to the drift velocity. If $\mathbf{u}_{\text{ion}}$ is the ion velocity, then

$$\left.\frac{\partial \mathbf{g}}{\partial t}\right|_{\text{coll}} = -\frac{\mathbf{g}_{eq}}{\tau} = -\tilde{n}_e m (\mathbf{v}_d - \mathbf{u}_{\text{ion}}). \qquad (10.117)$$

Eqs. (10.114) to (10.117) constitute the hydrodynamic equations of motion for the electron momentum density. The suppression of the drift velocity by the velocity of the ions is known as *phonon drag*.

10.7   SOUND PROPAGATION

We can also use the hydrodynamic approach to understand the propagation of sound waves in a metal. A sound wave arises from the collective motion of ions in a crystal. For electrons, we showed that the frequency for collective oscillations is given by $\omega_{\mathbf{p}}^2 = 4\pi n_e e^2/m$. Modifying this expression, so that the electric charge is replaced by the ion charge, $e \rightarrow eZ$ and $m \rightarrow M$ and $n_e \rightarrow Z n_i$, we find the equivalent ion "plasma" frequency is

$$\omega_{\text{ion}}^2 = \left(\frac{Z^3 m n_i}{n_e M}\right) \omega_{\mathbf{p}}^2. \qquad (10.118)$$

This conclusion is erroneous because the long-wavelength ion excitations should obey a linear k dispersion relationship, as opposed to the relatively constant dispersion relation for plasma oscillations.

To correct this problem, we must include both the electron and ion degrees of freedom in our hydrodynamic equations for the momentum density. If $\mathbf{g}_{\text{ion}}$ is the momentum density for the ions, then

$$\frac{\partial \mathbf{g}_{\text{ion}}}{\partial t} + eZ\,\mathbf{E}\tilde{n}_i = -\left(\frac{\partial \mathbf{g}_{\text{ion}}}{\partial t}\right)_{\text{coll}} \tag{10.119}$$

is the hydrodynamic equation for the ion degrees of freedom, with $\tilde{n}_i$ the ion density in the presence of the electric field. Because the charge on the ions is Ze, the coupling to the electric field has the opposite sign than in the electron problem. If we combine Eq. (10.119) with the equation of motion for $\mathbf{g}_e$, we obtain

$$\frac{\partial}{\partial t}(\mathbf{g}_e + \mathbf{g}_{\text{ion}}) + \nabla_{\mathbf{r}}P_e + e\mathbf{E}(Z\,\tilde{n}_i - \tilde{n}_e) = 0 \tag{10.120}$$

upon ignoring the quadratic terms in $\mathbf{v}_d$. We have set the collision terms equal to zero because once we have included the ion degrees of freedom, we can invoke momentum conservation.

To eliminate the particle densities from Eq. (10.120), we consider the time derivative of the electron number density

$$\begin{aligned}
\left.\frac{\partial \tilde{n}_e}{\partial t}\right|_{\text{coll}} &= 2\int \frac{d\mathbf{p}}{(2\pi\hbar)^3} \left.\frac{\partial f_{\mathbf{p}}}{\partial t}\right|_{\text{coll}} \\
&= \frac{\partial \tilde{n}_e}{\partial t} + 2\int \frac{d\mathbf{p}}{(2\pi\hbar)^3}[\nabla_{\mathbf{p}}\epsilon_{\mathbf{p}} \cdot \nabla_{\mathbf{r}}f_{\mathbf{p}} - \nabla_{\mathbf{r}}\epsilon_{\mathbf{p}} \cdot \nabla_{\mathbf{p}}f_{\mathbf{p}}] \\
&= \frac{\partial \tilde{n}_e}{\partial t} + \nabla \cdot \mathbf{J}(r,t), \tag{10.121}
\end{aligned}$$

where $\mathbf{J}(\mathbf{r},t)$ is the particle current

$$\begin{aligned}
\mathbf{J}(\mathbf{r},t) &= 2\int \frac{d\mathbf{p}}{(2\pi\hbar)^3}(\nabla_{\mathbf{p}}\epsilon_{\mathbf{p}})f_{\mathbf{p}} \\
&\simeq \mathbf{v}_d\,\tilde{n}_e. \tag{10.122}
\end{aligned}$$

Because the collision terms vanish in an isolated system, we have that

$$\frac{\partial \tilde{n}_e}{\partial t} + \nabla_{\mathbf{r}} \cdot (\mathbf{v}_d\,\tilde{n}_e) = 0. \tag{10.123}$$

In general, $\tilde{n}_e$ is a function of position because it involves an integral over the complete distribution function in the presence of the drift velocity. Nonetheless, we linearize about the equilibrium unperturbed value, n_e, and obtain

$$\frac{\partial \tilde{n}_e}{\partial t} + n_e \nabla_{\mathbf{r}} \cdot \mathbf{v}_d = 0, \tag{10.124}$$

and for the ion degrees of freedom,

$$\frac{\partial \tilde{n}_i}{\partial t} + n_i \nabla_{\mathbf{r}} \cdot \mathbf{u}_{\text{ion}} = 0. \tag{10.125}$$

If the electrons and ions are moving collectively, then $\mathbf{v}_d \sim \mathbf{u}_{\text{ion}}$. If we compare Eqs. (10.124) and (10.125) with the insight that $n_e = Zn_i$, we find that even in the drift-velocity distribution,

$$\frac{\partial \tilde{n}_e}{\partial t} = Z \frac{\partial \tilde{n}_i}{\partial t}, \tag{10.126}$$

or, equivalently, $\tilde{n}_e = Z\tilde{n}_i$. As a consequence, the last term vanishes in Eq. (10.120), and we obtain that

$$\frac{\partial}{\partial t}(\mathbf{g}_e + \mathbf{g}_{\text{ion}}) + \nabla_r P_e = 0. \tag{10.127}$$

Using the equilibrium form for $\mathbf{g}$, Eq. (10.84), and the approximation in Eq. (10.122), we reduce the momentum conservation constraint to

$$\frac{\partial}{\partial t}[(m\tilde{n}_e + M\tilde{n}_i)v_d] + \nabla_{\mathbf{r}} P_e = 0. \tag{10.128}$$

To eliminate the drift-velocity terms, we take the divergence of Eq. (10.128) and linearize, keeping only the $\nabla_{\mathbf{r}} v_d$ derivative:

$$(mn_e + Mn_i)\frac{\partial}{\partial t}\nabla_{\mathbf{r}} \cdot \mathbf{v}_d + \nabla_{\mathbf{r}}^2 P_e = 0. \tag{10.129}$$

Coupled with the equation for the second time derivative of the electron number density,

$$\frac{\partial^2 \tilde{n}_e}{\partial t^2} + n_e \frac{\partial}{\partial t}(\nabla_{\mathbf{r}} \cdot \mathbf{v}_d) = 0, \tag{10.130}$$

we arrive at the equation of motion,

$$\left(m + \frac{M}{Z}\right)\frac{\partial^2 \tilde{n}_e}{\partial t^2} - \nabla_r^2 P_e = 0. \tag{10.131}$$

In Eq. (10.131), we can ignore the electron mass relative to the ion mass. Let us calculate the gradient of the pressure. We note that $\nabla_r P_e = \partial P_e \, \partial \tilde{n}_e / \nabla_r \tilde{n}_e$. We showed previously that

$$\frac{\partial P_e}{\partial \tilde{n}_e} = \frac{2\mu_0}{3} = \frac{m v_F^2}{3}. \tag{10.132}$$

The equation of motion for the electron density, then, becomes

$$\frac{\partial^2 \tilde{n}_e}{\partial t^2} - \frac{Z}{M}\frac{m v_F^2}{3}\nabla_r^2 \tilde{n}_e = 0. \tag{10.133}$$

In Fourier space, we obtain

$$\left(-\omega^2 + \frac{Zm}{M}\frac{v_F^2}{3}k^2\right)\tilde{n}_e = 0 \tag{10.134}$$

or

$$\omega^2 = \frac{Zm}{M}\frac{v_F^2}{3}k^2 = v_{sound}^2 k^2, \tag{10.135}$$

which is quadratic in the wave vector k. The speed of sound in a metal is, then,

$$v_{sound} = \sqrt{\frac{Zm}{M}\frac{v_F^2}{3}}, \tag{10.136}$$

which is the Bohm-Staver result. It accurately describes the propagation of sound in a metal in the hydrodynamic limit. Plasma oscillations in general cannot be treated in an analogous manner because they are outside ($\omega\tau_{ee} \gg 1$) the hydrodynamic limit.

10.8   SUMMARY

Two key results were derived in this chapter: 1) the form of the linear interaction between electrons and phonons, and 2) the Boltzmann transport theory. Using the latter, we were able to derive the result we advertised initially in Chapter 7, namely, that the resistivity vanishes as T^5 in a metal where collisions with phonons dominate all scattering

processes. We derived this result using the relaxation-time approximation. An essential ingredient of the relaxation-time approximation is
the drift-velocity ansatz. As illustrated in Fig. (10.3), in this approximation the Fermi surface is translated by an amount proportional to the
drift velocity. Electrons in the translated Fermi surface are described
by the original Fermi-Dirac distribution with the momentum shifted
by $\mathbf{p} \rightarrow \mathbf{p} - m\mathbf{v}_d$.

PROBLEMS

1. Prove that $W_{\mathbf{q}}(\mathbf{k}, \mathbf{k}') = e^{\beta(\epsilon_{\mathbf{k}} - \epsilon_{\mathbf{k}'})} W_{\mathbf{q}}(\mathbf{k}', \mathbf{k})$.
2. Calculate explicitly the impurity contribution to the electrical re-
laxation rate $1/\tau$. Prove that the impurity contribution to the relaxation rate is proportional to the Fermi energy. A useful strategy
is to: 1) assume the electrons elastically scatter from the impurities
from one plane-wave state to another; 2) average over the random
position of the impurities using the random-phase approximation
in which an average of the form

$$\left\langle \sum_{i,j} e^{i(\mathbf{q}-\mathbf{q}')\cdot(\mathbf{R}_i - \mathbf{R}_j)} \right\rangle \qquad (10.137)$$

is nonzero only when $\mathbf{R}_i = \mathbf{R}_j$ and reduces to Vn_{imp} where n_{imp}
is the concentration of the impurities; 3) follow the same Boltz-
mann equation approach we used for the phonon problem, invoking again the relaxation-time approximation; 4) in the end,
evaluate a structure factor of the form $S_0(\mathbf{k}, -\mathbf{k} \cdot \mathbf{v}_d)$. Solve this
by a Taylor series expansion, retaining only the linear $\mathbf{k} \cdot \mathbf{v}_d$ term.
Be careful with the limits. It is from the structure factor that the
ϵ_F dependence emerges; 5) assume an impurity potential that is
of the screened Coulomb form. Again take the small-k limit.

REFERENCES

[AM1976] N. W. Ashcroft, N. D. Mermin, *Solid State Physics* (Holt,
Rinehart, and Winston, New York, 1976).

[G1963] J. Goldstone, *Nuovo Cimento*, **19**, 155 (1963).

[GF1987] M. Gurvitch, A. T. Fiory, *Phys. Rev. Lett.* **59**, 1337 (1987).

[M1981] G. D. Mahan, *Many-Particle Physics* (Plenum Press, New
York, 1981).

– 11 –

Superconductivity

In this chapter, we focus on the phenomenon of superconductivity and the Bardeen-Cooper-Schrieffer (BCS) theory behind it [BCS1957]. Superconductivity obtains when a finite fraction of the conduction electrons in a metal condense into a quantum state characterized by a unique quantum mechanical phase. The specific value of the quantum mechanical phase varies from one superconductor to another. The locking in of the phase of a number of electrons on the order of Avogadro's number ensures the rigidity of the superconducting state. For example, electrons in the condensate find it impossible to move individually. Rather, the whole condensate moves from one end of the sample to the other as a single unit. Likewise, electron-scattering events that tend to disrupt the condensate must disrupt the phase of a macroscopic number of electrons for the superconducting state to be destroyed. Hence, phase rigidity implies collective motion as well as collective destruction of a superconducting condensate. The only other physical phenomenon that arises from a similar condensation of a macroscopic number of particles into a phase-locked state is that of Bose-Einstein condensation. There is a crucial difference between these effects, however. The particles that constitute the condensate in superconductivity are Cooper pairs, which do not obey Bose statistics. In fact, it is the Pauli principle acting on the electrons comprising a Cooper pair that prevents the complete mapping of the superconducting problem onto a simple one of Bose condensation. As we will see, it is the Pauli principle that makes BCS theory work so well. What do we mean by this? In BCS theory, it is assumed that electrons form Cooper pairs, and the pairs are strongly overlapping. Such strong overlap would imply a strong correlation between pairs. In fact, it is the correlations between pairs that accounts for most of the observed properties of superconductors, for example the energy gap and the Meissner effect. In BCS theory, however, there is no explicit dynamical interaction between Cooper pairs. The only interaction, if it can be thought of in these terms, is that arising from the Pauli exclusion principle that precludes two Cooper pairs from occupying the same momentum state. That BCS theory works so well speaks volumes for the real nature of pair-pair correlations in metals. It would suggest that real pair-pair interactions in a metal arise primarily from the Pauli exclusion principle rather than from some additional dynamical interaction. It is primarily for this reason that the simple pairing hypothesis of BCS has had such profound success.

11.1   SUPERCONDUCTIVITY: PHENOMENOLOGY

At the outset, we lay plain the experimental facts that inspired the BCS theory of superconductivity.

(a)   Zero Resistance: The typical signature of superconductivity is the vanishing of the electrical resistance below some critical temperature, T_c. The superconducting state is a thermodynamically distinct state of matter. Below T_c, a current flows without any loss. Until the high T_c materials, Nb held the highest transition temperature at 9.26 K.

(b)   Meissner Effect: Another feature is the exclusion of magnetic fields, the Meissner effect. Materials in which the Meissner effect is complete are known as *Type I superconductors*. Consequently, the interior of a Type I superconductor is a perfect diamagnet. A magnetic field applied at the boundary of a Type I superconductor falls off exponentially with distance in the interior of the material as illustrated in Fig. (11.1). The penetration depth, λ_L, is defined as the distance over which the magnetic field decreases by the factor $1/e$. Below T_c, the field needed to destroy superconductivity increases to some critical value $H_c(T)$, as illustrated in Fig. (11.2). Because the magnetic field inside a

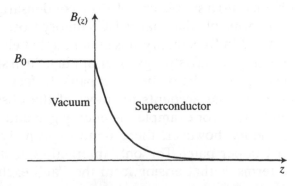

Figure 11.1 Fall-off of the magnetic field in a Type I superconductor.

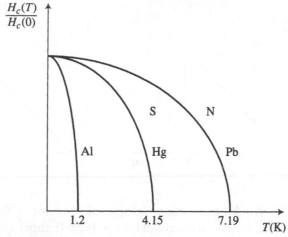

Figure 11.2 The dependence of the critical field as a function of temperature. The temperatures indicated on the horizontal axis represent the superconducting transition temperatures for a series of metals.

superconductor is zero,

$$B = 0 = B_{appl} + 4\pi M, \tag{11.1}$$

where B_{appl} is the applied field and M the magnetization. Solving this equation, we find that the magnetization is

$$M = -\frac{1}{4\pi} B_{appl}. \tag{11.2}$$

The negative value of M signals that the interior of a superconductor is diamagnetic. At any temperature less than T_c, the magnetization, M, should be a linear function of the applied field, H. A material having a magnetization of this form is called a *Type I superconductor*. The normal state is indicated with an N and the superconducting state with an S. Above H_c, $B \neq 0$ and the magnetization no longer obeys Eq. (11.2).

In some materials, superconductivity is observed up to an upper critical field H_{c_2}, but an incomplete Meissner effect is seen between a lower critical field H_{c_1} and H_{c_2}. The resultant magnetization is shown in Fig. (11.3b). Materials exhibiting a magnetization of this kind are known as *Type II superconductors*. Between H_{c_1} and H_{c_2}, the magnetic field penetrates the material but superconductivity is not destroyed. The field lines form a regular array known as the *Abrikosov vortex lattice* [A1957]. All high T_c cuprate superconductors are Type II.

We are concerned primarily with Type I materials. To understand the penetration depth in a superconductor, we resort to the London

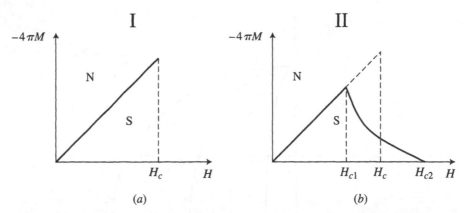

Figure 11.3 (a) Magnetization vs. applied field for a Type I superconductor. (b) Magnetization vs. applied field for a Type II superconductor. Between H_{c1} and H_{c2}, magnetic field lines penetrate the superconductor, but they do not destroy superconductivity. The field lines form a vortex lattice.

equations. First, we need the Maxwell equation for the curl of an electric field: $-\partial \mathbf{B}/\partial t = c\nabla \times \mathbf{E} = c\rho\nabla \times \mathbf{j}$, where ρ is the resistivity. In a perfect conductor, $\rho = 0$, and, as a consequence, $\partial \mathbf{B}/\partial t = 0$. In a superconductor, $\rho = 0$, as well. However, it is an experimental fact that $B = 0$ inside a superconductor. This result cannot be deduced from the Maxwell equations. It is the Meissner effect that sets superconductivity apart from materials that just display perfect conductivity. Inside a superconductor, expulsion of magnetic flux is mediated by the current that flows. London proposed in 1935 that everywhere in a superconductor

$$\mathbf{j} = -\text{const.}A, \tag{11.3}$$

where A is the vector potential and $\mathbf{B} = \nabla \times \mathbf{A}$. From this ansatz, London was able to show that a magnetic field decays exponentially inside a superconductor. Dimensionally, the constant has units of $1/(L\cdot\text{time})$. Let us write the constant as

$$\text{const.} = \frac{c}{4\pi\lambda_L^2}, \tag{11.4}$$

where c is the speed of light. If we take the curl of both sides of Eq. (11.3), we find that

$$\nabla \times \mathbf{j} = -\frac{c}{4\pi\lambda_L^2}\mathbf{B}. \tag{11.5}$$

From the Maxwell equation

$$\nabla \times \mathbf{B} = \frac{4\pi}{c}\mathbf{j}, \tag{11.6}$$

we find that

$$\nabla \times \nabla \times \mathbf{B} = \frac{4\pi}{c}\nabla \times \mathbf{j}, \tag{11.7}$$

which implies that

$$\nabla^2 \mathbf{B} = \frac{1}{\lambda_L^2}\mathbf{B}. \tag{11.8}$$

The solution to this equation,

$$B(r) = B(0)e^{-r/\lambda_L}, \tag{11.9}$$

is an exponentially decaying magnetic induction on a length scale λ_L. Exponential decay of the magnetic field into a superconductor is the Meissner effect. Let v_s, m^*, and e^*, respectively, be the velocity, mass, and charge of the current carriers in a superconductor. Then

$$m^*\dot{\mathbf{v}}_s = -e^*\mathbf{E}. \tag{11.10}$$

The current of these electrons is defined as $\mathbf{j} = -e^*\mathbf{v}_s n_s$, which, combined with Eq. (11.10), yields

$$\frac{\partial \mathbf{j}}{\partial t} = \frac{n_s e^{*2}}{m^*}E \tag{11.11}$$

for the time evolution of the current. Taking the curl of both sides, we find that

$$0 = \frac{\partial}{\partial t}(\nabla \times \mathbf{j}) - \frac{n_s e^2}{m^*}\nabla \times \mathbf{E} = \frac{\partial}{\partial t}\left(\nabla \times \mathbf{j} + \frac{n_s e^{*2}}{m^*c}\mathbf{B}\right). \tag{11.12}$$

Comparing Eq. (11.12) with Eq. (11.5), we obtain that the penetration depth is

$$\lambda_L = \left(\frac{m^*c^2}{4\pi n_s e^{*2}}\right)^{1/2}. \tag{11.13}$$

We see, then, that as the superconducting density increases, λ_L decreases.

We can justify the main assumption in the London approach by appealing to the theory of Ginsburg and Landau [GL1950]. The crucial ingredient in this phenomenological theory is that the difference in the free-energy density between the superconducting and normal states can be written as a functional of an order parameter, $\psi(\mathbf{r})$, for the superconducting state. Physically, $|\psi(\mathbf{r})|^2$ is proportional to the charge density in the superconducting state, n_s. Consequently, we can interpret $\psi(\mathbf{r})$ as the wave function of the superconducting state. In BCS theory, $\psi(\mathbf{r})$ plays the role of the center-of-mass wave function for a Cooper pair. Near T_c, the superfluid density is small; hence, $|\psi|^2 \ll n_e$. Consequently, Ginsburg and Landau expanded the free-energy density for the superconducting state in the vicinity of T_c as a power series in $|\psi|^2$,

$$F = F_N + \int d\mathbf{r} \left(\frac{\hbar^2}{2m^*}|\nabla\psi|^2 + a(T)|\psi(\mathbf{r})|^2 + b(T)|\psi(\mathbf{r})|^4 \right), \quad (11.14)$$

retaining the kinetic energy term, $|\nabla\psi|^2$, to account explicitly for spatial variations of the field $\psi(\mathbf{r})$. The free energy of the normal state is F_N. The coefficients $a(T)$ and $b(T)$ are real and temperature-dependent, and, for stability, $b(T) > 0$.

To find the ground state of the system, we minimize the free-energy density with respect to $\psi^*(\mathbf{r})$:

$$-\frac{\hbar^2}{2m^*}\nabla^2\psi(\mathbf{r}) + a(T)\psi(\mathbf{r}) + 2b(T)|\psi(\mathbf{r})|^2\psi(\mathbf{r}) = 0. \quad (11.15)$$

Because the free-energy density contains the term $|\nabla\psi|^2$, which is always positive, the free energy is minimized by demanding that $\nabla\psi(\mathbf{r}) = 0$, or, equivalently, that ψ be uniform in space. Consequently, the solution to the saddle point equation is either $\psi = 0$ or

$$|\psi_0|^2 = -\frac{a(T)}{2b(T)} = n_s, \quad (11.16)$$

which implies that $a(T) < 0$. Should $a(T)$ exceed zero, $\psi = 0$, and the system is in the normal state. Since superconductivity vanishes at T_c, we must have that $a(T_c) = 0$. A Taylor expansion of $a(T)$ around T_c to first order leads to the result that

$$a(T) = a_1(T - T_c) \quad (11.17)$$

with $a_1 > 0$.

We obtain the London conjecture by recalling that the current density in the presence of a vector potential, **A**, is

$$\mathbf{j} = \frac{e^*\hbar}{2im^*}(\psi^*\nabla\psi - \psi\nabla\psi^*) - \frac{e^{*2}}{m^*c}|\psi|^2\mathbf{A}. \qquad (11.18)$$

If we assume that the magnetic field is sufficiently small, so that the equilibrium value of ψ_0 is unchanged, then substitution of ψ_0 into Eq. (11.18) yields the London result

$$\mathbf{j} = -\frac{e^{*2}n_s}{m^*c}\mathbf{A}. \qquad (11.19)$$

Using Eq. (11.13), we find that this result is consistent with Eq. (11.4). Hence, from this simple phenomenological approach, we are able to justify the London ansatz for the current density.

(c)   Heat Capacity: In the superconducting state, the entropy decreases continuously but dramatically, signaling the formation of a highly ordered state. This is depicted in Figure (11.4a). As a result, the temperature derivative of the entropy must be steeper on the superconducting side than on the normal side of the transition. Consequently, the heat capacity is discontinuous at T_c, and superconductivity is a second-order phase transition. As shown in Fig. (11.4b), in the superconducting state, the heat capacity, c_s, falls off as $c_s \propto \exp -\Delta/k_B T$, where Δ is an energy scale. A heat capacity of this form is indicative of an energy gap in the excitation spectrum, with $\epsilon_p > \mu$. Let us verify this with the simple calculation:

$$c_V = \frac{\partial}{\partial T}\int\frac{\epsilon_p d\epsilon_p}{(e^{\beta(\epsilon_p-\mu)}+1)} \xrightarrow{T\to 0} \frac{\partial}{\partial T}\int \epsilon_p e^{-\beta(\epsilon_p-\mu)}d\epsilon_p. \qquad (11.20)$$

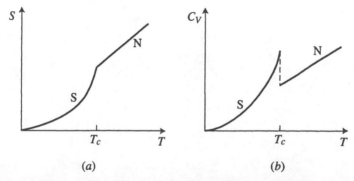

(a)   (b)

Figure 11.4 (a) Behavior of the entropy across the superconducting transition. (b) Behavior of the heat capacity in the normal and superconducting states.

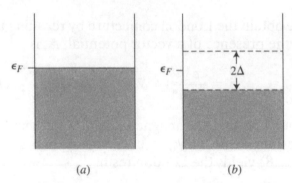

(a)   (b)

Figure 11.5 (a) Filled states in the normal state. (b) Formation of an energy gap in the superconducting state. The full gap is 2Δ, not Δ.

If $(\epsilon_p - \mu) \approx \Delta$, then $c_V \sim \exp(-\Delta(T = 0)/k_B T)$. Consequently, in the superconducting state, we adopt the picture for the energy gap shown in Fig. (11.5).

The formation of a gap at the Fermi level in the superconducting state results in a lowering of the ground state energy of the system. The gap is actually 2Δ, not Δ. Hence, $\epsilon_p - \mu$ accounts for only half the gap. Experimentally, the gap can be measured by tunneling or ultrasound attenuation experiments. Thermodynamically, the gap gives rise to a discontinuity in the heat capacity. That is, $c_s(T_c^-) - c_N(T_c^+) \neq 0$. Across the superconducting transition, both first derivatives of the free energy vanish. Hence, no latent heat is associated with the superconducting transition. Above T_c, $\Delta = 0$, and at $T = 0$, Δ has its largest value. A typical plot of $\Delta(T)/\Delta(T = 0)$ is shown in Fig. (11.6). A weak-coupling superconductor has a ratio of $2\Delta(T = 0)/k_B T_c$ in the range $1 \leftrightarrow 3$. Strong coupling corresponds to $2\Delta(T = 0)/k_B T_c > 4$. The basic energy scale for the creation of an electron-hole pair in a superconductor is 2Δ.

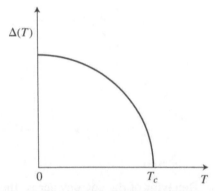

Figure 11.6 The behavior of the superconducting gap, Δ, as $T \rightarrow T_c$.

(d)   Microwave and Infrared Properties: As a result of the gap 2Δ, photons possessing energies less than 2Δ are not absorbed. All such photons are reflected. Perfect reflection occurs for $\omega < 2\Delta(T = 0)/\hbar$. When this condition is true, photons see a completely resistanceless surface. As $\omega > 2\Delta(T = 0)/\hbar$ at absolute zero, the resistance begins to approach that of the normal state. We estimate this energy by assuming a weak-coupling description is valid for the superconductor. Then $\Delta \sim 2k_B T_c$, and $\omega \sim 4k_B T_c/\hbar$. For a T_c of 5 K, $\omega \sim 10^{12}s$. This frequency is in the infrared. Infrared radiation can then penetrate a superconductor and scatter the electrons.

(e)   Ultrasonic Attenuation: No damping of an impinging beam of phonons is observed if $\omega_q < 2\Delta/\hbar$. As in the microwave absorption case, ω_q must exceed the energy needed to create an electron-hole pair.

(f)   Nuclear-Spin Lattice Relaxation: Consider a set of nuclei that have been forced to align with a magnetic field. The rate at which the equilibrium magnetization is recovered is the spin-lattice relaxation rate, $1/T_1$ (see Fig. 11.7). In a superconductor, $(1/T_1)_S > (1/T_1)_N$ just below T_c. That is, there is an enhancement in the relaxation rate that is brought on by the formation of the superconducting state. Hebel and Slichter [HS1959] were the first to see this effect experimentally. The peak in the relaxation rate just below T_c is known as the *Hebel-Slichter peak*.

(g)   Isotope effect: Experimentally, it is observed that, if the mass of the ions is changed isotopically, T_c changes accordingly:

$$T_c \propto \frac{1}{\sqrt{M}} \propto \omega_D. \qquad (11.21)$$

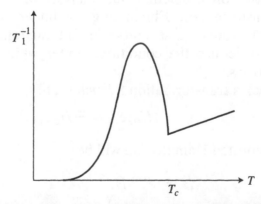

Figure 11.7 Behavior of the spin-lattice relaxation time at and below T_c. The enhancement in $1/T_1$ is a signature that the spins in a superconductor are acting in consort.

This observation implies that electron-phonon coupling is at the heart of superconductivity (in traditional superconductivity, at least).

11.2   ELECTRON-PHONON EFFECTIVE INTERACTION

The isotope dependence of T_c confirms that electron-phonon coupling is central to the mechanism of superconductivity. We now show how electron-phonon coupling can produce an attractive interaction between two electrons. The starting point for our analysis is an ordered band of electrons with kinetic energies ϵ_k coupled to the quantized vibrational modes of the crystal. As a result, our Hamiltonian $H = H_o + H_{e-ph}$ is a sum of an ordered part,

$$H_o = \sum_{\mathbf{q}} \hbar \omega_{\mathbf{q}} b_{\mathbf{q}}^{\dagger} b_{\mathbf{q}} + \sum_{\mathbf{k}} \epsilon_{\mathbf{k}} a_{\mathbf{k}}^{\dagger} a_{\mathbf{k}}, \tag{11.22}$$

and the electron-lattice interaction

$$H_{e-ph} = \sum_{\mathbf{k},\mathbf{q}} M_{\mathbf{q}} a_{\mathbf{k}+\mathbf{q}}^{\dagger} a_{\mathbf{k}} (b_{\mathbf{q}} + b_{-\mathbf{q}}^{\dagger}). \tag{11.23}$$

The electron-phonon coupling constant is defined in Section 9.3. We will assume that the electron-phonon interaction is weak and we will do perturbation theory to second order in this interaction. As a second-order perturbation, we anticipate that the electron-phonon interaction will provide a negative correction to the energy. Although the final result can be established quite straightforwardly (see Problem 11.3) using second-order perturbation theory, we will use a similarity transformation, as this method has numerous applications in electron-phonon problems. Our analysis will mirror the Schrieffer-Wolff transformation. Eliminating the linear electron-phonon interaction will produce a second-order interaction, from which we will be able to deduce the conditions under which phonons effectively bind electrons.

We seek a transformation S, such that

$$[H_o, S] = -H_{e-ph}. \tag{11.24}$$

The transformed Hamiltonian will be

$$\tilde{H} = e^{-S} H e^{S}$$

$$= H_o + \frac{1}{2}[H_{e-ph}, S] + \cdots. \tag{11.25}$$

We posit that S is of the form

$$S = \sum_{k,q} (Ab^\dagger_{-q} + Bb_q)M_q a^\dagger_{k+q} a_k, \tag{11.26}$$

where A and B are to be determined by the constraint in Eq. (11.24). To proceed, we need the commutators

$$[n_k, a^\dagger_{k_1} a_{k_2}] = a^\dagger_{k_1} a_{k_2}(\delta_{k_1 k} - \delta_{k_2 k})$$

$$[b^\dagger_q, b_q] = -1$$

$$[b^\dagger_q b_q, b_{q'}] = -b_q \delta_{qq'}. \tag{11.27}$$

We find, then, that

$$[H_o, S] = \sum_{q',k,q} \hbar\omega_{q'}[b^\dagger_{q'} b_{q'}, (Ab^\dagger_{-q} + Bb_q)a^\dagger_{k+q} a_k]M_q$$

$$+ \sum_{k',k,q} \epsilon_{k'} M_q[a^\dagger_{k'} a_{k'}, (Ab^\dagger_{-q} + Bb_q)a^\dagger_{k+q} a_k]$$

$$= \sum_{k,q} M_q b^\dagger_{-q} a^\dagger_{k+q} a_k(\hbar\omega_{-q} + \epsilon_{k+q} - \epsilon_k)A$$

$$+ \sum_{k,q} M_q b_q a^\dagger_{k+q} a_k(\epsilon_{k+q} - \epsilon_k - \hbar\omega_q)B. \tag{11.28}$$

This quantity must equal $-H_{e-ph}$. The A and B coefficients that satisfy this constraint are

$$A = -(\hbar\omega_{-q} + \epsilon_{k+q} - \epsilon_k)^{-1} \tag{11.29}$$

and

$$B = -(\epsilon_{k+q} - \epsilon_k - \hbar\omega_q)^{-1}. \tag{11.30}$$

Our unitary transformation is

$$S = \sum_{k,q} \left[\frac{b^\dagger_{-q}}{(\epsilon_k - \epsilon_{k+q} - \hbar\omega_{-q})} + \frac{b_q}{(\epsilon_k - \epsilon_{k+q} + \hbar\omega_q)} \right] M_q a^\dagger_{k+q} a_k. \tag{11.31}$$

Note the form of the energy denominators. The denominator of $b^\dagger_{-q}$ corresponds to the energy difference for emission of a phonon of energy $\hbar\omega_{-\mathbf{q}}$ and the denominator of $b_{\mathbf{q}}$ to absorption. Let us define $\Delta_{\pm}(\mathbf{k}, \mathbf{q}) = \epsilon_{\mathbf{k}} - \epsilon_{\mathbf{k}+\mathbf{q}} \pm \hbar\omega_{\pm\mathbf{q}}$. The effective Hamiltonian is given by

$$\tilde{H} = H_o + \frac{1}{2} \sum_{\mathbf{k},\mathbf{q},\mathbf{k}',\mathbf{q}'} \left[M_{\mathbf{q}} M_{\mathbf{q}'} (b^{\dagger}_{-\mathbf{q}} + b_{\mathbf{q}}) a^{\dagger}_{\mathbf{k}+\mathbf{q}} a_{\mathbf{k}}, \right.$$

$$\left. \left(\frac{b^{\dagger}_{-\mathbf{q}'}}{\Delta_{-}(\mathbf{k}', \mathbf{q}')} + \frac{b_{\mathbf{q}'}}{\Delta_{+}(\mathbf{k}', \mathbf{q}')} \right) a^{\dagger}_{\mathbf{k}'+\mathbf{q}'} a_{\mathbf{k}'} \right]. \tag{11.32}$$

We need to evaluate a commutator of the form

$$[(b^{\dagger}_{-\mathbf{q}} + b_{\mathbf{q}}) a^{\dagger}_{\mathbf{k}+\mathbf{q}} a_{\mathbf{k}}, b^{\dagger}_{-\mathbf{q}'} a^{\dagger}_{\mathbf{k}'+\mathbf{q}'} a_{\mathbf{k}'}]. \tag{11.33}$$

Specifically, we are interested in the two-electron terms that are produced from this commutator. If we were to evaluate the electron commutator part, we would obtain only a one-body potential. Such one-body potentials are not of interest here, as they do not affect the interaction between two electrons. The commutator involving the phonon operators produces an effective two-body potential that, interestingly enough, is negative in a narrow energy scale around the Fermi energy. To see this, we note that the phonon part of the commutator in Eq. (11.33) yields the constraint $\delta_{\mathbf{q}+\mathbf{q}'}$. Consequently, our effective Hamiltonian is

$$\tilde{H} = H_o + \frac{1}{2} \sum_{\mathbf{k},\mathbf{k}',\mathbf{q}} M_{\mathbf{q}} M_{-\mathbf{q}} \left[\frac{a^{\dagger}_{\mathbf{k}+\mathbf{q}} a_{\mathbf{k}} a^{\dagger}_{\mathbf{k}'-\mathbf{q}} a_{\mathbf{k}'}}{\Delta_{-}(\mathbf{k}', -\mathbf{q})} - \frac{a^{\dagger}_{\mathbf{k}+\mathbf{q}} a_{\mathbf{k}} a^{\dagger}_{\mathbf{k}'-\mathbf{q}} a_{\mathbf{k}'}}{\Delta_{+}(\mathbf{k}', -\mathbf{q})} \right]$$

$$= H_o + \sum_{\mathbf{k},\mathbf{k}',\mathbf{q}} |M_{\mathbf{q}}|^2 a^{\dagger}_{\mathbf{k}+\mathbf{q}} a^{\dagger}_{\mathbf{k}'-\mathbf{q}} a_{\mathbf{k}'} a_{\mathbf{k}} \frac{\hbar\omega_{\mathbf{q}}}{(\Delta\epsilon_{\mathbf{k}',\mathbf{q}})^2 - (\hbar\omega_{\mathbf{q}})^2}, \tag{11.34}$$

with $\Delta\epsilon_{\mathbf{k}',\mathbf{q}} = \epsilon_{\mathbf{k}'} - \epsilon_{\mathbf{k}'-\mathbf{q}}$. In general, $\epsilon_{\mathbf{k}'}$ and $\epsilon_{\mathbf{k}'-\mathbf{q}}$ vastly exceed $\hbar\omega_{\mathbf{q}}$. However, if $|\epsilon_{\mathbf{k}} - \epsilon_{\mathbf{k}-\mathbf{q}}| < \hbar\omega_{\mathbf{q}}$, then the 2-electron potential in Eq. (11.34) is negative, and the electrons experience a net attraction. Note that we have shown that a net attraction exists in k-space, not real space. The k-space attraction allows two electrons to bind together, so that they have a total energy that is lower than their noninteracting counterparts. This attraction is the basis for pairing in BCS superconductivity. Diagrammatically, this interaction is depicted in Fig. (11.8), where the wavy line represents the electron-phonon interaction

$$V(\mathbf{k}', \mathbf{q}) = \frac{|M_{\mathbf{q}}|^2 \hbar\omega_{\mathbf{q}}}{(\epsilon_{\mathbf{k}'} - \epsilon_{\mathbf{k}'-\mathbf{q}})^2 - (\hbar\omega_{\mathbf{q}})^2}. \tag{11.35}$$

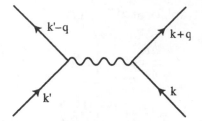

Figure 11.8 Electron-phonon interaction giving rise to BCS pairing near the Fermi surface. The wavy line represents the interaction in Eq. (11.35).

11.3   MODEL INTERACTION

In a metal, of course, we must consider the electron–electron repulsion. We can represent this potential as

$$V_{ee}^o = \frac{V_e(\mathbf{q})}{\varepsilon(\mathbf{q}, \omega)}, \tag{11.36}$$

with $\varepsilon(\mathbf{q}, \omega)$ the dynamic dielectric screening function. There is, of course, another source of electron–electron interaction, the attraction induced by phonons. The full interaction between electrons is of the form

$$V_{ee} = \frac{V_e(\mathbf{q})}{\varepsilon(\mathbf{q})} + \frac{\hbar\omega_\mathbf{q}|M_\mathbf{q}|^2}{(\Delta\epsilon_{\mathbf{k},\mathbf{q}})^2 - (\hbar\omega_\mathbf{q})^2}, \tag{11.37}$$

where we have taken the static limit of the screening function. Now $\varepsilon(\mathbf{q}) > 0$, unless the static compressibility is negative, as discussed in Chapter 8. We exclude this situation here as a negative compressibility occurs in the dilute regime where nonperturbative treatments of the Coulomb interaction are required. Typically, $V_e(\mathbf{q})$ is much greater than the phonon part of the potential. However, when $\Delta\epsilon_{\mathbf{k},\mathbf{q}} \simeq \omega_\mathbf{q}$, the phonon part dominates, and a net attraction is introduced. We see, then, that on different energy scales, either the phonon or the electron part can dominate. The largest phonon frequency that is relevant is $\hbar\omega_D$, the Debye energy. Hence, for $\Delta\epsilon_{\mathbf{k},\mathbf{q}} \le \hbar\omega_D$, a net attraction is produced that dominates the electron repulsion. When both electrons have an energy close to the Fermi energy, then $\Delta\epsilon_{\mathbf{k},\mathbf{q}} \approx 0$, and an attractive interaction obtains. This is an essential feature of the BCS theory.

Before we leave this section, it is instructive to consider the time Fourier transform of the interaction potential, V_{ee}. Let $\hbar\omega = \Delta\epsilon_{\mathbf{k},\mathbf{q}}$.

The Fourier transform of V_{ee} is

$$V_{ee}(t) = \int_{-\infty}^{\infty} \frac{d\omega}{2\pi} e^{-i\omega t} \left[V_{ee}^{o} + \frac{\hbar\omega_{\mathbf{q}}|M_{\mathbf{q}}|^2}{(\hbar\omega)^2 - (\hbar\omega_{\mathbf{q}})^2} \right]$$

$$= V_{ee}^{o}\delta(t) + |M_{\mathbf{q}}|^2 \frac{\hbar\omega_{\mathbf{q}}}{2\pi} \int_{-\infty}^{\infty} \frac{e^{-i\omega t}}{(\hbar\omega - \hbar\omega_{\mathbf{q}})(\hbar\omega + \hbar\omega_{\mathbf{q}})} d\omega$$

$$= V_{ee}^{o}\delta(t) - |M_{\mathbf{q}}|^2 \sin \omega_{\mathbf{q}} t. \qquad (11.38)$$

The repulsive part acts only at $t = 0$, while the attraction is an oscillatory function of time. The total potential is shown in Fig. (11.9). We find that the electron potential is attractive only over a time interval where the phonon attraction dominates. This is a key point; V_{ee} in the time domain is not always attractive. Electron repulsions and the phonon-mediated attraction act on different timescales and, hence, pair-binding of electrons is possible.

Let us evaluate the two-body scattering amplitude,

$$A_s = \langle \mathbf{p}_{4\sigma_1} \mathbf{p}_{3\sigma_2} | V_{ee} | \mathbf{p}_{1\sigma_1} \mathbf{p}_{2\sigma_2} \rangle, \qquad (11.39)$$

where

$$V_{ee} = \sum_{\mathbf{k},\mathbf{k}',\mathbf{q}} V(\mathbf{k}, \mathbf{k}' - \mathbf{q}) a_{\mathbf{k}+\mathbf{q}}^{\dagger} a_{\mathbf{k}'-\mathbf{q}}^{\dagger} a_{\mathbf{k}'} a_{\mathbf{k}}. \qquad (11.40)$$

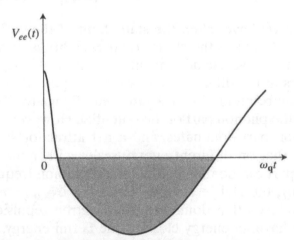

Figure 11.9 Fourier transform of the full electron interaction including the contribution from phonons. The delta function contribution from the electron repulsions has been smeared slightly in the vicinity of $t = 0$.

As discussed in Chapter 4, any two-body amplitude of this form separates into a difference of direct and exchange Coulomb integrals. For a general momentum-dependent potential, the direct and exchange terms will enter with different combinations of the momenta and spin. For example, the exchange term will be of the form $V(\mathbf{p}_1, \mathbf{p}_3)$ with $\sigma_1 = \sigma_2$, whereas the direct term will enter with no restriction on the spins and will depend on $V(\mathbf{p}_1, \mathbf{p}_4)$. Let us assume that the net attractive potential is a constant of the form

$$V_{ee} = \begin{cases} -V_o, & |\Delta \epsilon_{k_F, \mathbf{q}}| < \hbar \omega_D \\ 0, & |\Delta \epsilon_{k_F, \mathbf{q}}| > \hbar \omega_D \end{cases}. \qquad (11.41)$$

Because the potential is momentum independent, the direct and exchange integrals are equal, and the scattering amplitude vanishes when $\sigma_1 = \sigma_2$. The only nonzero contribution arises from $\sigma_1 \neq \sigma_2$. As a consequence, for a constant interaction, the scattering amplitude is nonzero only if the electrons are locked into a singlet state. In this case, phonons induce a net attraction between electrons of opposite spin. The corresponding matrix element is of the form

$$\langle \mathbf{p}_{4\uparrow} \mathbf{p}_{3\downarrow} | V_{ee} | \mathbf{p}_{1\uparrow} \mathbf{p}_{2\downarrow} \rangle = -V_0 \delta_{\mathbf{p}_1 + \mathbf{p}_2, \mathbf{p}_3 + \mathbf{p}_4} \qquad (11.42)$$

for scattering in the vicinity of the Fermi surface. Two particles locked into a singlet state will give rise to a gap at the Fermi level.

11.4 COOPER PAIRS

Consider now the somewhat artificial problem of a full Fermi sea containing N noninteracting electrons with two additional interacting electrons outside the sea. As a result of the Pauli exclusion principle, the momentum of the electrons outside the Fermi sea must exceed p_F. We take the potential of interaction to be the constant singlet pair potential derived in the previous section, Eq. (11.42). The spin wave function is, hence, antisymmetric with respect to spin. The corresponding spatial part must be symmetric to satisfy the overall antisymmetry requirement. The eigenvalue equation for our subsystem of two particles is

$$\left[-\frac{\hbar^2}{2m} (\nabla_1^2 + \nabla_2^2) + V(\mathbf{r}_1 - \mathbf{r}_2) - E \right] \Psi(\mathbf{r}_1, \mathbf{r}_2; \uparrow_1, \downarrow_2) = 0. \qquad (11.43)$$

In general, the two-body wave function is a composite state

$$\Psi(\mathbf{r}_1, \mathbf{r}_2; \uparrow_1, \downarrow_2) = \psi(\mathbf{r}_1, \mathbf{r}_2) \chi_S(\uparrow_1, \downarrow_2), \qquad (11.44)$$

where $\chi_S(\uparrow_1, \downarrow_2)$ represents the singlet spin state and ψ contains the spatial dependence. For two particles, we can express the spatial wave function in terms of the relative coordinate, $\mathbf{r} = \mathbf{r}_1 - \mathbf{r}_2$, and a center of mass, $\mathbf{R} = (\mathbf{r}_1 + \mathbf{r}_2)/2$, as

$$\psi(\mathbf{r}_1, \mathbf{r}_2) = \varphi(\mathbf{r}) e^{i\mathbf{Q}\cdot\mathbf{R}/\hbar}. \qquad (11.45)$$

Likewise, the momenta of interest are the center of mass, $\mathbf{Q} = \mathbf{p}_1 + \mathbf{p}_2$, and the relative momentum, $\mathbf{q} = (\mathbf{p}_1 - \mathbf{p}_2)/2$. We expand $\varphi(\mathbf{r})$ in a Fourier series as

$$\varphi(\mathbf{r}) = \sum_{\mathbf{k}}' \frac{e^{i\mathbf{k}\cdot\mathbf{r}/\hbar}}{\sqrt{V}} \alpha_{\mathbf{k}}, \qquad (11.46)$$

in which the prime indicates the restriction $\mathbf{k} > \mathbf{k}_F$ is restricted over all states whose energy exceeds ϵ_F. Because $\mathbf{k} \cdot \mathbf{r} = \mathbf{k} \cdot \mathbf{r}_1 - \mathbf{k} \cdot \mathbf{r}_2$, we see that the pair state has momenta $(\mathbf{k}, -\mathbf{k})$. Note, if $\mathbf{k}_1 + \mathbf{k}_2 = 0$, then the center-of-mass motion drops out of the problem.

We focus first on the $\mathbf{Q} = 0$ solution. To this end, we define the Fourier components of the interaction potential

$$V_{\mathbf{k}\mathbf{k}'} = \int \frac{d\mathbf{r}}{V} e^{-i(\mathbf{k}-\mathbf{k}')\cdot\mathbf{r}/\hbar} V(\mathbf{r}) \qquad (11.47)$$

and introduce the center-of-mass Schrödinger equation

$$\left(-\frac{\hbar^2}{2\mu}\nabla^2 + V(\mathbf{r})\right)\varphi(\mathbf{r}) = E\varphi(\mathbf{r}), \qquad (11.48)$$

where μ is the reduced mass, $\mu = m/2$. Substituting the Fourier representation of $\varphi(\mathbf{r})$, multiplying by $\exp(-i\mathbf{k} \cdot \mathbf{r}/\hbar)$, and integrating, we obtain

$$(E - 2\epsilon_{\mathbf{k}})\alpha_{\mathbf{k}} = \sum_{\mathbf{k}'} V_{\mathbf{k}\mathbf{k}'}\alpha_{\mathbf{k}'} \qquad (11.49)$$

as our new eigenvalue equation. Here, $\epsilon_{\mathbf{k}} = k^2/2m$. The matrix element $V_{\mathbf{k}\mathbf{k}'}$ is equivalent to

$$V_{\mathbf{k}\mathbf{k}'} = \langle \mathbf{k}, -\mathbf{k} | V_{ee} | \mathbf{k}', -\mathbf{k}' \rangle. \qquad (11.50)$$

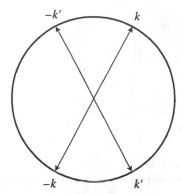

Figure 11.10 Scattering between a pair of electron states across the Fermi surface.

A typical scattering process in $V_{\mathbf{kk'}}$ is shown in Fig. (11.10). If we now introduce the approximation that

$$V_{\mathbf{kk'}} = \begin{cases} -V_0 & k,k' > p_F \\ 0 & \text{otherwise,} \end{cases} \tag{11.51}$$

the eigenvalue equation can be recast as

$$1 = -V_0 \sum_{k>k_F} \frac{1}{E - 2\epsilon_{\mathbf{k}}}$$

$$= -V_0 \phi(E). \tag{11.52}$$

This equation is satisfied as long as $\phi(E) = -1/V_0$. The poles of $\phi(E)$ occur at $E = 2\epsilon_{\mathbf{k}}$, the total energy of the pair, which is bounded from below by $2\epsilon_F$. In a finite system, $\epsilon_{\mathbf{k}}$ takes on discrete values because $\mathbf{k}$ is quantized. As E approaches $2\epsilon_{\mathbf{k}}$ from below, $\phi(E)$ approaches $-\infty$. Just above $2\epsilon_{\mathbf{k}}$, $\phi(E)$ is $\sim +\infty$. For all $E < 2\epsilon_F$, $\phi(E)$ is negative. Hence, a bound state forms when $\phi(E)$ crosses $-1/V_0$ for $E < 2\epsilon_F$. The intersection of $-1/V_0$ with $\phi(E)$ is illustrated graphically in Fig. (11.11). The existence of such a solution is the Cooper pair problem [C1956]. To find the precise energy of the bound state, we convert the sum in Eq. (11.52) to an integral,

$$1 = V_0 \sum_{\epsilon(\mathbf{k})}' \frac{1}{2\epsilon_{\mathbf{k}} - E} \rightarrow V_0 \int_{\epsilon_F}^{\epsilon_F + \hbar\omega_D} \frac{N(x)dx}{2x - E} \tag{11.53}$$

by introducing the density of states, $N(x)$. If we assume that $N(x)$ does not change significantly in the narrow range of integration, we can set

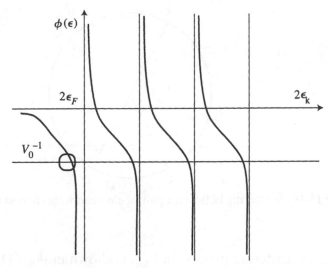

Figure 11.11 A plot of $\phi(E)$ versus ϵ_k in the Cooper-pair problem. The intersection of $\phi(E)$ with the straight line $-V_0^{-1}$ determines the bound-state solutions for the pair.

$N(x) \approx N(\epsilon_F)$. Under these assumptions, the integral yields

$$-\frac{2}{V_0 N(\epsilon_F)} = \ln\left[\frac{\epsilon_F - \frac{E}{2}}{\epsilon_F - \frac{E}{2} + \hbar\omega_D}\right], \qquad (11.54)$$

which implies that

$$\left(\epsilon_F - \frac{E}{2} + \hbar\omega_D\right)\exp\left(-\frac{2}{V_0 N(\epsilon_F)}\right) = \epsilon_F - \frac{E}{2}. \qquad (11.55)$$

This linear equation can be solved immediately for the eigenenergy E:

$$E = 2\epsilon_F - \frac{2\hbar\omega_D \exp\left(-2/V_0 N(\epsilon_F)\right)}{1 - \exp\left(-2/V_0 N(\epsilon_F)\right)}. \qquad (11.56)$$

In the limit that $2 \gg V_0 N(\epsilon_F)$, the exponential in the denominator can be expanded. The pair-binding energy

$$E \simeq 2\epsilon_F - 2\hbar\omega_D \exp\left(-\frac{2}{V_0 N(\epsilon_F)}\right) \qquad (11.57)$$

in the weak-coupling limit results. Either of these expressions indicates that the Cooper pair is bound with an energy $< 2\epsilon_F$ whenever V_0 is nonzero and positive. This is a profound result. It implies that two electrons directly below the Fermi surface can lower their energy by being excited into a Cooper pair with momentum $(\mathbf{k}, -\mathbf{k})$ just above the Fermi surface, provided that an attractive interaction of the form in Eq. (11.41) exists. This is known as the *Cooper instability*.

Further, we can estimate how the pair-binding energy depends on the center of mass of the Cooper pair. At the onset, we suspect that this quantity might scale as Q^2. We will show that this is not the case. To proceed, we extend the pair-binding criterion to the case in which $Q \neq 0$. For nonzero Q, the bare energy of the pair becomes $\epsilon_{\mathbf{q}+\mathbf{Q}/2} + \epsilon_{-\mathbf{q}+\mathbf{Q}/2}$. Consequently, the pair-binding condition becomes

$$1 = -V_0 \sum_{\mathbf{q}} \frac{1}{E - \epsilon_{\mathbf{q}+\mathbf{Q}/2} - \epsilon_{-\mathbf{q}+\mathbf{Q}/2}}. \tag{11.58}$$

For small Q, we can rewrite Eq. (11.58) as

$$1 = -V_0 \int_{\epsilon_F + v_F Q/2}^{\epsilon_F + v_F Q/2 + \hbar\omega_D} \frac{N(\epsilon_{\mathbf{q}})d\epsilon_{\mathbf{q}}}{E - 2\epsilon_{\mathbf{q}}}, \tag{11.59}$$

dropping terms of $O(Q^2)$. The center of mass simply shifts the zero of the Fermi energy. The new pair-binding energy,

$$E = 2\epsilon_F + Qv_F - \frac{2\hbar\omega_D}{\exp(2/V_0 N(\epsilon_F)) - 1}, \tag{11.60}$$

is a linear function of the center-of-mass momentum. Translation of the center of mass strongly reduces the binding energy and could eventually break up the pair. To show this, we set $\epsilon_F = 0$ and evaluate the value of Q at which the Cooper pair loses most of its binding energy. We must solve, then,

$$Qv_F = \frac{2\hbar\omega_D}{\exp(2/V_0 N(\epsilon_F)) - 1}$$

$$\approx k_B T_c. \tag{11.61}$$

Equivalently, when $Q/\hbar \sim k_B T_c/\hbar v_F \sim 10^4 cm^{-1}$, which is roughly the reciprocal of the Pippard coherence length, $\xi \sim 10^{-4} cm$. This is the effective radius of gyration of a Cooper pair, an enormous distance when compared to interatomic spacings. Such a large coherence length is a typical feature of phonon-pairing mechanisms.

11.5   FERMI-LIQUID THEORY

Of course, our problem is somewhat artificial in that we have ignored all interactions between the electrons save for the pair just above the Fermi surface. It is certainly reasonable to expect the simple picture of the Cooper instability to break down once we consider repulsive interactions among all of the electrons. That is, when electrons are interacting, we cannot *a priori* regard the noninteracting eigenstates as a valid description of our system (as we have done in the Cooper problem). We then are led to the question, is the instability real? We can answer this question by appealing to Fermi-liquid theory. It is now well accepted that the normal state of a metal is well described by Landau-Fermi-liquid theory. In this account, it is claimed that the dominant effect of electron interactions in a metal is to renormalize the effective mass of the electron. The observed shift is on the order of 10 to 50%. Another essential claim of Fermi-liquid theory is that there is a one-to-one correspondence between the excited states of the normal state of a metal with those of a noninteracting electron gas. The elementary excitations in Fermi-liquid theory are called *quasi particles*. A quasi particle is a composite particle with a lifetime. The lifetime stems from collisions with other quasi particles. When the lifetime (τ) of a quasi particle is infinite, the state with such a particle is an eigenstate of the system. However, the minimum constraint that must hold for a quasi-particle state to be an eigenstate of a system is that $\hbar/\tau \ll \tilde{\epsilon}_{\mathbf{p}}$, where $\tilde{\epsilon}_{\mathbf{p}}$ is the energy of the quasi particle. We will see below that as the energy of a quasi particle approaches the Fermi level, its lifetime goes to infinity. The stability of quasi particles at the Fermi level is a crucial tenet of Fermi-liquid theory.

The vanishing of the scattering rate of electrons in the vicinity of the Fermi level can be shown as follows. Consider a near $T = 0$ distribution in which all but one of the electrons is below the Fermi surface. Let ϵ_1 be the energy of the electron above the Fermi surface. For an electron with this energy to scatter, it must interact with some electron with energy $\epsilon_2 < \epsilon_F$. The Pauli exclusion principle requires that after the scattering event, the electrons must occupy two states above the Fermi surface. Let the energy of these states be ϵ_3 and ϵ_4. Energy conservation requires that $\epsilon_1 + \epsilon_2 = \epsilon_3 + \epsilon_4$. If $\epsilon_1 = \epsilon_F$, then all the other states must be at the Fermi level as well to satisfy energy conservation. As a consequence, for electrons at the Fermi level, the number of states into which they can be scattered is zero. Hence, the scattering rate, which is proportional to the density of scattering states, must vanish. As a result, the lifetime of electrons or quasi particles at the Fermi level is infinite.

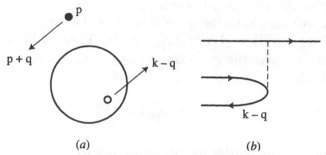

(a)   *(b)*

Figure 11.12 *(a)* The scattering of an excited quasi particle against particles in the ground state. *(b)* A diagrammatic representation of the momentum space scattering.

In the event that ϵ_1 moves away from the Fermi level, there is a window of states of width $\epsilon_1 - \epsilon_F$, from which we can choose the other three electronic states. Once we have chosen two of them from the narrow window $\epsilon_1 - \epsilon_F$, the energy of the last state cannot be chosen freely as a result of energy conservation. Consequently, there are $\epsilon_1 - \epsilon_F$ choices for ϵ_2 and $\epsilon_1 - \epsilon_F$ choices for ϵ_3; ϵ_4 is now fixed. As a result, the scattering rate should scale as $(\epsilon_1 - \epsilon_F)^2$.

To see this more rigorously, we calculate the scattering rate for two particles in momentum states $\mathbf{k}$ and $\mathbf{p}$ scattering to states $\mathbf{k} - \mathbf{q}$ and $\mathbf{p} + \mathbf{q}$. A diagrammatic depiction of this interaction is shown in Fig. (11.12). The scattering rate for this process is

$$\frac{1}{\tau} = \frac{2\pi}{\hbar} \sum_{\mathbf{q},\mathbf{k},\sigma} |V(q)|^2 n_{\mathbf{k}\sigma}(1 - n_{\mathbf{k}-\mathbf{q}\sigma})(1 - n_{\mathbf{p}+\mathbf{q}})\delta(\hbar\omega_{\mathbf{pq}} + \hbar\omega_{-\mathbf{qk}}),$$

$$(11.62)$$

where $V(q)$ is the Fourier transform of the Coulomb potential and $\hbar\omega_{\mathbf{pq}} = \epsilon_{\mathbf{p}+\mathbf{q}} - \epsilon_{\mathbf{p}}$. From energy conservation, it follows that $|\mathbf{p} + \mathbf{q}| < p$ and $|\mathbf{p} + \mathbf{q}| > p_F$. These conditions are summarized by the constraint $p_F^2 < |\mathbf{p} + \mathbf{q}|^2 < p^2$, or, equivalently,

$$p_F^2 - p^2 < 2pq \cos\theta + q^2 < 0. \qquad (11.63)$$

As in our analysis of screening, we can simplify the sum over $\mathbf{k}$, using the definition of the screening function for free particles. At $T = 0$, we obtain

$$\frac{1}{\tau} = \frac{N}{2\hbar^2} \sum_{\mathbf{q}} |V(q)|^2 S_0(q, \hbar\omega_{\mathbf{pq}})(1 - n_{\mathbf{p}+\mathbf{q}}). \qquad (11.64)$$

Because we are interested primarily in $p - p_F \approx 0$, we can take q to be small. We then take the $q \to 0$ limit of the screened Coulomb potential, $V(q \to 0) = 4\pi e^2 / V \kappa_{TF}^2$, in the Thomas-Fermi approximation. Also, we take the $q \to 0$ form of the structure function derived in Chapter 8:

$$S(q, \omega_{pq}) \propto \frac{\omega_{pq}}{q}. \tag{11.65}$$

Consequently, the scattering rate simplifies to

$$\frac{1}{\tau} \propto \frac{1}{\epsilon_F} \int \sin\theta d\theta \int q \omega_{pq} dq. \tag{11.66}$$

We can replace the q-integral with one over ω_{pq} by solving $\hbar \omega_{pq} = (2pq \cos\theta + q^2)/2m$. The scattering rate is proportional to

$$\frac{1}{\tau} \propto \frac{1}{\epsilon_F} \int_0^{\epsilon_p - \epsilon_F} d\omega \int_{-1}^{1} \omega \frac{1 - x p_F}{\sqrt{x^2 p_F^2 + \omega}} dx$$

$$\propto \frac{1}{\epsilon_F} \int_0^{\epsilon_p - \epsilon_F} \omega d\omega \propto \frac{(\epsilon_p - \epsilon_F)^2}{\epsilon_F}. \tag{11.67}$$

We see clearly, then, that the scattering rate scales as $(\epsilon_p - \epsilon_F)^2$ and, hence, vanishes for states at the Fermi level. Setting $\epsilon_p - \epsilon_F \propto k_B T$, we find that the scattering rate is proportional to T^2. As mentioned in the previous chapter, the resistivity [GF1987] in the normal state of the cuprate high-T_c materials is linear in temperature down to T_c. In light of the Fermi-liquid result, this behavior strongly suggests that the normal state of these materials is a non–Fermi liquid.

The quadratic dependence of the quasi particle scattering rate on temperature is a central prediction of Fermi-liquid theory. At room temperature, $(k_B T)^2 / \epsilon_F$ is on the order of $10^{-4} eV$. This gives rise to a scattering lifetime that is on the order of $10^{-10} s$ at room temperature. A typical relaxation time in a metal is four orders of magnitude shorter (see Chapter 10). Hence, electron-electron interactions are not the dominant scattering mechanism for electrons in the vicinity of the Fermi level. It is for this reason that the noninteracting picture works so well and is a good approximation for the normal state properties of a metal, at least at room temperature. Of course, at sufficiently low temperatures, the importance of electron-electron interactions increases tremendously, superconductivity being a case in point.

The energy of a particular quasi particle is the energy required to add or subtract that particle from a state in the system. To illustrate this concept, we consider an interacting system described by some distribution function, $n^0_{\mathbf{p}\sigma}$, generally taken to be the Fermi-Dirac distribution. Suppose a particle is added or subtracted from the system ever so slowly, so that the system remains in the ground state. This process represents an adiabatic change in the particle number. Nonetheless, this process will change the particle distribution to some new quantity, $n_{\mathbf{p}\sigma}$, and in turn the ground state energy will change. Let $\delta n_{\mathbf{p}\sigma}$ represent the resultant change in the particle distribution function. We can calculate the change in the ground state energy,

$$\delta E = \frac{1}{V} \sum_{\mathbf{p}\sigma} \epsilon_{\mathbf{p}\sigma} \delta n_{\mathbf{p}\sigma} + \frac{1}{2V^2} \sum_{\mathbf{p}\sigma,\mathbf{p}'\sigma'} f_{\mathbf{p}\sigma,\mathbf{p}'\sigma'} \delta n_{\mathbf{p}\sigma} \delta n_{\mathbf{p}'\sigma'} + \cdots,$$

(11.68)

by expanding in powers of the fluctuation, $\delta n_{\mathbf{p}\sigma}$. The Landau parameter $f_{\mathbf{p}\sigma,\mathbf{p}'\sigma'}$ describes how quasi particles interact [L1956]. This quantity is invariant with respect to interchange of $\mathbf{p}$ and $\mathbf{p}'$. We identify the energy of a quasi particle,

$$\tilde{\epsilon}_{\mathbf{p}\sigma} = \epsilon_{\mathbf{p}\sigma} + \frac{1}{V} \sum_{\mathbf{p}'\sigma'} f_{\mathbf{p}\sigma,\mathbf{p}'\sigma'} \delta n_{\mathbf{p}'\sigma'} + \cdots,$$

(11.69)

by taking the variation of the new ground state energy with respect to $\delta n_{\mathbf{p}\sigma}$. We see clearly, then, that the quasi-particle energy is itself a function of the distribution function. More explicitly, quasi particles obey a Fermi-Dirac distribution function with $\epsilon_{\mathbf{p}} - \mu$ replaced with $\tilde{\epsilon}_{\mathbf{p}\sigma} - \mu$. At the Fermi surface, quasi particles move with a velocity

$$v_F = \left(\frac{\partial \epsilon_{\mathbf{p}\sigma}}{\partial \mathbf{p}}\right)_{p=p_F},$$

(11.70)

which is equal to the Fermi velocity. Because $v_F = p_F/m^*$, the formal definition of the *effective mass* of a quasi particle is

$$\frac{1}{m^*} = \frac{1}{p_F} \left(\frac{\partial \epsilon_{\mathbf{p}\sigma}}{\partial \mathbf{p}}\right)_{p=p_F}.$$

(11.71)

Consequently, in the vicinity of the Fermi surface, the quasi-particle energy

$$\epsilon_{\mathbf{p}\sigma} = \epsilon_F + v_F(p - p_F)$$

(11.72)

is linear in the displacement momentum $p - p_F$. As most of our focus will be on processes in the vicinity of the Fermi surface, this expression should suffice to describe the energy of a quasi particle.

It is a crucial assumption of the BCS pairing theory of superconductivity that the normal state properties are described by noninteracting quasi particles with infinite lifetimes [L1964, L1965]. This assumption has been borne out experimentally in the normal state of a wide range of Type I, as well as Type II, materials. Hence, the instability observed by Cooper is at the heart of the $T = 0$ properties of metals.

11.6 PAIR AMPLITUDE

At this point, it is customary to write down the BCS wave function and go on our merry way to elucidate the superconducting transition. However, we want to take a step back and explore the expectation value of the pairing amplitude for a Cooper pair. In so doing, we will be able to swindle BCS theory. Thus far, we have established that two electrons above the Fermi surface can lower their energy by forming a bound singlet Cooper pair just below ϵ_F, if the temperature is sufficiently low. This tells us that we should be able to define an order parameter whose expectation value should discern if the Cooper instability has occurred. The appropriate pairing operator for a $(\mathbf{p} \uparrow, -\mathbf{p} \downarrow)$ pair is

$$b_{\mathbf{p}}^{\dagger} = a_{\mathbf{p}\uparrow}^{\dagger} a_{-\mathbf{p}\downarrow}^{\dagger}, \tag{11.73}$$

whereas

$$b_{\mathbf{p}} = a_{-\mathbf{p}\downarrow} a_{\mathbf{p}\uparrow} \tag{11.74}$$

is the corresponding pair annihilation operator. Let $|N\rangle$ represent the ground state of an N-particle system. The Cooper instability suggests that $\langle N - 2|b_{\mathbf{p}}|N\rangle \neq 0$ in the superconducting state. Let us define

$$\alpha_{\mathbf{p}} = \langle N - 2|b_{\mathbf{p}}|N\rangle, \tag{11.75}$$

which is generally referred to as the *pair amplitude*. What we will show is that below a certain temperature, $\alpha_{\mathbf{p}}$ grows exponentially. In the normal state, $\alpha_{\mathbf{p}}$ equals 1 or 0. In BCS theory, $\alpha_{\mathbf{p}}$ is closely related to the order parameter.

We start by showing that $\alpha_{\mathbf{p}}$ is the general expansion coefficient for a pair state. To proceed, we write the general singlet pair state as

$$|\psi\rangle = \sum_{\mathbf{p}>p_F} \eta_{\mathbf{p}} b_{\mathbf{p}}^{\dagger}|0\rangle = \sum_{\mathbf{p}>p_F} \eta_{\mathbf{p}}|\mathbf{p} \uparrow -\mathbf{p} \downarrow\rangle, \tag{11.76}$$

where $\eta_{\mathbf{p}}$ is an expansion coefficient. To determine $\eta_{\mathbf{p}}$, we note that

$$\langle 0|b_{\mathbf{p}}|\psi\rangle = \sum_{\mathbf{p}'} \eta_{\mathbf{p}'}\langle 0|b_{\mathbf{p}}b_{\mathbf{p}'}^\dagger|0\rangle$$

$$= \sum_{\mathbf{p}'} \eta_{\mathbf{p}'}\delta_{\mathbf{pp}'} = \eta_{\mathbf{p}}. \tag{11.77}$$

Because $|\psi\rangle$ simply differs from $|0\rangle$ by two particles, $\eta_{\mathbf{p}} = \alpha_{\mathbf{p}}$.

The pair amplitude is the expansion coefficient for the general pairing state. Let us calculate the time evolution of $\alpha_{\mathbf{p}}$. To do this, we write the BCS Hamiltonian as

$$H = H_0 + H_{\text{int}}, \tag{11.78}$$

where

$$H_0 = \sum_{\mathbf{p},\sigma} \epsilon_{\mathbf{p}} a_{\mathbf{p}\sigma}^\dagger a_{\mathbf{p}\sigma}$$

$$H_{\text{int}} = \sum_{\mathbf{pp}'} V_{\mathbf{pp}'} b_{\mathbf{p}}^\dagger b_{\mathbf{p}'}. \tag{11.79}$$

In the Schrödinger picture, the time dependence of $\alpha_{\mathbf{p}}$ is carried in the wave functions. As a consequence,

$$i\hbar\frac{\partial}{\partial t}\alpha_{\mathbf{p}}(t) = i\hbar\frac{\partial}{\partial t}\langle 0(t)|b_{\mathbf{p}}|\psi(t)\rangle$$

$$= \langle 0(t)|[b_{\mathbf{p}}, H]|\psi(t)\rangle. \tag{11.80}$$

To evaluate $[b_{\mathbf{p}}, H]$, we will find the commutators useful:

$$[b_{\mathbf{p}}, n_{\mathbf{p}'\sigma}] = b_{\mathbf{p}}(\delta_{\mathbf{pp}'}\delta_{\sigma\uparrow} + \delta_{\mathbf{p}-\mathbf{p}'}\delta_{\sigma\downarrow})$$

$$[b_{\mathbf{p}}, b_{\mathbf{p}'}^\dagger] = \delta_{\mathbf{pp}'}(1 - n_{\mathbf{p}\uparrow} - n_{-\mathbf{p}\downarrow}). \tag{11.81}$$

We see explicitly that the occupancy factors, $n_{\mathbf{p}\uparrow} + n_{-\mathbf{p}\downarrow}$, render the pair-creation and annihilation operators noncommutating. These terms preserve the Pauli principle between electrons forming the pair. We can now evaluate $[b_{\mathbf{p}}, H]$:

$$[b_{\mathbf{p}}, H] = \sum_{\mathbf{p}'\sigma}[b_{\mathbf{p}}, \epsilon_{\mathbf{p}'} n_{\mathbf{p}'\sigma}] + {\sum_{\mathbf{p}'\mathbf{p}''}}' V_{\mathbf{p}'\mathbf{p}''}[b_{\mathbf{p}}, b_{\mathbf{p}'}^\dagger b_{\mathbf{p}''}]$$

$$= 2\epsilon_{\mathbf{p}} b_{\mathbf{p}} + {\sum_{\mathbf{p}'}}' V_{\mathbf{pp}'}(1 - n_{-\mathbf{p}\downarrow} - n_{\mathbf{p}\uparrow})b_{\mathbf{p}'}. \tag{11.82}$$

The prime on the sum indicates a restricted sum over the thin momentum shell of width $2\hbar\omega_D$ around ϵ_F. We can gain physical insight into this result by noting that $(1 - n_{-\mathbf{p}\downarrow} - n_{\mathbf{p}\uparrow}) = (1 - n_{\mathbf{p}\uparrow})(1 - n_{-\mathbf{p}\downarrow}) - n_{\mathbf{p}\uparrow}n_{-\mathbf{p}\downarrow}$. If we use this result with Eq. (11.82), we obtain

$$\left(i\hbar\frac{\partial}{\partial t} - 2\epsilon_{\mathbf{p}}\right)\alpha_{\mathbf{p}}(t) = [(1 - n_{\mathbf{p}\uparrow})(1 - n_{-\mathbf{p}\downarrow}) - n_{\mathbf{p}\uparrow}n_{-\mathbf{p}\downarrow}]\sum_{\mathbf{p}'}{}' V_{\mathbf{p}\mathbf{p}'}\alpha_{\mathbf{p}'}(t)$$

(11.83)

as the time evolution of the pair amplitude. The probability that the particles forming the pair lie outside the Fermi surface is determined by the product $(1 - n_{\mathbf{p}\uparrow})(1 - n_{-\mathbf{p}\downarrow})$, whereas $n_{\mathbf{p}\uparrow}n_{-\mathbf{p}\downarrow}$ is the probability that they lie inside. This term can be ignored if we are interested strictly in pair formation between two momentum states just outside the Fermi surface, as in the Cooper problem. Our effective equation of motion is

$$\left(i\hbar\frac{\partial}{\partial t} - 2\epsilon_{\mathbf{p}}\right)\alpha_{\mathbf{p}}(t) = (1 - n_{\mathbf{p}\uparrow})(1 - n_{-\mathbf{p}\downarrow})\sum_{\mathbf{p}'}{}' V_{\mathbf{p}\mathbf{p}'}\alpha_{\mathbf{p}'}(t) \quad (11.84)$$

in this limit. Because $i\hbar\partial/\partial t \rightarrow E$, the above expression is precisely the Schrödinger equation we solved previously to determine the Cooper bound state energy. The general pair amplitude equation, Eq. (11.83), dictates, however, that all particles be involved in the pairing process whether or not they lie inside or outside the Fermi surface. This is the essence of BCS theory. Cooper included only the particles outside ϵ_F. BCS simply included the $n_{\mathbf{p}\uparrow}n_{-\mathbf{p}\downarrow}$ term in addition. This simple change made all the difference.

11.6.1   Instability: Superconducting State

We turn to the evaluation of the pair amplitude [KB1962]. We want to show that below some characteristic temperature, T_c, the pair amplitude grows exponentially. We seek a solution of the form

$$\alpha_{\mathbf{p}}(t) = e^{-i\frac{zt}{\hbar}}\alpha_{\mathbf{p}}(t = 0). \quad (11.85)$$

Substituting this expression into the pairing equation, we obtain

$$(z - 2\epsilon_{\mathbf{p}})\alpha_{\mathbf{p}}(0) = (1 - n_{\mathbf{p}\uparrow} - n_{-\mathbf{p}\downarrow})\sum_{\mathbf{p}'}{}' V_{\mathbf{p}\mathbf{p}'}\alpha_{\mathbf{p}'}(0) \quad (11.86)$$

as the Fourier transform of the pair amplitude equation.

For the model attractive potential in Eq. (11.51), the pair amplitude evolution equation becomes

$$(z - 2\epsilon_{\mathbf{p}})\alpha_{\mathbf{p}}(0) = -V_0(1 - n_{\mathbf{p}\uparrow} - n_{-\mathbf{p}\downarrow}){\sum_{\mathbf{p}'}}'\alpha_{\mathbf{p}'}(0). \quad (11.87)$$

As before, we sum both sides of this expression to obtain

$${\sum_{\mathbf{p}}}'\alpha_{\mathbf{p}}(0) = -V_0{\sum_{\mathbf{p}}}'\frac{(1 - n_{\mathbf{p}\uparrow} - n_{-\mathbf{p}\downarrow})}{z - 2\epsilon_{\mathbf{p}}}{\sum_{\mathbf{p}'}}'\alpha_{\mathbf{p}'}(0), \quad (11.88)$$

which implies that

$$1 = -V_0{\sum_{\mathbf{p}}}'\frac{(1 - n_{\mathbf{p}\uparrow} - n_{-\mathbf{p}\downarrow})}{z - 2\epsilon_{\mathbf{p}}}. \quad (11.89)$$

This is almost the expression we had previously, except we are now including pairing between all particles. To obtain a solution, we plot the intersection of the left- and right-hand sides of the above equation. As before, we plot the RHS in the vicinity of the discrete energies $\epsilon_{\mathbf{p}}$.

As is evident, a solution exists only if $2\epsilon_F - 2\hbar\omega_D \leq \mathrm{Re}z \leq 2\epsilon_F + 2\hbar\omega_D$. To solve our pair condition, we convert the sum to an integral and note that $n_{\mathbf{p}} = n_{-\mathbf{p}}$, for free particles. Also, because $\epsilon_{\mathbf{p}\uparrow} = \epsilon_{\mathbf{p}\downarrow}$, we have immediately that

$$n_{\mathbf{p}\uparrow} = n_{-\mathbf{p}\downarrow} = (e^{\beta(\epsilon_{\mathbf{p}}-\mu)} + 1)^{-1} \quad (11.90)$$

and that

$$1 = -V_0 N(\epsilon_F)\int_{-\hbar\omega_D}^{\hbar\omega_D} d\epsilon\,\frac{\tanh\beta\epsilon/2}{z - 2\mu - 2\epsilon}, \quad (11.91)$$

where $N(\epsilon_F)$ is the density of states at the Fermi level. Exponential growth of the pairing amplitude will occur if z has a positive imaginary part. Let $z = 2\mu + x + iy$, where x and y are real. Consider the first case in which $y = 0$. We obtain in this limit that

$$1 = -V_0 N(\epsilon_F)\int_{-\hbar\omega_D}^{\hbar\omega_D} d\epsilon\,\tanh\frac{\beta\epsilon}{2}\left[P\left(\frac{1}{x - 2\epsilon}\right) - i\pi\delta(x - 2\epsilon)\right]. \quad (11.92)$$

To satisfy this equation, the imaginary part must vanish. This obtains only if $|x|/2 > \hbar\omega_D$, or, equivalently, $\mathrm{Re}z > 2\epsilon_F + 2\hbar\omega_D$. As illustrated graphically in Fig. (11.13), the real part of the integrand is positive in this energy range. Consequently, $y = 0$ is not permissible.

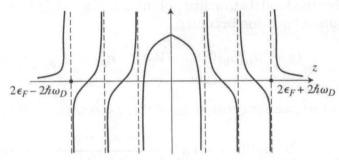

Figure 11.13 The right-hand side of the pair amplitude equation as a function of the complex energy, z.

For a solution to exist on the energy shell, $y \neq 0$. We write our pair constraint as

$$1 = -V_0 N(\epsilon_F) \int_{-\hbar\omega_D}^{\hbar\omega_D} \frac{(x - 2\epsilon - iy)}{(x - 2\epsilon)^2 + y^2} \tanh \frac{\beta\epsilon}{2} d\epsilon, \quad (11.93)$$

which is only true if the imaginary part,

$$0 = \int_{-\hbar\omega_D}^{\hbar\omega_D} \frac{\tanh \beta\epsilon/2 \, d\epsilon}{(x - 2\epsilon)^2 + y^2}, \quad (11.94)$$

vanishes. Because $\tanh(x) = -\tanh(-x)$, we rewrite this integrand as

$$0 = \frac{1}{2} \int_{-\hbar\omega_D}^{\hbar\omega_D} d\epsilon \tanh \frac{\beta\epsilon}{2} \left[\frac{1}{(x - 2\epsilon)^2 + y^2} - \frac{1}{(x + 2\epsilon)^2 + y^2} \right]$$

$$= \frac{1}{2} \int_{-\hbar\omega_D}^{\hbar\omega_D} d\epsilon \tanh \frac{\beta\epsilon}{2} \left[\frac{8\epsilon x}{((x - 2\epsilon)^2 + y^2)((x + 2\epsilon)^2 + y^2)} \right]. \quad (11.95)$$

It is now clear that $x = 0$ because $\epsilon \tanh \beta\epsilon/2$ is positive everywhere. Consequently, our solution must be of the form $z = 2\mu + iy$. We must solve, then,

$$1 = -N(\epsilon_F) V_0 \int_{-\hbar\omega_D}^{\hbar\omega_D} d\epsilon \frac{(-iy - 2\epsilon) \tanh \beta\epsilon/2}{4\epsilon^2 + y^2}. \quad (11.96)$$

The imaginary part of this integral is odd in ϵ and, hence, vanishes exactly. We are left with the single equation

$$1 = 4N(\epsilon_F) V_0 \int_0^{\hbar\omega_D} d\epsilon \frac{\epsilon \tanh \beta\epsilon/2}{4\epsilon^2 + y^2}. \quad (11.97)$$

Here again, the integrand is positive. As y increases, the integrand decreases. The same thing is true as T increases. Hence, for T greater than some temperature, no solution exists. This defines the critical temperature, T_c. As $T \to 0$, we find that

$$
1 = \frac{N(\epsilon_F) V_0}{2} \int_0^{(2\hbar\omega_D)^2} \frac{da}{a + y^2}
$$

$$
= \frac{N(\epsilon_F) V_0}{2} \ln\left(\frac{(2\hbar\omega_D)^2 + y^2}{y^2}\right). \tag{11.98}
$$

Exponentiation of both sides of this equation,

$$
y^2 = (2\hbar\omega_D)^2 \left(\exp\left(\frac{2}{N(\epsilon_F) V_0}\right) - 1\right)^{-1}
$$

$$
\approx (2\hbar\omega_D)^2 \exp\left(-\frac{2}{N(\epsilon_F) V_0}\right), \tag{11.99}
$$

results in the familiar equation for y, namely,

$$
y = \pm 2\hbar\omega_D \exp\left(-\frac{1}{N(\epsilon_F) V_0}\right). \tag{11.100}
$$

Choosing the positive solution, we conclude that as $T \to 0$, the pair amplitude grows as

$$
\alpha_{\mathbf{p}}(t) \simeq e^{-izt} = e^{-2i\mu t} e^{2\hbar\omega_D t \exp(-1/N(\epsilon_F) V_0)}. \tag{11.101}
$$

The existence of an exponentially growing pair amplitude is the signature of an instability in the N-particle ground state. We have essentially derived BCS theory by focusing solely on the pair amplitude, $\alpha_{\mathbf{p}}$. To obtain an exact expression for T_c, we consider our pair-binding equation in the limit that $y = 0$. In this limit, the temperature that satisfies Eq. (11.97) is maximized. Hence, the $y = 0$ solution can be used to estimate T_c. Let us define $\beta_c = 1/k_B T_c$. We find that

$$
1 = N(\epsilon_F) V_0 \int_0^{\hbar\omega_D \beta_c/2} \frac{\tanh x}{x} dx. \tag{11.102}
$$

This integral must be done numerically. The transition temperature consistent with Eq. (11.102) is

$$
k_B T_c = 1.14 \hbar\omega_D \exp\left(-\frac{1}{N(\epsilon_F) V_0}\right). \tag{11.103}
$$

We will show that the parameter y plays the role of the energy gap in BCS theory.

11.7 BCS GROUND STATE

Thus far, we have shown how the pairing hypothesis leads to an instability in the ground state for $T < T_c$. Our focus now is on the many-particle wave function describing this state. The starting point for our analysis is the many-body state for N noninteracting electrons:

$$|\Psi^0(N)\rangle = \prod_{\mathbf{p} < \mathbf{p}_F} b_{\mathbf{p}}^\dagger |0\rangle. \tag{11.104}$$

Because each momentum state is doubly occupied, we can construct such a state by successively operating on the vacuum with the pairing operator $b_{\mathbf{p}}^\dagger$. Relaxing the restriction $\mathbf{p} < \mathbf{p}_F$, we recast $|\Psi^0(N)\rangle$ in the form

$$|\Psi^0(N)\rangle = \prod_{\mathbf{p}} (u_{\mathbf{p}}^0 + v_{\mathbf{p}}^0 b_{\mathbf{p}}^\dagger)|0\rangle, \tag{11.105}$$

where

$$u_{\mathbf{p}}^0 = \left\{ \begin{matrix} 0, & \mathbf{p} < \mathbf{p}_F \\ 1, & \mathbf{p} > \mathbf{p}_F \end{matrix} \right\} \tag{11.106}$$

$$v_{\mathbf{p}}^0 = \left\{ \begin{matrix} 1, & \mathbf{p} < \mathbf{p}_F \\ 0, & \mathbf{p} > \mathbf{p}_F \end{matrix} \right\}. \tag{11.107}$$

This representation of $|\Psi^0\rangle$ is identical to Eq. (11.104). Note also that because the $a_{\mathbf{p}}^\dagger$'s anticommute, $|\Psi^0(N)\rangle$ is properly antisymmetrized.

BCS postulated that the general electron superconducting ground state can be written as

$$|\Psi_{\mathrm{BCS}}\rangle = \prod_{\mathbf{p}} (u_{\mathbf{p}} + v_{\mathbf{p}} b_{\mathbf{p}}^\dagger)|0\rangle, \tag{11.108}$$

where the coefficients $u_{\mathbf{p}}$ and $v_{\mathbf{p}}$ are to be determined variationally and are referred to as *coherence factors*. The normalization condition $\langle \Psi_{\mathrm{BCS}} | \Psi_{\mathrm{BCS}} \rangle = 1$ imposes the constraint

$$1 = \langle 0| \prod_{\mathbf{p}, \mathbf{p}'} (u_{\mathbf{p}}^* + v_{\mathbf{p}}^* b_{\mathbf{p}})(u_{\mathbf{p}'} + v_{\mathbf{p}'} b_{\mathbf{p}'}^\dagger)|0\rangle. \tag{11.109}$$

Products of the form appearing in Eq. (11.109) can be simplified using the commutation relation in Eq. (11.81) and by noting that $b_{\mathbf{p}}b_{\mathbf{p}}^{\dagger}|0\rangle = 1$. We find that only the diagonal $(p = p')$ survives

$$\langle\Psi_{\text{BCS}}|\Psi_{\text{BCS}}\rangle = \prod_{\mathbf{p}}(|u_{\mathbf{p}}|^2 + |v_{\mathbf{p}}|^2), \qquad (11.110)$$

implying that

$$|u_{\mathbf{p}}|^2 + |v_{\mathbf{p}}|^2 = 1. \qquad (11.111)$$

If we expand the product in the BCS ground state, it becomes clear that $|\Psi_{\text{BCS}}\rangle$ is a sum of all $2m$-particle states,

$$|\Psi_{\text{BCS}}\rangle = \sum_{m} A_{2m}|\Psi_{2m}\rangle, \qquad (11.112)$$

where m is an integer, A_{2m} is an expansion coefficient, and $|\Psi_{2m}\rangle$ is the ground state for $2m$ particles. We can relate $|\Psi_{2m}\rangle$ to $|\Psi_{\text{BCS}}\rangle$ by writing the expansion coefficients [S1964] as $A_{2m} = |A_{2m}|e^{2im\phi}$, so that

$$|\Psi_{\text{BCS}}^{\phi}\rangle = \sum_{m} e^{2im\phi}|A_{2m}||\Psi_{2m}\rangle. \qquad (11.113)$$

That is, we associate with each $2m$-particle state an overall phase, $2m\phi$. To this end, we rewrite the BCS ground state

$$|\Psi_{\text{BCS}}^{\phi}\rangle = \prod_{\mathbf{p}}(u_{\mathbf{p}} + e^{2i\phi}v_{\mathbf{p}}b_{\mathbf{p}}^{\dagger})|0\rangle \qquad (11.114)$$

by multiplying each pair-creation operator by $\exp(2i\phi)$. Multiplying Eq. (11.113) by $\exp(-i2m\phi)$ and integrating over ϕ, we obtain a concise relation

$$|\Psi_{2m}\rangle = \frac{1}{2\pi|A_{2m}|}\int_{0}^{2\pi} d\phi\, e^{-i2m\phi}\prod_{\mathbf{p}}(u_{\mathbf{p}} + e^{2i\phi}v_{\mathbf{p}}b_{\mathbf{p}}^{\dagger})|0\rangle \qquad (11.115)$$

between each $2m$-particle state and the BCS ground state. That only states with $2m$ particles survive on the right-hand side of Eq. (11.115) follows from the fact that

$$\frac{1}{2\pi}\int_{0}^{2\pi} e^{i\phi(N-N')}d\phi = \delta_{NN'}. \qquad (11.116)$$

Because the $2m$-particle states are mutually orthogonal, when we multiply Eq. (11.115) by $\langle\Psi_{2m}|$, the left-hand side becomes unity, allowing us to identify the square of the expansion coefficient as

$$
|A_{2m}|^2 = \int_0^{2\pi} \int_0^{2\pi} \frac{d\phi \, d\phi'}{2\pi \, 2\pi} e^{i2m(\phi-\phi')} \langle 0| \prod_{\mathbf{p},\mathbf{p'}} (u_{\mathbf{p}}^* + e^{-2i\phi'} v_{\mathbf{p}}^* b_{\mathbf{p}})
$$

$$
\times (u_{\mathbf{p'}} + e^{2i\phi} v_{\mathbf{p'}} b_{\mathbf{p'}}^\dagger)|0\rangle
$$

$$
= \frac{1}{2\pi} \int_0^{2\pi} d\phi \, e^{i2m\phi} \prod_{\mathbf{p}} (|u_{\mathbf{p}}|^2 + e^{2i\phi} |v_{\mathbf{p}}|^2). \tag{11.117}
$$

Equation (11.117) should be construed as the probability distribution for $2m$ particles in the BCS state. This quantity is strongly peaked at the average number of particles in the BCS state and has a width proportional to the square root of the average number of particles in the system (see Problem 11.8).

Our rewriting of the BCS state as a linear superposition of all possible $2m$-particle states, all with the same phase, ϕ, suggests that, in a superconducting state, a conjugacy relationship exists between the phase and the particle number. Indeed, such a relationship exists. However, such a relationship cannot exist if one of the variables is bounded from below. Without loss of generality, we can treat the particle number as being a continuous variable from $-\infty < N < \infty$. We can establish the conjugacy relationship straightforwardly by acting $-i\partial/\partial\phi$ on $|\psi_{BCS}^\phi\rangle$ in Eq. (11.115) and then integrating by parts. The result

$$
\frac{1}{2\pi|A_{2m}|} \int_0^{2\pi} d\phi \, e^{-i2m\phi} \frac{-i\partial}{\partial\phi} \prod_{\mathbf{p}} (u_{\mathbf{p}} + e^{2i\phi} v_{\mathbf{p}} b_{\mathbf{p}}^\dagger)|0\rangle = 2m|\psi_{2m}\rangle \tag{11.118}
$$

implies that the particle number and the phase are related by

$$
N \leftrightarrow -i\frac{\partial}{\partial\phi}, \tag{11.119}
$$

as long as we regard the particle number as a continuous variable. Similarly, by applying $i\partial/\partial(2m)$ to Eq. (11.113), we establish analogously that

$$
i\frac{\partial}{\partial N} \leftrightarrow \phi. \tag{11.120}
$$

Hence, the phase and the particle number are conjugate dynamical variables as long as we consider the particle number to be a continuous variable. As such, the particle and phase must satisfy the Heisenberg uncertainty relationship $\Delta\phi\Delta N \geq 2\pi$ for two canonically conjugate variables. Consequently, complete certainty in the phase implies infinite uncertainty in the particle number, as is the case in a superconductor. In addition, the phase and the particle number operators should also satisfy Hamilton's equations

$$i\hbar\dot{\hat{N}} = [\hat{H}, \hat{N}] = i\frac{\partial\hat{H}}{\partial\hat{\phi}}$$

$$i\hbar\dot{\hat{\phi}} = [\hat{H}, \hat{\phi}] = -i\frac{\partial\hat{H}}{\partial\hat{N}} \qquad (11.121)$$

for two canonically conjugate variables. We will use these equations in the last section of this chapter, where we formulate the Josephson effect.

11.8   PAIR FLUCTUATIONS

We focus now on computing $\langle b_{\mathbf{p}} \rangle$ for a BCS wave function. In the BCS ground state, the average of the pair operator

$$\langle b_{\mathbf{p}}\rangle_{\mathrm{BCS}} = \langle 0| \prod_{\mathbf{p}',\mathbf{p}''} (u_{\mathbf{p}'}^* + v_{\mathbf{p}'}^* b_{\mathbf{p}'}) b_{\mathbf{p}} (u_{\mathbf{p}''} + v_{\mathbf{p}''} b_{\mathbf{p}''}^\dagger)|0\rangle$$

$$= \langle 0| u_{\mathbf{p}}^* v_{\mathbf{p}} b_{\mathbf{p}} b_{\mathbf{p}}^\dagger|0\rangle = u_{\mathbf{p}}^* v_{\mathbf{p}} \qquad (11.122)$$

is nonzero. However, for a normal state, $u_{\mathbf{p}}$ and $v_{\mathbf{p}}$ are never nonzero simultaneously (see Eqs. 11.106 and 11.107). Hence, $\langle b_{\mathbf{p}} \rangle = 0$ in the normal state. Although $\langle b_{\mathbf{p}} \rangle$ is not identical in form to the pair amplitude studied in the previous sections, it contains the same information. This state of affairs obtains because the BCS ground state is made up of states with different numbers of electrons. A nonzero value of $\langle b_{\mathbf{p}} \rangle$ for a particular ground state indicates that that state favors pair formation.

It is reasonable, then, to define

$$\chi = \frac{1}{V} \sum_{\mathbf{p}} b_{\mathbf{p}} \qquad (11.123)$$

as the effective order parameter for superconductivity. The average value of χ,

$$\langle \chi \rangle_{\mathrm{BCS}} = \frac{1}{V} \sum_{\mathbf{p}} u_{\mathbf{p}}^* v_{\mathbf{p}}, \qquad (11.124)$$

has a definite value in the superconducting state. To prove this, it is sufficient to show that the second-order fluctuations of the order parameter vanish. We proceed by evaluating the second moment

$$\langle \chi^2 \rangle_{BCS} = \frac{1}{V^2} \left\langle \sum_{\mathbf{p},\mathbf{p}'} b_{\mathbf{p}} b_{\mathbf{p}'} \right\rangle. \tag{11.125}$$

When we insert $|\Psi_{BCS}\rangle$, the only nonzero terms will come from the states whose momenta coincide with the indices in Eq. (11.125). Consequently, $\langle \chi^2 \rangle_{BCS}$ is a sum of

$$\frac{1}{V^2} \langle 0 | \sum_{\mathbf{p},\mathbf{p}',\mathbf{p} \neq \mathbf{p}'} (u_{\mathbf{p}}^* + v_{\mathbf{p}}^* b_{\mathbf{p}})(u_{\mathbf{p}'}^* + v_{\mathbf{p}'}^* b_{\mathbf{p}'}) b_{\mathbf{p}} b_{\mathbf{p}'}$$

$$\times (u_{\mathbf{p}} + v_{\mathbf{p}} b_{\mathbf{p}}^\dagger)(u_{\mathbf{p}'} + v_{\mathbf{p}'} b_{\mathbf{p}'}^\dagger) | 0 \rangle \tag{11.126}$$

when $\mathbf{p} \neq \mathbf{p}'$ and

$$\frac{1}{V^2} \langle 0 | \sum_{\mathbf{p}} (u_{\mathbf{p}}^* + v_{\mathbf{p}}^* b_{\mathbf{p}}) b_{\mathbf{p}}^2 (u_{\mathbf{p}} + v_{\mathbf{p}} b_{\mathbf{p}}^\dagger) | 0 \rangle \tag{11.127}$$

for $\mathbf{p} = \mathbf{p}'$. The second expression vanishes because $\langle b_{\mathbf{p}}^2 \rangle = 0$. Hence, $\langle \chi^2 \rangle_{BCS}$ is off-diagonal in momentum space, and Eq. (11.126) reduces to

$$\langle \chi^2 \rangle_{BCS} = \frac{1}{V^2} \sum_{\mathbf{p},\mathbf{p}',\mathbf{p} \neq \mathbf{p}'} u_{\mathbf{p}}^* v_{\mathbf{p}} u_{\mathbf{p}'}^* v_{\mathbf{p}'}$$

$$= \frac{1}{V^2} \sum_{\mathbf{p},\mathbf{p}'} \langle b_{\mathbf{p}} \rangle \langle b_{\mathbf{p}'} \rangle (1 - \delta_{\mathbf{p}\mathbf{p}'})$$

$$= \langle \chi \rangle_{BCS}^2 - \frac{1}{V^2} \sum_{\mathbf{p}} \langle b_{\mathbf{p}} \rangle^2. \tag{11.128}$$

The second-order fluctuation in the order parameter is

$$\langle \delta \chi^2 \rangle = \langle \chi^2 \rangle_{BCS} - \langle \chi \rangle_{BCS}^2 = -\frac{1}{V} \left[\frac{1}{V} \sum_{\mathbf{p}} \langle b_{\mathbf{p}} \rangle^2 \right]. \tag{11.129}$$

Once we determine the coefficients $u_{\mathbf{p}}$ and $v_{\mathbf{p}}$, we will be able to show that $\langle \delta \chi^2 \rangle \propto V^{-1}$ and, hence, vanishes in the thermodynamic limit. On this basis, we will argue that $\langle \chi \rangle$ takes on a definite value in the superconducting state and, as advertised, is the order parameter for the superconducting transition.

11.9   GROUND STATE ENERGY

We determine the coefficients $u_{\mathbf{p}}$ and $v_{\mathbf{p}}$ by minimizing the ground state energy subject to the normalization constraint (Eq. 11.111) and a particle number constraint on the ground state energy. We impose that the average number of particles in the BCS ground state is the desired value (supposedly even) by introducing a Lagrange multiplier into the average value of the Hamiltonian:

$$E' = \langle H \rangle - \mu \langle N \rangle, \tag{11.130}$$

where μ is the chemical potential. To simplify this calculation, we rescale the single-particle energies, such that $\epsilon_{\mathbf{p}} \equiv p^2/2m - \mu$. The energy E' is now

$$E' = \sum_{\mathbf{p},\sigma} \epsilon_{\mathbf{p}} \langle n_{\mathbf{p}\sigma} \rangle + \sum_{\mathbf{p},\mathbf{p}'} V_{\mathbf{p}\mathbf{p}'} \langle b_{\mathbf{p}}^{\dagger} b_{\mathbf{p}'} \rangle. \tag{11.131}$$

Let us first compute $\langle n_{\mathbf{p}\sigma} \rangle$:

$$\langle n_{\mathbf{p}\sigma} \rangle = \langle 0 | \prod_{\mathbf{p}'\mathbf{p}''} (u_{\mathbf{p}'}^* + v_{\mathbf{p}'}^* b_{\mathbf{p}'}) a_{\mathbf{p}\sigma}^{\dagger} a_{\mathbf{p}\sigma} (u_{\mathbf{p}''} + v_{\mathbf{p}''} b_{\mathbf{p}''}^{\dagger}) | 0 \rangle$$

$$= \prod_{\mathbf{p}',\mathbf{p}''} |v_{\mathbf{p}'}|^2 \langle 0 | b_{\mathbf{p}'} n_{\mathbf{p}\sigma} b_{\mathbf{p}'}^{\dagger} | 0 \rangle (\delta_{\mathbf{p},\mathbf{p}'} \delta_{\sigma,\uparrow} + \delta_{\mathbf{p},-\mathbf{p}''} \delta_{\sigma,\downarrow})$$

$$= \begin{cases} |v_{\mathbf{p}}|^2 & \sigma = \uparrow \\ |v_{-\mathbf{p}}|^2 & \sigma = \downarrow \end{cases}.$$

Reflection symmetry requires that the expansion coefficients be invariant under $\mathbf{p} \to -\mathbf{p}$.

The dominant term in $\langle b_{\mathbf{p}}^{\dagger} b_{\mathbf{p}'} \rangle$ arises from $\mathbf{p} \neq \mathbf{p}'$ because $\langle b_{\mathbf{p}}^2 \rangle = 0$. In this case, the average can be factorized: $\langle b_{\mathbf{p}}^{\dagger} b_{\mathbf{p}'} \rangle = \langle b_{\mathbf{p}}^{\dagger} \rangle \langle b_{\mathbf{p}'} \rangle = u_{\mathbf{p}} u_{\mathbf{p}'}^* v_{\mathbf{p}}^* v_{\mathbf{p}'}$. Consequently, the average value of the ground state energy is

$$E' = 2 \sum_{\mathbf{p}} |v_{\mathbf{p}}|^2 \epsilon_{\mathbf{p}} + \sum_{\mathbf{p},\mathbf{p}'} V_{\mathbf{p}\mathbf{p}'} u_{\mathbf{p}} u_{\mathbf{p}'}^* v_{\mathbf{p}}^* v_{\mathbf{p}'}. \tag{11.132}$$

We need to vary E',

$$\delta E' = 0 = 2 \sum_{\mathbf{p}} \epsilon_{\mathbf{p}} \delta v_{\mathbf{p}}^* v_{\mathbf{p}} + \sum_{\mathbf{p},\mathbf{p}'} V_{\mathbf{p}\mathbf{p}'} \delta v_{\mathbf{p}}^* (u_{\mathbf{p}} u_{\mathbf{p}'}^* v_{\mathbf{p}'})$$

$$+ \sum_{\mathbf{p},\mathbf{p}'} V_{\mathbf{p}\mathbf{p}'} \delta u_{\mathbf{p}'}^* (u_{\mathbf{p}} v_{\mathbf{p}}^* v_{\mathbf{p}'}), \tag{11.133}$$

with respect to $v_{\mathbf{p}}^*$ and $u_{\mathbf{p}}^*$ to find the minimum energy. To eliminate $\delta u_{\mathbf{p}'}^*$, we note that the variation of the normalization constraint can be rewritten as

$$u_{\mathbf{p}} \delta u_{\mathbf{p}}^* + v_{\mathbf{p}} \delta v_{\mathbf{p}}^* = 0, \tag{11.134}$$

or, equivalently,

$$\delta u_{\mathbf{p}}^* = -\frac{v_{\mathbf{p}}}{u_{\mathbf{p}}} \delta v_{\mathbf{p}}^*. \tag{11.135}$$

Substitution of this expression into Eq. (11.133) yields

$$0 = \sum_{\mathbf{p}} \delta v_{\mathbf{p}}^* \left[2\epsilon_{\mathbf{p}} v_{\mathbf{p}} + \sum_{\mathbf{p}'} \left[V_{\mathbf{p}\mathbf{p}'} (u_{\mathbf{p}} u_{\mathbf{p}'}^* v_{\mathbf{p}'}) - V_{\mathbf{p}'\mathbf{p}} u_{\mathbf{p}'} v_{\mathbf{p}'}^* \frac{v_{\mathbf{p}}^2}{u_{\mathbf{p}}} \right] \right], \tag{11.136}$$

which implies that

$$0 = 2\epsilon_{\mathbf{p}} v_{\mathbf{p}} - \Delta_{\mathbf{p}} u_{\mathbf{p}} + \Delta_{\mathbf{p}}^* \frac{v_{\mathbf{p}}^2}{u_{\mathbf{p}}}, \tag{11.137}$$

where we have defined

$$\Delta_{\mathbf{p}} = -\sum_{\mathbf{p}'} V_{\mathbf{p}\mathbf{p}'} u_{\mathbf{p}'}^* v_{\mathbf{p}'}. \tag{11.138}$$

As we will see, $\Delta_{\mathbf{p}}$ will play the role of the energy gap. In fact, we can see this immediately by dividing Eq. (11.137) by $u_{\mathbf{p}}$,

$$0 = 2\epsilon_{\mathbf{p}} \frac{v_{\mathbf{p}}}{u_{\mathbf{p}}} + \Delta_{\mathbf{p}}^* \left(\frac{v_{\mathbf{p}}}{u_{\mathbf{p}}}\right)^2 - \Delta_{\mathbf{p}}, \tag{11.139}$$

and solving this quadratic equation for $v_{\mathbf{p}}/u_{\mathbf{p}}$,

$$\frac{v_{\mathbf{p}}}{u_{\mathbf{p}}} = \frac{-\epsilon_{\mathbf{p}} \pm \sqrt{\epsilon_{\mathbf{p}}^2 + |\Delta_{\mathbf{p}}|^2}}{\Delta_{\mathbf{p}}^*}. \tag{11.140}$$

Let us define

$$\varepsilon_{\mathbf{p}} \equiv \sqrt{\epsilon_{\mathbf{p}}^2 + |\Delta_{\mathbf{p}}|^2},\qquad(11.141)$$

which will turn out to be the quasi-particle energy in the superconducting state. At this point we can see, at least heuristically, that $\Delta_{\mathbf{p}}$ does introduce a gap into the single-particle spectrum, $\epsilon_{\mathbf{p}}$. We can simplify Eq. (11.140) further by noting that

$$\frac{v_{\mathbf{p}}}{u_{\mathbf{p}}} = \frac{\epsilon_{\mathbf{p}}^2 - \varepsilon_{\mathbf{p}}^2}{(-\epsilon_{\mathbf{p}} \mp \varepsilon_{\mathbf{p}})\Delta_{\mathbf{p}}^*}$$

$$= \frac{\Delta_{\mathbf{p}}}{\epsilon_{\mathbf{p}} \pm \varepsilon_{\mathbf{p}}}.\qquad(11.142)$$

We assess which sign is appropriate by noting that in the free system, $v_{\mathbf{p}}/u_{\mathbf{p}} \to 0$ when $\epsilon_{\mathbf{p}} > \epsilon_F$. Consequently, we should choose the $+$ sign. Coupled with the normalization constraint,

$$1 = |u_{\mathbf{p}}|^2 + |u_{\mathbf{p}}|^2 \frac{|\Delta_{\mathbf{p}}|^2}{(\epsilon_{\mathbf{p}} + \varepsilon_{\mathbf{p}})^2} = 2|u_{\mathbf{p}}|^2 \left[\frac{\varepsilon_{\mathbf{p}}(\epsilon_{\mathbf{p}} + \varepsilon_{\mathbf{p}})}{(\epsilon_{\mathbf{p}} + \varepsilon_{\mathbf{p}})^2} \right],\quad(11.143)$$

we determine the magnitude of $u_{\mathbf{p}}$ to be

$$|u_{\mathbf{p}}|^2 = \frac{\epsilon_{\mathbf{p}} + \varepsilon_{\mathbf{p}}}{2\varepsilon_{\mathbf{p}}}.\qquad(11.144)$$

There is clearly a phase degree of freedom associated with the definition of $u_{\mathbf{p}}$. To make life simple, we choose $u_{\mathbf{p}}$ to be positive and real. We find, then, that

$$u_{\mathbf{p}} = \sqrt{\frac{\epsilon_{\mathbf{p}} + \varepsilon_{\mathbf{p}}}{2\varepsilon_{\mathbf{p}}}}\qquad(11.145)$$

and

$$v_{\mathbf{p}} = \frac{\Delta_{\mathbf{p}}}{\epsilon_{\mathbf{p}} + \varepsilon_{\mathbf{p}}} \sqrt{\frac{\epsilon_{\mathbf{p}} + \varepsilon_{\mathbf{p}}}{2\varepsilon_{\mathbf{p}}}}$$

$$= \sqrt{\frac{\Delta_{\mathbf{p}}}{\Delta_{\mathbf{p}}^*} \left(\frac{\varepsilon_{\mathbf{p}} - \epsilon_{\mathbf{p}}}{2\varepsilon_{\mathbf{p}}} \right)}.\qquad(11.146)$$

For the simple attractive interaction chosen, Δ_p is real if motion of the center of mass is ignored and v_p simplifies to

$$v_p = \sqrt{\frac{\varepsilon_p - \epsilon_p}{2\varepsilon_p}}. \tag{11.147}$$

Returning to our original problem of determining the ground state energy, we reduce the average energy

$$
\begin{aligned}
E_s' &= 2\sum_p \epsilon_p |v_p|^2 + \sum_{p,p'} V_{pp'} u_{p'}^* v_{p'} u_p v_p^* \\
&= \sum_p (2\epsilon_p |v_p|^2 - \Delta_p u_p v_p^*) \\
&= -\sum_p \frac{(\varepsilon_p - \epsilon_p)^2}{2\varepsilon_p} \tag{11.148}
\end{aligned}
$$

to a single sum using the form of the coherence factors just derived. In the absence of the attractive interaction, we had that

$$E_n' = \sum_p 2\epsilon_p \theta \,(\mathbf{p} - \mathbf{p}_F) \tag{11.149}$$

with θ the Heaviside step function. Because we have redefined $\epsilon_p \equiv p^2/2m - \mu$, $\epsilon_p < 0$ for $p < p_F$. Consequently, the energy difference between the superconducting and normal states is given by the integral

$$
\begin{aligned}
\delta E &= -N \,(\epsilon_F - \mu) \left[\int_{-\hbar\omega_D}^{\hbar\omega_D} \frac{(\varepsilon_p - \epsilon)^2}{2\varepsilon_p} d\epsilon + \int_{-\hbar\omega_D}^{0} 2\epsilon\, d\epsilon \right] \\
&= -N \,(\epsilon_F - \mu) \int_0^{\hbar\omega_D} \frac{(\varepsilon_p - \epsilon)^2}{\varepsilon_p} d\epsilon. \tag{11.150}
\end{aligned}
$$

To obtain the final result, we substitute the expression for ε_p and perform the integral over ϵ. In obtaining the final result, it is expedient to introduce the new variable $\epsilon = \Delta \sinh y$. Consequently,

$$
\begin{aligned}
(\varepsilon_p - \epsilon)^2 &= (\Delta\sqrt{\sinh^2 y + 1} - \Delta \sinh y)^2 \\
&= (\Delta \cosh y - \Delta \sinh y)^2 = \Delta^2 e^{-2y}. \tag{11.151}
\end{aligned}
$$

As a result,

$$\delta E = -N \,(\epsilon_F - \mu)\Delta^2 \int_0^{\sinh^{-1}(\hbar\omega_D/\Delta)} e^{-2y}\, dy. \tag{11.152}$$

In general, $\hbar\omega_D/\Delta \gg 1$, and the upper limit can then be extended to ∞. The y-integration yields $1/2$, and the energy difference is

$$\delta E = \frac{-N(\epsilon_F - \mu)\Delta^2}{2}. \tag{11.153}$$

That is, the ground state energy of the superconducting state is lower than that of the corresponding normal state. The condensation energy in a superconductor is δE.

11.10   CRITICAL MAGNETIC FIELD

We showed in the previous section that the condensation energy in a superconductor is $2\delta E = -N(\epsilon_F - \mu)\Delta^2$. The negative sign ensures that the superconducting state is lower in energy than the normal state. The condensation energy is related to the critical field, H_c, as follows.

Consider placing a superconductor in a magnetic field with the cylindrical geometry shown in Fig. (11.14). The energy per unit volume produced by the magnetic field is $B^2/8\pi$. The current in the magnetic coils is held constant by the field H. Hence, there must be a contribution from the interaction of the vector field, $\mathbf{A}(\mathbf{r})$, with the current produced by the magnetic field. Ampere's law states that

$$\nabla \times \mathbf{H} = \frac{4\pi}{c}\mathbf{J}, \tag{11.154}$$

where $\mathbf{J}$ is the current in the coils that produce the magnetic field. The resultant contribution to the energy density is

$$\begin{aligned}
W &= -\int \mathbf{A} \cdot \frac{\nabla \times \mathbf{H}}{4\pi}\frac{d\mathbf{r}}{V} \\
&= -\int \frac{\nabla \times \mathbf{A}}{4\pi} \cdot \mathbf{H}\frac{d\mathbf{r}}{V} \\
&= -\int \mathbf{B} \cdot \mathbf{H}\frac{d\mathbf{r}}{4\pi V}. \tag{11.155}
\end{aligned}$$

In a uniform system, $\mathbf{B} = \mathbf{H}$, and the total internal energy due to the field is $H^2/8\pi - H^2/4\pi = -H^2/8\pi$. In the normal state, this energy must be included in the total ground state energy. As a consequence, the energy of the normal state in the presence of a magnetic field is given by $E_0^n = E_n' - H^2/8\pi$. Because the field does not penetrate a superconductor, the internal energy of the superconductor remains

Figure 11.14 Cylindrical piece of a superconductor surrounded by a magnetic coil.

unchanged from E_s'. Superconductivity will persist until $E_n' - H^2/8\pi > E_s'$. Consequently,

$$\frac{H^2}{8\pi} > E_n' - E_s' = -\delta E \tag{11.156}$$

is the criterion for the critical value of the field. We find that the critical field,

$$\begin{aligned} H_c &= \sqrt{-8\pi\delta E} \\ &= \sqrt{4\pi N \left(\epsilon_F - \mu \right)}\Delta, \end{aligned} \tag{11.157}$$

is directly proportional to the gap parameter, Δ.

11.11 ENERGY GAP

The physical motivation for introducing $\Delta_{\mathbf{p}}$,

$$\begin{aligned} \Delta_{\mathbf{p}} &= -\sum_{\mathbf{p'}} V_{\mathbf{pp'}} u_{\mathbf{p'}}^* v_{\mathbf{p'}} \\ &= -\sum_{\mathbf{p'}} V_{\mathbf{pp'}} \frac{\Delta_{\mathbf{p'}}}{2\varepsilon_{\mathbf{p'}}} \end{aligned} \tag{11.158}$$

as the gap is that this quantity breaks the continuous single-particle spectrum $\epsilon_{\mathbf{p}}$ into two branches. Specializing to the model potential that is nonzero only over a narrow momentum shell around the Fermi energy, we rewrite the gap equation as

$$\Delta_{\mathbf{p}} = V_0 {\sum_{\mathbf{p'}}}' \frac{\Delta_{\mathbf{p'}}}{2\sqrt{\epsilon_{\mathbf{p'}}^2 + \Delta_{\mathbf{p'}}^2}}. \tag{11.159}$$

The prime on the sum indicates the restriction $|\epsilon_{\mathbf{p}} - \epsilon_{\mathbf{p}_F}| < \hbar\omega_D$. Within the momentum shell, $\Delta_{\mathbf{p}}$ is relatively independent of $\mathbf{p}$ only if the Fermi surface is isotropic. For s-wave pairing, the Fermi surface is entirely isotropic, and the gap can be taken to be independent of $\mathbf{p}$. For higher angular momentum pairing states such as d-wave, the Fermi surface has nodes, and the momentum dependence of the gap cannot be ignored. For our purposes, we will assume here that $\Delta_{\mathbf{p}}$ is independent of $\mathbf{p}$, at least within the thin momentum shell around the Fermi surface. An immediate solution to the gap equation is $\Delta = 0$. If $\Delta = 0$, then $\langle b_{\mathbf{p}} \rangle = 0$ and only the normal phase is possible.

To obtain a nontrivial solution to the gap equation, we convert the sum to an integral:

$$1 = N(\epsilon_F - \mu) V_0 \int_{-\hbar\omega_D}^{\hbar\omega_D} \frac{d\epsilon}{2\sqrt{\epsilon^2 + \Delta^2}}$$

$$= N(\epsilon_F - \mu) V_0 \int_0^{\hbar\omega_D/\Delta} \frac{d\epsilon}{\sqrt{\epsilon^2 + 1}}$$

$$= N(\epsilon_F - \mu) V_0 \sinh^{-1} \frac{\hbar\omega_D}{\Delta}. \tag{11.160}$$

Consequently, the gap is given by

$$\Delta = \frac{\hbar\omega_D}{\sinh\left(1/N(\epsilon_F - \mu) V_0\right)}. \tag{11.161}$$

Typically, $N(\epsilon_F - \mu) V_0 \ll 1$. In this limit, the gap takes on the familiar form

$$\Delta(T = 0) \simeq 2\hbar\omega_D \exp\left(-\frac{1}{N(0)V_0}\right). \tag{11.162}$$

In the Cooper problem, we showed that the binding energy of a pair is identical to $\Delta(T = 0)$, except $1/(N(0)V_0) \to 2/(N(0)V_0)$. In the full theory, as remarked earlier, pairing of particles below the Fermi energy is included, as well. As a consequence, the exact exponent is $2/2N(0)V_0 = 1/(N(0)V_0)$. In this sense, pairing of particles below the Fermi surface effectively doubles the density of states at the Fermi level. Comparing the gap with T_c (Eq. 11.103), we find that

$$\frac{2\Delta}{k_B T_c} = \frac{4\hbar\omega_D}{1.14\hbar\omega_D} = 3.52, \tag{11.163}$$

which is the crucial ratio in BCS theory. It determines whether a superconductor is in the weak coupling (with respect to the phonon interaction) limit. Strong coupling corresponds to $2\Delta/k_B T_c > 3.5$.

To evaluate the gap at finite temperature, we include the Fermi distribution weight factor, $1 - n_{\mathbf{p}\uparrow} - n_{-\mathbf{p}\downarrow}$, which accounts for pairing among all the particles. Because $n_{\mathbf{p}\uparrow} = n_{-\mathbf{p}\downarrow}$, we can write the finite temperature gap equation as

$$\Delta_{\mathbf{p}} = -\sum_{\mathbf{p}'} V_{\mathbf{p}\mathbf{p}'} \frac{(1 - 2n_{\mathbf{p}'})}{2\varepsilon_{\mathbf{p}'}} \Delta_{\mathbf{p}'}. \tag{11.164}$$

For an isotropic system and the BCS model potential, we obtain

$$1 = V_0 N(0) \int_{-\hbar\omega_D}^{\hbar\omega_D} \frac{\tanh \beta \sqrt{\epsilon^2 + \Delta^2}/2}{2\sqrt{\epsilon^2 + \Delta^2}} d\epsilon \tag{11.165}$$

as the defining equation for the finite temperature gap in the excitation spectrum. This equation must be solved numerically. However, the approximate solution is $\Delta(T) = \Delta(T = 0)\sqrt{1 - T/T_c}$. This behavior is illustrated in Fig. (11.15).

Recalling that the second-order fluctuation in the order parameter,

$$\langle \delta\chi^2 \rangle = -\frac{1}{V^2} \sum_{\mathbf{p}} u_{\mathbf{p}}^{*2} v_{\mathbf{p}}^2, \tag{11.166}$$

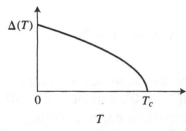

Figure 11.15 Asymptotic behavior of the gap for the model BCS interaction. Asymptotically, the gap scales as $\Delta(T) = \Delta(T = 0)(1 - \frac{T}{T_c})^{1/2}$.

is determined by the same combination of coherence factors as is the gap, we see immediately that if the gap is well behaved, then $\langle \delta \chi^2 \rangle$ vanishes in the thermodynamic limit. Consequently, $\langle b_{\mathbf{p}} \rangle$ takes on a definite value in the superconducting state.

11.12   QUASI-PARTICLE EXCITATIONS

To be able to determine the energy required to add an extra particle to a superconductor, we first determine the normalized creation operator for adding an extra electron to $|\Psi_{BCS}\rangle$. We note at the outset that

$$
\begin{aligned}
a^{\dagger}_{\mathbf{p}'\uparrow}|\Psi_{BCS}\rangle &= a^{\dagger}_{\mathbf{p}'\uparrow} \prod_{\mathbf{p}} (u_{\mathbf{p}} + v_{\mathbf{p}} b^{\dagger}_{\mathbf{p}})|0\rangle \\
&= a^{\dagger}_{\mathbf{p}'\uparrow} u_{\mathbf{p}'} \prod_{\mathbf{p} \neq \mathbf{p}'} (u_{\mathbf{p}} + v_{\mathbf{p}} b^{\dagger}_{\mathbf{p}})|0\rangle \\
&= u_{\mathbf{p}'}|\psi_{\mathbf{p}'\uparrow}\rangle
\end{aligned}
\tag{11.167}
$$

and

$$
\begin{aligned}
a_{-\mathbf{p}'\downarrow}|\Psi_{BCS}\rangle &= a_{-\mathbf{p}'\downarrow} v_{\mathbf{p}'} a^{\dagger}_{\mathbf{p}'\uparrow} a^{\dagger}_{-\mathbf{p}'\downarrow} \prod_{\mathbf{p} \neq \mathbf{p}'} (u_{\mathbf{p}} + v_{\mathbf{p}} b^{\dagger}_{\mathbf{p}})|0\rangle \\
&= -v_{\mathbf{p}'}|\psi_{\mathbf{p}'\uparrow}\rangle,
\end{aligned}
\tag{11.168}
$$

where

$$
|\Psi_{\mathbf{p}'\uparrow}\rangle = a^{\dagger}_{\mathbf{p}'\uparrow} \prod_{\mathbf{p} \neq \mathbf{p}'} (u_{\mathbf{p}} + v_{\mathbf{p}} b^{\dagger}_{\mathbf{p}})|0\rangle
\tag{11.169}
$$

is the normalized wave function when an extra electron with momentum $\mathbf{p}$ and spin $\uparrow$ is added to the BCS ground state. Consequently, $a^{\dagger}_{\mathbf{p}\uparrow}$ and $a_{-\mathbf{p}\downarrow}$ are not the operators that should be used to construct the excited states because the resultant states are not normalized. However, the linear combination of the electron operators,

$$
\gamma^{\dagger}_{\mathbf{p}\uparrow} = u_{\mathbf{p}} a^{\dagger}_{\mathbf{p}\uparrow} - v_{\mathbf{p}} a_{-\mathbf{p}\downarrow},
\tag{11.170}
$$

when acting on $|\Psi_{BCS}\rangle$,

$$
\begin{aligned}
\gamma^{\dagger}_{\mathbf{p}\uparrow}|\Psi\rangle_{BCS} &= (u^*_{\mathbf{p}} a^{\dagger}_{\mathbf{p}\uparrow} - v^*_{\mathbf{p}} a_{-\mathbf{p}\downarrow}) \prod_{\mathbf{p}'} (u_{\mathbf{p}'} + v_{\mathbf{p}'} b^{\dagger}_{\mathbf{p}'})|0\rangle \\
&= (|u_{\mathbf{p}}|^2 + |v_{\mathbf{p}}|^2)|\psi_{\mathbf{p}\uparrow}\rangle \\
&= |\psi_{\mathbf{p}\uparrow}\rangle
\end{aligned}
\tag{11.171}
$$

does produce the normalized state $|\psi_{\mathbf{p}\uparrow}\rangle$. The corresponding creation operator for the state $|\psi_{-\mathbf{p}\downarrow}\rangle$ is

$$\gamma^\dagger_{-\mathbf{p}\downarrow} = u^*_\mathbf{p} a^\dagger_{-\mathbf{p}\downarrow} + v^*_\mathbf{p} a_{\mathbf{p}\uparrow}. \tag{11.172}$$

Because the $\gamma_\mathbf{p}$'s are linear combinations of the $a_\mathbf{p}$'s, they must obey the standard Fermi-Dirac anticommutation relationships, namely,

$$[\gamma_{\mathbf{p}\sigma}, \gamma^\dagger_{\mathbf{p}'\sigma'}]_+ = \delta_{\mathbf{p}\mathbf{p}'}\delta_{\sigma\sigma'}$$

$$[\gamma_{\mathbf{p}\sigma}, \gamma^\dagger_{\mathbf{p}'\sigma'}]_+ = [\gamma^\dagger_{\mathbf{p}\sigma}, \gamma^\dagger_{\mathbf{p}'\sigma'}]_+ = 0. \tag{11.173}$$

The anticommutation relationships guarantee that

$$\gamma_{\mathbf{p}\uparrow}|\Psi_{\text{BCS}}\rangle = \gamma_{-\mathbf{p}\downarrow}|\Psi_{\text{BCS}}\rangle = 0. \tag{11.174}$$

That is, no quasi particles exist in the ground state of a superconductor.

Because the $\gamma^\dagger_\mathbf{p}$'s are linear combinations of an electron and a hole operator with antiparallel spins, the excitations created by these operators do not carry a well-defined charge. That is, creation of a quasi particle in a superconductor does not conserve charge. Quasi particles have a well-defined spin, however, because adding an electron with spin $\hbar/2$ or a hole with spin $-\hbar/2$ creates an eigenstate of S_z with spin $\hbar/2$. Cooper pairs, on the other hand, carry charge $2e$ but are spinless in the singlet channel. Consequently, the quantities that carry well-defined spin and charge in a standard singlet BCS superconductor are distinct entities. It is in this sense that spin-charge separation obtains in singlet BCS superconductors. We show now that the energy dispersion relation for the spin carrier in a singlet superconductor is $\varepsilon_\mathbf{p}$. To prove this assertion, we start with the equations of motion for $a_{\mathbf{p}\uparrow}$,

$$i\hbar\dot{a}_{\mathbf{p}\uparrow} = \epsilon_\mathbf{p} a^\dagger_{\mathbf{p}\uparrow} - \sum_{\mathbf{p}_1,\mathbf{p}_2} V_{\mathbf{p}_1\mathbf{p}_2}[a_{\mathbf{p}\uparrow}, b^\dagger_{\mathbf{p}_1} b_{\mathbf{p}_2}]$$

$$= \epsilon_\mathbf{p} a^\dagger_{\mathbf{p}\uparrow} + \sum_{\mathbf{p}_1} V_{\mathbf{p}\mathbf{p}_1} b_{\mathbf{p}_1} a^\dagger_{-\mathbf{p}\downarrow}, \tag{11.175}$$

and for $a^\dagger_{-\mathbf{p}\downarrow}$,

$$i\hbar\dot{a}^\dagger_{-\mathbf{p}\downarrow} = -\epsilon_\mathbf{p} a^\dagger_{-\mathbf{p}\downarrow} + \sum_{\mathbf{p}_2} V_{\mathbf{p}\mathbf{p}_2} b^\dagger_{\mathbf{p}_2} a_{\mathbf{p}\uparrow}. \tag{11.176}$$

To simplify these equations of motion, we replace $\sum_{\mathbf{p}_1} V_{\mathbf{p}\mathbf{p}_1} b_{\mathbf{p}_1}$ by its average value: $-\Delta_\mathbf{p} = \sum_{\mathbf{p}_1} V_{\mathbf{p}\mathbf{p}_1}\langle b_{\mathbf{p}_1}\rangle$. As a consequence, the effective

Hamiltonian

$$H_{\text{eff}} = \sum_{\mathbf{p}\sigma} \epsilon_{\mathbf{p}} n_{\mathbf{p}\sigma} + \sum_{\mathbf{p}} \Delta_{\mathbf{p}}(b_{\mathbf{p}}^{\dagger} + b_{\mathbf{p}}) \tag{11.177}$$

is linear in the pairing operator $b_{\mathbf{p}}^{\dagger}$, as opposed to quadratic, as in the full Hamiltonian. As a result, particle number is not conserved in the reduced Hamiltonian. Nonetheless, the mean-field treatment of the BCS Hamiltonian provides an accurate description of the superconducting state. This state of affairs occurs partly because the BCS ground state does not have a fixed number of particles.

The mean-field closure leads to a coupled set of equations

$$i\hbar \dot{a}_{\mathbf{p}\uparrow} = \epsilon_{\mathbf{p}} a_{\mathbf{p}\uparrow} - \Delta_{\mathbf{p}} a_{-\mathbf{p}\downarrow}^{\dagger}$$
$$i\hbar \dot{a}_{-\mathbf{p}\downarrow}^{\dagger} = -\epsilon_{\mathbf{p}} a_{-\mathbf{p}\downarrow}^{\dagger} - \Delta_{\mathbf{p}} a_{\mathbf{p}\uparrow} \tag{11.178}$$

linear in the gap parameter, $\Delta_{\mathbf{p}}$. For economy of notation, we rewrite this set of equations in matrix form:

$$i\hbar \frac{\partial}{\partial t} \begin{pmatrix} a_{\mathbf{p}\uparrow} \\ a_{-\mathbf{p}\downarrow}^{\dagger} \end{pmatrix} = \begin{pmatrix} \epsilon_{\mathbf{p}} & -\Delta_{\mathbf{p}} \\ -\Delta_{\mathbf{p}} & -\epsilon_{\mathbf{p}} \end{pmatrix} \begin{pmatrix} a_{\mathbf{p}\uparrow} \\ a_{-\mathbf{p}\downarrow}^{\dagger} \end{pmatrix} = H_{\text{eff}} \begin{pmatrix} a_{\mathbf{p}\uparrow} \\ a_{-\mathbf{p}\downarrow}^{\dagger} \end{pmatrix}. \tag{11.179}$$

To aid in diagonalizing the reduced Hamiltonian matrix, we introduce the angle $\theta_{\mathbf{p}}$, such that

$$\cos \theta_{\mathbf{p}} = \frac{\epsilon_{\mathbf{p}}}{\varepsilon_{\mathbf{p}}} \tag{11.180}$$

and

$$\sin \theta_{\mathbf{p}} = \frac{\Delta_{\mathbf{p}}}{\varepsilon_{\mathbf{p}}}. \tag{11.181}$$

The utility of this transformation is immediate because the BCS coefficients

$$u_{\mathbf{p}} = \frac{1}{\sqrt{2}} \left(1 + \frac{\epsilon_{\mathbf{p}}}{\varepsilon_{\mathbf{p}}} \right)^{1/2} = \frac{1}{\sqrt{2}} (1 + \cos \theta_{\mathbf{p}})^{1/2} \tag{11.182}$$

and

$$v_{\mathbf{p}} = \frac{1}{\sqrt{2}} \left(1 - \frac{\epsilon_{\mathbf{p}}}{\varepsilon_{\mathbf{p}}} \right)^{1/2} = \frac{1}{\sqrt{2}} (1 - \cos \theta_{\mathbf{p}})^{1/2} \tag{11.183}$$

form the eigenvector basis of the reduced Hamiltonian matrix,

$$\mathbf{H}_{\text{eff}} = \varepsilon_{\mathbf{p}} \begin{pmatrix} \cos \theta_{\mathbf{p}} & -\sin \theta_{\mathbf{p}} \\ -\sin \theta_{\mathbf{p}} & -\cos \theta_{\mathbf{p}} \end{pmatrix}. \tag{11.184}$$

The eigenvalues of this matrix are ± 1. Consequently, the eigenvectors are of the form

$$\begin{pmatrix} \sin \theta_{\mathbf{p}} \\ \cos \theta_{\mathbf{p}} \pm 1 \end{pmatrix}. \tag{11.185}$$

The ratio of the coefficients in the eigenvectors is

$$\frac{\sin \theta_{\mathbf{p}}}{(\cos \theta_{\mathbf{p}} \pm 1)} = \pm \left(\tan \frac{\theta_{\mathbf{p}}}{2} \right)^{\pm 1}. \tag{11.186}$$

Because

$$\tan \frac{\theta_{\mathbf{p}}}{2} = \pm \left(\frac{\sqrt{1 - \cos \theta_{\mathbf{p}}}}{\sqrt{1 + \cos \theta_{\mathbf{p}}}} \right), \tag{11.187}$$

the eigenvectors of $\mathbf{H}_{\text{eff}}$ can be written as

$$\mathbf{e}_1 = \begin{pmatrix} v_{\mathbf{p}} \\ u_{\mathbf{p}} \end{pmatrix} = \begin{pmatrix} \sin(\theta_{\mathbf{p}}/2) \\ \cos(\theta_{\mathbf{p}}/2) \end{pmatrix} \tag{11.188}$$

and

$$\mathbf{e}_2 = \begin{pmatrix} u_{\mathbf{p}} \\ -v_{\mathbf{p}} \end{pmatrix} = \begin{pmatrix} \cos(\theta_{\mathbf{p}}/2) \\ -\sin(\theta_{\mathbf{p}}/2) \end{pmatrix}. \tag{11.189}$$

It is now clear that $\mathbf{H}_{\text{eff}}\mathbf{e}_1 = -\varepsilon_{\mathbf{p}}\mathbf{e}_1$ and $\mathbf{H}_{\text{eff}}\mathbf{e}_2 = \varepsilon_{\mathbf{p}}\mathbf{e}_2$. Simply put, $\pm\varepsilon_{\mathbf{p}}$ are the two eigenenergies of $\mathbf{H}_{\text{eff}}$. To lay plain the utility of $\mathbf{e}_1$ and $\mathbf{e}_2$, we multiply the equations of motion by $\mathbf{e}_1$ and $\mathbf{e}_2$. The result for $\mathbf{e}_2$,

$$i\hbar \frac{\partial}{\partial t}(a_{\mathbf{p}\uparrow}u_{\mathbf{p}} - v_{\mathbf{p}}a^{\dagger}_{-\mathbf{p}\downarrow}) = \mathbf{e}_2^{\dagger}\mathbf{H}_{\text{eff}}\begin{pmatrix} a_{\mathbf{p}\uparrow} \\ a^{\dagger}_{-\mathbf{p}\downarrow} \end{pmatrix} = \varepsilon_{\mathbf{p}}\mathbf{e}_2^{\dagger}\begin{pmatrix} a_{\mathbf{p}\uparrow} \\ a^{\dagger}_{-\mathbf{p}\downarrow} \end{pmatrix}, \tag{11.190}$$

demonstrates that

$$i\hbar\dot{\gamma}_{\mathbf{p}\uparrow} = \varepsilon_{\mathbf{p}}\gamma_{\mathbf{p}\uparrow}. \tag{11.191}$$

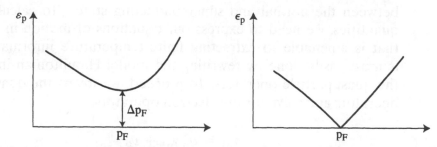

Figure 11.16 Quasi-particle spectrum in the vicinity of the Fermi surface in the (*a*) superconducting and (*b*) normal states. The value of ϵ_p at the Fermi level is chosen to vanish. Consequently, the minimum excitation energy required at the Fermi surface is the gap energy.

Similarly for $\mathbf{e}_1$, we have that

$$i\hbar\frac{\partial}{\partial t}(v_{\mathbf{p}}^{*}a_{\mathbf{p}\uparrow} + u_{\mathbf{p}}^{*}a_{-\mathbf{p}\downarrow}^{\dagger}) = -\varepsilon_{\mathbf{p}}\mathbf{e}_{1}^{\dagger}\begin{pmatrix}a_{\mathbf{p}\uparrow}\\ a_{-\mathbf{p}\downarrow}^{\dagger}\end{pmatrix} \tag{11.192}$$

and

$$i\hbar\dot{\gamma}_{-\mathbf{p}\downarrow}^{\dagger} = -\varepsilon_{\mathbf{p}}\gamma_{-\mathbf{p}\downarrow}^{\dagger}. \tag{11.193}$$

These equations of motion imply that the diagonalized form for the reduced Hamiltonian is

$$\mathbf{H}_{\text{eff}} = \sum_{\mathbf{p}}\varepsilon_{\mathbf{p}}(\gamma_{\mathbf{p}\uparrow}^{\dagger}\gamma_{\mathbf{p}\uparrow} + \gamma_{-\mathbf{p}\downarrow}^{\dagger}\gamma_{-\mathbf{p}\downarrow}). \tag{11.194}$$

Consequently, creating any quasi particle, regardless of its spin and momentum, costs an energy $\varepsilon_{\mathbf{p}}$. To reiterate, the quasi particles of BCS theory are not electrons. Rather, they are spinful entities with no well-defined charge, in contrast to singlet Cooper pairs, which have charge 2e but no spin. In this sense, spin-charge separation can be thought to obtain in a BCS superconductor. A plot of the quasi-particle energies is shown in Fig. (11.16). As is evident, adding or subtracting a particle costs at least the minimum energy $\Delta_{\mathbf{p}}$.

11.13   THERMODYNAMICS

We are now in a position to calculate the thermodynamic properties of a superconductor. Of special interest are the heat capacity, C_v, and the Helmholtz free energy, the former of which has a discontinuity between the normal and superconducting states. To calculate these quantities, we need to express our equations of motion in a manner that is amenable to extracting finite-temperature information. This is most easily done by rewriting the model Hamiltonian in terms of the quasi-particle operators. To proceed, we invert the quasi-particle operators and solve for the electron operators,

$$a_{\mathbf{p}\uparrow} = u_{\mathbf{p}}^* \gamma_{\mathbf{p}\uparrow} + v_{\mathbf{p}} \gamma_{-\mathbf{p}\downarrow}^\dagger \tag{11.195}$$

and

$$a_{-\mathbf{p}\downarrow} = -v_{\mathbf{p}} \gamma_{\mathbf{p}\uparrow}^\dagger + u_{\mathbf{p}}^* \gamma_{-\mathbf{p}\downarrow}. \tag{11.196}$$

We will show that the thermal overage

$$\langle \gamma_{\mathbf{p}\sigma}^\dagger \gamma_{\mathbf{p}\sigma} \rangle = n_{\mathbf{p}\sigma}, \tag{11.197}$$

where $n_{\mathbf{p}\sigma}$ is the Fermi-Dirac distribution function evaluated at the quasi-particle energy, $\varepsilon_{\mathbf{p}}$. We will assume that the entropy associated with such quasi-particle excitations is given by the noninteracting form

$$S = -k_B \sum_{\mathbf{p},\sigma} [n_{\mathbf{p}\sigma} \ln n_{\mathbf{p}\sigma} + (1 - n_{\mathbf{p}\sigma}) \ln(1 - n_{\mathbf{p}\sigma})] \tag{11.198}$$

presented in Chapter 1.

We evaluate first the Helmholtz free energy, $F = E' - TS$, where $E' = \langle H - \mu N \rangle$. We need to express the internal energy at the mean-field level,

$$E' = \sum_{\mathbf{p},\sigma} \epsilon_{\mathbf{p}} \langle a_{\mathbf{p}\sigma}^\dagger a_{\mathbf{p}\sigma} \rangle + \sum_{\mathbf{p},\mathbf{p}'} V_{\mathbf{p}\mathbf{p}'} \langle b_{\mathbf{p}}^\dagger \rangle \langle b_{\mathbf{p}'} \rangle, \tag{11.199}$$

in terms of the $\gamma_{\mathbf{p}}$'s. The thermal average in these expressions is to be interpreted as a trace over all states of the system. An arbitrary excitation above the ground state is represented by

$$|\text{excited state}\rangle = \prod_i \gamma_{\mathbf{p}_i \sigma_i}^\dagger |\Psi_{\text{BCS}}\rangle \tag{11.200}$$

and, hence, contains a definite number of excited quasi particles. As a consequence, any average of the form $\langle \gamma\gamma \rangle$ identically vanishes.

From Eqs. (11.195) and (11.196), it follows that

$$\sum_{\mathbf{p},\sigma} \langle a_{\mathbf{p}\sigma}^{\dagger} a_{\mathbf{p}\sigma} \rangle = \sum_{\mathbf{p}} \langle (u_{\mathbf{p}} \gamma_{\mathbf{p}\uparrow}^{\dagger} + v_{\mathbf{p}}^* \gamma_{-\mathbf{p}\downarrow})(u_{\mathbf{p}}^* \gamma_{\mathbf{p}\uparrow} + v_{\mathbf{p}} \gamma_{-\mathbf{p}\downarrow}^{\dagger}) \rangle$$

$$+ \sum_{\mathbf{p}} \langle (-v_{\mathbf{p}}^* \gamma_{\mathbf{p}\uparrow} + u_{\mathbf{p}} \gamma_{-\mathbf{p}\downarrow}^{\dagger})(-v_{\mathbf{p}} \gamma_{\mathbf{p}\uparrow}^{\dagger} + u_{\mathbf{p}}^* \gamma_{-\mathbf{p}\downarrow}) \rangle$$

$$= \sum_{\mathbf{p}} (|u_{\mathbf{p}}|^2 - |v_{\mathbf{p}}|^2)(n_{\mathbf{p}\uparrow} + n_{-\mathbf{p}\downarrow}) + 2|v_{\mathbf{p}}|^2. \quad (11.201)$$

In the ground state, we showed previously (see Eq. 11.122) that $\langle b_{\mathbf{p}} \rangle = u_{\mathbf{p}}^* v_{\mathbf{p}}$. At finite temperature, the average of the pair operator

$$\langle b_{\mathbf{p}} \rangle = \langle a_{-\mathbf{p}\downarrow} a_{\mathbf{p}\uparrow} \rangle$$

$$= \langle (-v_{\mathbf{p}} \gamma_{\mathbf{p}\uparrow}^{\dagger} + u_{\mathbf{p}}^* \gamma_{-\mathbf{p}\downarrow})(u_{\mathbf{p}}^* \gamma_{\mathbf{p}\uparrow} + v_{\mathbf{p}} \gamma_{-\mathbf{p}\downarrow}^{\dagger}) \rangle$$

$$= v_{\mathbf{p}} u_{\mathbf{p}}^* (1 - n_{-\mathbf{p}\downarrow} - n_{\mathbf{p}\uparrow}) \quad (11.202)$$

contains the thermal factors that reflect the pairing of electrons above and below the Fermi level. Consequently,

$$\sum_{\mathbf{p},\mathbf{p}'} V_{\mathbf{p}\mathbf{p}'} \langle b_{\mathbf{p}}^{\dagger} \rangle \langle b_{\mathbf{p}'} \rangle = \sum_{\mathbf{p},\mathbf{p}'} V_{\mathbf{p}\mathbf{p}'} u_{\mathbf{p}} u_{\mathbf{p}'}^* v_{\mathbf{p}}^* v_{\mathbf{p}'} (1 - n_{\mathbf{p}\uparrow} - n_{-\mathbf{p}\downarrow})$$

$$\times (1 - n_{-\mathbf{p}'\downarrow} - n_{\mathbf{p}'\uparrow}), \quad (11.203)$$

and the total Helmholtz energy is

$$F' = \sum_{\mathbf{p}} \epsilon_{\mathbf{p}} [(|u_{\mathbf{p}}|^2 - |v_{\mathbf{p}}|^2)(n_{\mathbf{p}\uparrow} + n_{-\mathbf{p}\downarrow}) + 2|v_{\mathbf{p}}|^2]$$

$$+ \sum_{\mathbf{p},\mathbf{p}'} V_{\mathbf{p}\mathbf{p}'} (1 - n_{-\mathbf{p}\downarrow} - n_{\mathbf{p}\uparrow})(1 - n_{-\mathbf{p}'\downarrow} - n_{\mathbf{p}'\uparrow}) u_{\mathbf{p}} v_{\mathbf{p}}^* u_{\mathbf{p}'}^* v_{\mathbf{p}'}$$

$$+ k_B T \sum_{\mathbf{p},\sigma} [n_{\mathbf{p}\sigma} \ln n_{\mathbf{p}\sigma} + (1 - n_{\mathbf{p}\sigma}) \ln (1 - n_{\mathbf{p}\sigma})]. \quad (11.204)$$

Minimizing F' with respect to $u_{\mathbf{p}}$ and $v_{\mathbf{p}}$, we obtain the standard result that $u_{\mathbf{p}} = ((1 + \cos \theta_{\mathbf{p}})/2)^{1/2}$ and $v_{\mathbf{p}} = ((1 - \cos \theta_{\mathbf{p}})/2)^{1/2}$ with $\cos \theta_{\mathbf{p}} = \epsilon_{\mathbf{p}}/\varepsilon_{\mathbf{p}}$. In this case, however, the gap is temperature dependent through

$$\Delta_{\mathbf{p}} = -\sum_{\mathbf{p}'} V_{\mathbf{p}\mathbf{p}'} \langle b_{\mathbf{p}'} \rangle$$

$$= -\sum_{\mathbf{p}'} V_{\mathbf{p}\mathbf{p}'} (1 - n_{\mathbf{p}'\uparrow} - n_{-\mathbf{p}'\downarrow}) u_{\mathbf{p}'}^* v_{\mathbf{p}'}, \quad (11.205)$$

where $\sin \theta_{\mathbf{p}} = \Delta_{\mathbf{p}}/\varepsilon_{\mathbf{p}}$. Varying F' with respect to $n_{\mathbf{p}\uparrow}$, we find that

$$\frac{\partial F'}{\partial n_{\mathbf{p}\uparrow}} = 0 = \epsilon_{\mathbf{p}}(|u_{\mathbf{p}}|^2 - |v_{\mathbf{p}}|^2) + k_B T \ln\left(\frac{n_{\mathbf{p}\uparrow}}{1 - n_{\mathbf{p}\uparrow}}\right) + \Delta_{\mathbf{p}} u_{\mathbf{p}} v_{\mathbf{p}}^*,$$
(11.206)

holding the coherence factors $u_{\mathbf{p}}$ and $v_{\mathbf{p}}$ fixed. The condition that is now imposed on $n_{\mathbf{p}\uparrow}$, namely,

$$k_B T \ln (n_{\mathbf{p}\uparrow}^{-1} - 1) = \epsilon_{\mathbf{p}} \cos \theta_{\mathbf{p}} + 2\Delta_{\mathbf{p}} u_{\mathbf{p}} v_{\mathbf{p}}^*$$

$$= \epsilon_{\mathbf{p}} \cos \theta_{\mathbf{p}} + \Delta_{\mathbf{p}} \sin \theta_{\mathbf{p}}$$

$$= \varepsilon_{\mathbf{p}},$$
(11.207)

is simply the requirement that $n_{\mathbf{p}\uparrow}$ is the Fermi-Dirac distribution evaluated at the quasi energy $\varepsilon_{\mathbf{p}}$. The Fermi-Dirac distribution is the most likely distribution of the quasi-particle excitations at finite temperature.

The free energy calculation is now straightforward if we use the mean-field value of the gap, Eq. (11.205):

$$E' = \sum_{\mathbf{p}} \epsilon_{\mathbf{p}}[(|u_{\mathbf{p}}|^2 - |v_{\mathbf{p}}|^2)(n_{\mathbf{p}\uparrow} + n_{\mathbf{p}\downarrow}) + 2|v_{\mathbf{p}}|^2] + \sum_{\mathbf{p}} \Delta_{\mathbf{p}} \frac{\sin \theta_{\mathbf{p}}}{2}(1 - 2n_{\mathbf{p}})$$

$$= \sum_{\mathbf{p}} \left[2\epsilon_{\mathbf{p}}\left(\cos \theta_{\mathbf{p}} n_{\mathbf{p}} + \sin^2 \frac{\theta_{\mathbf{p}}}{2}\right) + \Delta_{\mathbf{p}} \frac{\sin \theta_{\mathbf{p}}}{2}(1 - 2n_{\mathbf{p}})\right].$$
(11.208)

Coupled with the fact that $\ln (1/n_{\mathbf{p}\sigma} - 1) = \varepsilon_{\mathbf{p}}/k_B T$, we simplify the entropy term to

$$TS = -2k_B T \sum_{\mathbf{p}} [\beta \varepsilon_{\mathbf{p}} n_{\mathbf{p}} + \ln (1 + e^{-\beta \varepsilon_{\mathbf{p}}})].$$
(11.209)

The total free energy is, then,

$$F' = \sum_{\mathbf{p}} \left[2\epsilon_{\mathbf{p}}\left(\cos \theta_{\mathbf{p}} n_{\mathbf{p}} + \sin^2 \frac{\theta_{\mathbf{p}}}{2}\right) + \Delta_{\mathbf{p}} \frac{\sin \theta_{\mathbf{p}}}{2}(1 - 2n_{\mathbf{p}})\right]$$

$$+ 2k_B T \sum_{\mathbf{p}} [\beta \varepsilon_{\mathbf{p}} n_{\mathbf{p}} + \ln (1 + e^{-\beta \varepsilon_{\mathbf{p}}})].$$
(11.210)

The evaluation of this expression is left as a homework problem (see Problem 11.10).

The heat capacity,

$$C_v = T \left(\frac{\partial S}{\partial T} \right)$$

$$= -k_B T \sum_{\mathbf{p},\sigma} \frac{\partial n_{\mathbf{p}\sigma}}{\partial T} \ln \left(\frac{n_{\mathbf{p}\sigma}}{1 - n_{\mathbf{p}\sigma}} \right)$$

$$= \sum_{\mathbf{p},\sigma} \varepsilon_{\mathbf{p}} \frac{\partial n_{\mathbf{p}\sigma}}{\partial T}, \tag{11.211}$$

contains the temperature derivative

$$\frac{\partial n_{\mathbf{p}}}{\partial T} = \frac{n_{\mathbf{p}}^2}{k_B T} \left[\frac{\varepsilon_{\mathbf{p}}}{T} - \frac{\partial \varepsilon_{\mathbf{p}}}{\partial T} \right] e^{\beta \varepsilon_{\mathbf{p}}}, \tag{11.212}$$

which is compounded by the implicit temperature dependence of the gap, $\Delta_{\mathbf{p}}$. From the definition of $\varepsilon_{\mathbf{p}}$, it follows that

$$\frac{\partial \varepsilon_{\mathbf{p}}}{\partial T} = \frac{\partial \varepsilon_{\mathbf{p}}}{\partial \Delta_{\mathbf{p}}} \frac{\partial \Delta_{\mathbf{p}}}{\partial T} = \frac{\Delta_{\mathbf{p}}}{\varepsilon_{\mathbf{p}}} \frac{\partial \Delta_{\mathbf{p}}}{\partial T} = \frac{1}{2\varepsilon_{\mathbf{p}}} \frac{\partial \Delta_{\mathbf{p}}^2}{\partial T}. \tag{11.213}$$

Because $\partial \varepsilon_{\mathbf{p}} / \partial T$ vanishes in the normal state, we write the difference between the heat capacity in the superconducting and normal states at T_c as

$$(C_v^S - C_v^N)_{T=T_c} = -\beta \sum_{\mathbf{p}} n_{\mathbf{p}}^2 \frac{\partial \Delta_{\mathbf{p}}^2}{\partial T} e^{\beta \varepsilon_{\mathbf{p}}}$$

$$= \sum_{\mathbf{p}} \frac{\partial n_{\mathbf{p}}}{\partial \varepsilon_{\mathbf{p}}} \frac{\partial \Delta_{\mathbf{p}}^2}{\partial T} \Big|_{T=T_c}. \tag{11.214}$$

In the vicinity of $T = T_c$, the gap is well approximated by $\Delta^2 \sim |1 - T/T_c|$. We find, then, that $\partial \Delta^2 / \partial T \sim 1/T_c$, and the discontinuity in the heat capacity is identically

$$C_v^S - C_v^N = \frac{N(\epsilon_F)}{T_c} \tag{11.215}$$

at $T = T_c$. The positive sign associated with this difference indicates that the heat capacity in the superconducting state just below T_c exceeds C_v^N.

The second result we seek is the exponential fall-off of C_v^S just below T_c. To obtain this behavior, we analyze Eq. (11.211) in the limit that

$T \to 0$. In this limit, $\partial \Delta_{\mathbf{p}}^2 / \partial T \to 0, n_{\mathbf{p}} \to e^{-\beta \varepsilon_{\mathbf{p}}}$, and the heat capacity is dominated by

$$\lim_{T \to 0} C_v = 2\beta^2 k_B \sum_{\mathbf{p}} \varepsilon_{\mathbf{p}}^2 n_{\mathbf{p}} (1 - n_{\mathbf{p}})$$

$$\simeq 2\beta^2 k_B N(\epsilon_F) \Delta^2 (T = 0) \int_0^\infty d\epsilon \, e^{-\beta \sqrt{\Delta^2(T=0) + \epsilon^2}}$$

$$\simeq 2\beta^2 k_B N(\epsilon_F) \Delta_0^2 e^{-\beta \Delta_0} \int_0^\infty d\epsilon \, e^{-\beta \epsilon^2 / 2\Delta_0}$$

$$= N(\epsilon_F) \sqrt{\frac{\pi}{2}} k_B \beta^{3/2} \Delta_0^{5/2} e^{-\beta \Delta_0}, \tag{11.216}$$

which decays exponentially, as advertised with $\Delta_0 = \Delta(T = 0)$. The exponential fall-off is dominated by the gap at $T = 0$. In simplifying the integrals leading to this result, we used the fact that the integrals are largest when $\Delta \gg \epsilon$.

11.14   EXPERIMENTAL APPLICATIONS

One of the key successes of the BCS theory is the ease by which physical observables, such as the phonon attenuation rate, can be calculated. In fact, it was the calculation of the spin-lattice relaxation time using the BCS theory and its subsequent experimental confirmation that led to the wide acceptance of BCS theory. In this section, we perform three of these calculations explicitly: 1) electromagnetic absorption, 2) ultrasonic attenuation, and 3) the spin-lattice relaxation rate. Our derivation is an expanded version of the treatment by Schrieffer [S1964]. Appearing in each of these calculations are particular combinations of the coherence factors, $u_{\mathbf{p}}$ and $v_{\mathbf{p}}$. We will see that each combination of the coherence factors can generate completely different physics. That the physical properties are so intimately connected with the coherence factors is a hallmark of the BCS pairing hypothesis. In each of these calculations, we will treat the coherence factors as being real. As we have seen, for the simple model potential we have chosen, Eq. (11.41), the coherence factors are in fact real.

11.14.1   Electromagnetic Absorption

In a BCS superconductor, absorption of electromagnetic radiation is suppressed for frequencies less than $\Delta / \hbar$. We have now the machinery to derive this result explicitly. In the electromagnetic problem, the perturbation term is obtained by replacing the momentum with

$\mathbf{p} - e\mathbf{A}(\mathbf{r}, t)/c$, where $\mathbf{A}(\mathbf{r}, t)$ is the vector potential. To lowest order in $\mathbf{A}(\mathbf{r}, t)$, the perturbation is of the form $\mathbf{p} \cdot \mathbf{A} + \mathbf{A} \cdot \mathbf{p}$. We will require that the vector potential satisfy the transversality condition, $\nabla \cdot \mathbf{A} = 0$. As a consequence, the second-quantized form of the perturbation for absorption of an electromagnetic wave with frequency $\omega_{\mathbf{q}}$ is

$$H_{\text{abs}} = -\frac{e}{mcV} \sum_{\mathbf{p}, \sigma} \mathbf{A} \cdot \mathbf{p} a_{\mathbf{p}+\mathbf{q}\sigma}^{\dagger} a_{\mathbf{p}\sigma}. \qquad (11.217)$$

We must compute a matrix element of the form

$$\langle f | H_{\text{abs}} | i \rangle = -\frac{e}{Vmc} \sum_{\mathbf{p}\sigma} \mathbf{A} \cdot \mathbf{p} M_{if}^{\sigma}, \qquad (11.218)$$

where

$$M_{if}^{\sigma} = \langle f | a_{\mathbf{p}+\mathbf{q}\sigma}^{\dagger} a_{\mathbf{p}\sigma} | i \rangle \qquad (11.219)$$

and $|i\rangle$ and $|f\rangle$ are the initial and final quasi-particle states, respectively. As in the calculation of the heat capacity, we rewrite this matrix element,

$$\sum_{\sigma} M_{if}^{\sigma} = \langle f | (u_{\mathbf{p}'} \gamma_{\mathbf{p}'\uparrow}^{\dagger} + v_{\mathbf{p}'} \gamma_{-\mathbf{p}'\downarrow})(u_{\mathbf{p}} \gamma_{\mathbf{p}\uparrow} + v_{\mathbf{p}} \gamma_{-\mathbf{p}\downarrow}^{\dagger}) | i \rangle$$
$$+ \langle f | (-v_{-\mathbf{p}'} \gamma_{-\mathbf{p}'\uparrow} + u_{-\mathbf{p}'} \gamma_{\mathbf{p}'\downarrow}^{\dagger})(-v_{-\mathbf{p}} \gamma_{-\mathbf{p}\uparrow}^{\dagger} + u_{-\mathbf{p}} \gamma_{\mathbf{p}\downarrow}^{\dagger}) | i \rangle,$$

in terms of the quasi-particle operators. There are a number of distinct kinds of quasi-particle excitations that arise from the terms above. These are illustrated in Fig. (11.17).

At $T = 0$, there are no quasi particles in the ground state. Hence, only the last process shown in Fig. (11.17), which creates two quasi particles, survives at $T = 0$. We will specialize to $T = 0$, as in this case the essential physics is easily unearthed. We find that

$$M_{if}^{\uparrow} + M_{if}^{\downarrow} = u_{\mathbf{p}+\mathbf{q}} v_{\mathbf{p}} \langle f | \gamma_{\mathbf{p}+\mathbf{q}\uparrow}^{\dagger} \gamma_{-\mathbf{p}\downarrow}^{\dagger} | i \rangle$$
$$- u_{-(\mathbf{p}+\mathbf{q})} v_{-\mathbf{p}} \langle f | \gamma_{\mathbf{p}+\mathbf{q}\downarrow}^{\dagger} \gamma_{-\mathbf{p}\uparrow}^{\dagger} | i \rangle \qquad (11.220)$$

and

$$\langle f | H_{\text{abs}} | i \rangle = -\frac{e}{mcV} \sum_{\mathbf{p}} (\mathbf{A} \cdot \mathbf{p}) \langle f | u_{\mathbf{p}+\mathbf{q}} v_{-\mathbf{p}} \gamma_{\mathbf{p}+\mathbf{q}\uparrow}^{\dagger} \gamma_{-\mathbf{p}\downarrow}^{\dagger} $$
$$- u_{-(\mathbf{p}+\mathbf{q})} v_{-\mathbf{p}} \gamma_{\mathbf{p}+\mathbf{q}\downarrow}^{\dagger} \gamma_{-\mathbf{p}\uparrow}^{\dagger} | i \rangle. \qquad (11.221)$$

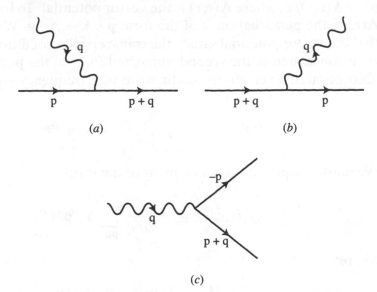

Figure 11.17 Quasi-particle excitations arising from the matrix element for electromagnetic absorption: (a) $\gamma_{\mathbf{p+q}}^{\dagger}\gamma_{\mathbf{p}}^{\dagger}$, (b) $\gamma_{\mathbf{p}}^{\dagger}\gamma_{\mathbf{p}}$, and (c) $\gamma_{\mathbf{p+q}}^{\dagger}\gamma_{-\mathbf{p}}^{\dagger}$. The wavy line represents the phonon interaction.

To simplify this expression, we note that if we shift the momentum in the second term by $\mathbf{p} \to -(\mathbf{p} + \mathbf{q})$, then $\mathbf{p} \cdot \mathbf{A} \to -\mathbf{A} \cdot (\mathbf{p} + \mathbf{q}) = -\mathbf{A} \cdot \mathbf{p}$. The last equality follows because the vector potential is transverse to the electromagnetic field. We find that

$$\langle f|H_{\text{abs}}|i\rangle = \frac{-e}{mcV} \sum_{\mathbf{p}} (\mathbf{A} \cdot \mathbf{p})(u_{\mathbf{p+q}}v_{\mathbf{p}} - u_{\mathbf{p}}v_{\mathbf{p+q}})\langle f|\gamma_{\mathbf{p+q}\uparrow}^{\dagger}\gamma_{-\mathbf{p}\downarrow}^{\dagger}|i\rangle. \tag{11.222}$$

In the matrix element in Eq. (11.222), we write the final state as $|f\rangle = |\mathbf{p'}\uparrow, \mathbf{p''}\downarrow\rangle$.

The quasi-particle matrix element reduces to

$$\langle f|\gamma_{\mathbf{p+q}\uparrow}^{\dagger}\gamma_{-\mathbf{p}\downarrow}^{\dagger}|i\rangle = \delta_{\mathbf{p''},-\mathbf{p}}\delta_{\mathbf{p'},\mathbf{p+q}}. \tag{11.223}$$

The total absorption rate per unit volume,

$$\Gamma_s = \frac{2\pi}{\hbar}\left(\frac{e}{mcV}\right)^2 \sum_{\mathbf{p},\mathbf{p'}} (\mathbf{A} \cdot \mathbf{p})^2 \delta_{\mathbf{p+p'},\mathbf{q}}(u_{\mathbf{p'}}v_{\mathbf{p}} - u_{\mathbf{p}}v_{\mathbf{p'}})^2$$

$$\times \delta(\varepsilon_{\mathbf{p}} + \varepsilon_{\mathbf{p'}} - \hbar\omega_{\mathbf{q}}), \tag{11.224}$$

involves the coherence factors

$$(u_{\mathbf{p}'}v_{\mathbf{p}} - u_{\mathbf{p}}v_{\mathbf{p}'})^2 = \left[\sqrt{\frac{\varepsilon_{\mathbf{p}'} + \epsilon_{\mathbf{p}'}}{2\varepsilon_{\mathbf{p}'}}}\sqrt{\frac{\varepsilon_{\mathbf{p}} - \epsilon_{\mathbf{p}}}{2\varepsilon_{\mathbf{p}}}} - \sqrt{\frac{\varepsilon_{\mathbf{p}'} - \epsilon_{\mathbf{p}'}}{2\varepsilon_{\mathbf{p}'}}}\sqrt{\frac{\varepsilon_{\mathbf{p}} + \epsilon_{\mathbf{p}}}{2\varepsilon_{\mathbf{p}}}}\right]^2$$

$$= \frac{1}{2\varepsilon_{\mathbf{p}}\varepsilon_{\mathbf{p}'}}(\varepsilon_{\mathbf{p}}\varepsilon_{\mathbf{p}'} - \epsilon_{\mathbf{p}}\epsilon_{\mathbf{p}'} - \Delta^2). \tag{11.225}$$

We convert the sums to integrals,

$$\frac{1}{V}\sum_{\mathbf{p}} \rightarrow N(\epsilon_F)\int d\epsilon_{\mathbf{p}} \int \frac{d\Omega_{\mathbf{p}}}{4\pi}, \tag{11.226}$$

where $d\Omega_{\mathbf{p}}$ represents an integral over the solid angle. The absorption rate is now

$$\Gamma_s = N(\epsilon_F)^2 \frac{2\pi}{\hbar}\int d\epsilon d\epsilon' \delta(\varepsilon + \varepsilon' - \hbar\omega_{\mathbf{q}})\frac{\varepsilon\varepsilon' - \epsilon\epsilon' - \Delta^2}{2\varepsilon\varepsilon'}$$

$$\times \int \frac{d\Omega\, d\Omega'}{4\pi\, 4\pi}\delta(\mathbf{p} + \mathbf{p}' - \mathbf{q})\left(\frac{e}{mc}\mathbf{p}\cdot\mathbf{A}\right)^2. \tag{11.227}$$

We note first that symmetry about the Fermi surface dictates that we must obtain the same result under the transformation $\epsilon \leftrightarrow -\epsilon$. As a consequence, the $\epsilon\epsilon'$ terms vanish. If we are interested primarily in the contribution from the vicinity of the Fermi surface, then we can set $|\mathbf{p}| = |\mathbf{p}'| = p_F$. The angular terms are trivial and yield $(p_F Ae)^2/3(mc)^2$. The remaining integration,

$$\Gamma_s = N(\epsilon_F)^2 \frac{8\pi}{3\hbar}\left(\frac{p_F Ae}{mc}\right)^2 \int_0^\infty d\epsilon$$

$$\int_0^\infty d\epsilon' \delta(\varepsilon + \varepsilon' - \hbar\omega_{\mathbf{q}})\frac{\varepsilon\varepsilon' - \Delta^2}{2\varepsilon\varepsilon'}, \tag{11.228}$$

is over the free-particle energies. The factor of 4 arises from the conversion of the lower limit of integration from $-\infty$ to zero. Because $\Gamma_s > 0$, the integral must be positive for all ε and ε'. We see immediately that $\varepsilon\varepsilon' > \Delta^2$, or, equivalently, that $\hbar\omega_{\mathbf{q}} > 2\Delta$. Thus, Γ_s must vanish then for $\hbar\omega_{\mathbf{q}} < 2\Delta$. This is the principal result of this section.

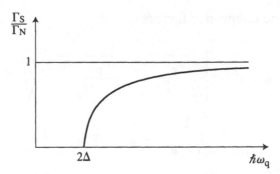

Figure 11.18 The ratio of the electromagnetic absorption in the superconducting state versus that in the normal state. The x-axis is the frequency, and Δ is the gap energy.

We can obtain a more quantitative prediction by changing variables and noting that because $\varepsilon = (\epsilon^2 + \Delta^2)^{1/2}$, $\varepsilon d\varepsilon = \epsilon d\epsilon$. This variable change introduces a lower limit cut-off

$$\Gamma_s = \text{const.} \int_\Delta^\infty d\varepsilon \int_\Delta^\infty d\varepsilon' \delta(\varepsilon + \varepsilon' - \hbar\omega_q) \frac{(\varepsilon\varepsilon' - \Delta^2)}{(\varepsilon'^2 - \Delta^2)^{1/2}(\varepsilon^2 - \Delta^2)^{1/2}}$$

$$(11.229)$$

on the range of integration on ε and ε'. The δ-fcn. constraint, $\varepsilon' - \hbar\omega_q = -\varepsilon$, allows us to rewrite the ε-integral as

$$\Gamma_s = \text{const.} \int_\Delta^{\hbar\omega_q - \Delta} d\varepsilon \frac{\varepsilon(\hbar\omega_q - \varepsilon) - \Delta^2}{(\varepsilon^2 - \Delta^2)^{1/2}((\hbar\omega_q - \varepsilon)^2 - \Delta^2)^{1/2}}, \quad (11.230)$$

where we approximated the upper limit by the lower bound $\hbar\omega_q - \Delta$. When $\Delta = 0$, the integrand reduces to a constant and $\Gamma_s \to \text{const.}\hbar\omega_q$. If we refer to the $\Delta = 0$ value of Γ_s as Γ_N and plot Γ_s/Γ_N, we obtain the result shown in Fig. (11.18). The form we have derived here for Γ_s agrees well with experiment.

11.14.2   Ultrasonic Attenuation

Consider now the problem of a beam of phonons impinging on a superconductor. We have advertised that a superconductor should be transparent to phonon absorption unless the energy of the impinging phonon beam exceeds 2Δ. To show this, we rewrite the electron-phonon

Hamiltonian, Eq. (11.23), in terms of the quasi-particle operators,

$$
H_{e-ph} = \sum_{\mathbf{p},\mathbf{q}}' M_{\mathbf{q}}(b_{\mathbf{q}} + b^\dagger_{-\mathbf{q}})[(u_{\mathbf{p}+\mathbf{q}}\gamma^\dagger_{\mathbf{p}+\mathbf{q}\uparrow} + v_{\mathbf{p}+\mathbf{q}}\gamma_{-(\mathbf{p}+\mathbf{q})\downarrow})
$$
$$
\times (u_{\mathbf{p}}\gamma_{\mathbf{p}\uparrow} + v_{\mathbf{p}}\gamma^\dagger_{-\mathbf{p}\downarrow}) + (u_{-(\mathbf{p}+\mathbf{q})}\gamma^\dagger_{\mathbf{p}+\mathbf{q}\downarrow} - v_{-(\mathbf{p}+\mathbf{q})}\gamma_{-(\mathbf{p}+\mathbf{q})\uparrow})
$$
$$
\times (u_{-\mathbf{p}}\gamma_{\mathbf{p}\downarrow} - v_{-\mathbf{p}}\gamma^\dagger_{-\mathbf{p}\uparrow})],
\tag{11.231}
$$

using Eqs. (11.195) and (11.196). There are two distinct kinds of terms that arise from the quasi-particle phonon coupling: 1) single quasi-particle excitations and 2) quasi-particle pair creation and annihilation. The single quasi-particle creation and annihilation processes,

$$
H^{qp}_{e-ph} = \sum_{\mathbf{p},\mathbf{q}}' M_{\mathbf{q}}(b_{\mathbf{q}} + b^\dagger_{-\mathbf{q}})[u_{\mathbf{p}+\mathbf{q}}u_{\mathbf{p}}\gamma^\dagger_{\mathbf{p}+\mathbf{q}\uparrow}\gamma_{\mathbf{p}\uparrow}
$$
$$
+ v_{\mathbf{p}+\mathbf{q}}v_{\mathbf{p}}\gamma_{-(\mathbf{p}+\mathbf{q})\downarrow}\gamma^\dagger_{-\mathbf{p}\downarrow} + u_{-(\mathbf{p}+\mathbf{q})}u_{-\mathbf{p}}\gamma^\dagger_{\mathbf{p}+\mathbf{q}\downarrow}\gamma_{\mathbf{p}\downarrow}
$$
$$
+ v_{-(\mathbf{p}+\mathbf{q})}v_{-\mathbf{p}}\gamma_{-(\mathbf{p}+\mathbf{q})\uparrow}\gamma^\dagger_{-\mathbf{p}\uparrow}],
\tag{11.232}
$$

arise from a conjugate $\gamma_{\mathbf{p}}$, $\gamma^\dagger_{\mathbf{p}}$, pair. The additional scattering channel,

$$
H^{pair}_{e-ph} = \sum_{\mathbf{p},\mathbf{q}} M_{\mathbf{q}}(b_{\mathbf{q}} + b^\dagger_{-\mathbf{q}})[u_{\mathbf{p}+\mathbf{q}}v_{\mathbf{p}}\gamma^\dagger_{\mathbf{p}+\mathbf{q}\uparrow}\gamma^\dagger_{-\mathbf{p}\downarrow}
$$
$$
- u_{-(\mathbf{p}+\mathbf{q})}v_{-\mathbf{p}}\gamma^\dagger_{\mathbf{p}+\mathbf{q}\downarrow}\gamma^\dagger_{-\mathbf{p}\uparrow} + v_{\mathbf{p}+\mathbf{q}}u_{\mathbf{p}}\gamma_{-(\mathbf{p}+\mathbf{q})\downarrow}\gamma_{\mathbf{p}\uparrow}
$$
$$
- v_{-(\mathbf{p}+\mathbf{q})}u_{-\mathbf{p}}\gamma_{-(\mathbf{p}+\mathbf{q})\uparrow}\gamma_{\mathbf{p}\downarrow}],
\tag{11.233}
$$

is mediated by quasi-particle pair creation and annihilation.

Both types of processes are expected to occur when phonons are excited in a superconductor. However, as in the electromagnetic absorption problem, pair creation and annihilation is expected only when $\hbar\omega_{\mathbf{q}} > 2\Delta$. Because our goal is to show that absorption of phonons is attenuated at frequencies less than 2Δ, we can ignore the pair-scattering terms. We specialize, then, to the $\hbar\omega_{\mathbf{q}} < \Delta$ case in which only single-particle scattering contributes to the transition amplitude. The two representative classes of emission and absorption terms are shown in Figs. (11.19a) and (11.19b). Each term contributes identically for $\downarrow$ and $\uparrow$ electron states. First, consider emission. The operator describing emission of a phonon in a single quasi-particle exchange is

$$
V^{qp}_{emiss} = M_{\mathbf{q}}b^\dagger_{-\mathbf{q}}(u_{\mathbf{p}+\mathbf{q}}u_{\mathbf{p}}\gamma^\dagger_{\mathbf{p}+\mathbf{q}\sigma}\gamma_{\mathbf{p}\sigma} + v_{-(\mathbf{p}+\mathbf{q})}v_{-\mathbf{p}}\gamma_{\mathbf{p}\sigma}\gamma^\dagger_{\mathbf{p}+\mathbf{q}\sigma}),
$$
$$
\tag{11.234}
$$

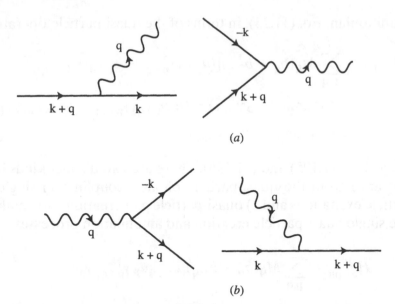

Figure 11.19 (a) Emission and (b) absorption processes contributing to the ultra-sonic attenuation. The wavy line represents the quasi particle–phonon interaction.

where we have set $-\mathbf{p} \to \mathbf{p}$ and $-\mathbf{p} - \mathbf{q} \to \mathbf{p}$ in the second term in Eq. (11.232). From the anticommutation relation for the γ's, we can rewrite the emission term as

$$V^{qp}_{\text{emiss}} = M_{\mathbf{q}} b^{\dagger}_{-\mathbf{q}}[(u_{\mathbf{p}+\mathbf{q}} u_{\mathbf{p}} - v_{-(\mathbf{p}+\mathbf{q})} v_{-\mathbf{p}}) \gamma^{\dagger}_{\mathbf{p}+\mathbf{q}\sigma} \gamma_{\mathbf{p}\sigma}$$
$$+ v_{-(\mathbf{p}+\mathbf{q})} v_{-\mathbf{p}} \delta_{\mathbf{p},\mathbf{p}+\mathbf{q}}]. \qquad (11.235)$$

The last term requires that $\mathbf{p} = \mathbf{p} + \mathbf{q}$ or, equivalently, $\mathbf{q} = 0$. However, $M_{\mathbf{q}}$ is linear in $\mathbf{q}$ and, hence, vanishes when $\mathbf{q} = 0$. In the emission process, the final composite electron-phonon state must contain the phonon with momentum $\mathbf{q}$ and the quasi particle with momentum $\mathbf{p} + \mathbf{q}$ and spin σ. The quasi particle with momentum $\mathbf{p}$ is absent from this state. The initial state is in general some determinantal state containing any number of quasi particles but in particular the quasi particle with momentum $\mathbf{p}$ and spin σ. Let $|\mathbf{p} + \mathbf{q}\sigma, \mathbf{q}\rangle$ and $|\mathbf{p}\sigma\rangle$ be the final and initial states, respectively. The emission matrix element simplifies to

$$\langle \mathbf{p} + \mathbf{q}\sigma, \mathbf{q}|V^{qp}_{\text{emiss}}|\mathbf{p}\sigma\rangle = M_{\mathbf{q}}(u_{\mathbf{p}+\mathbf{q}} u_{\mathbf{p}} - v_{-(\mathbf{p}+\mathbf{q})} v_{-\mathbf{p}})$$
$$\times \sqrt{N_{\mathbf{q}} + 1} \sqrt{n_{\mathbf{p}}(1 - n_{\mathbf{p}+\mathbf{q}})}, \qquad (11.236)$$

where $n_{\mathbf{p}}$ is the occupation of quasi particles in momentum state $\mathbf{p}$. Consequently, when both $\uparrow$ and $\downarrow$ spins are included, the Fermi golden rule rate for emission is

$$\Gamma^{qp}_{\text{emiss}} = \frac{2}{\hbar} \sum_{\mathbf{p}} (N_{\mathbf{q}} + 1)|M_{\mathbf{q}}|^2 (u_{\mathbf{p}+\mathbf{q}} u_{\mathbf{p}} - v_{\mathbf{p}+\mathbf{q}} v_{\mathbf{p}})^2$$

$$\times n_{\mathbf{p}}(1 - n_{\mathbf{p}+\mathbf{q}}) 2\pi \delta(\hbar \omega_{\mathbf{q}} + \varepsilon_{\mathbf{p}} - \varepsilon_{\mathbf{p}+\mathbf{q}}). \quad (11.237)$$

Analogous arguments can be applied to absorption of a phonon. There are two principal differences in the form of the transition rate. First, the factors of $N_{\mathbf{q}} + 1$ are replaced by $N_{\mathbf{q}}$. This change arises because in the case of absorption, the final state has one less phonon. Second, the initial state and final states are of the form $|\mathbf{p} + \mathbf{q}\sigma\rangle$ and $|\mathbf{p}\sigma, \mathbf{q}\rangle$, respectively. The rate of absorption between all such initial and final states

$$\Gamma^{qp}_{\text{abs}} = \frac{2}{\hbar} \sum_{\mathbf{p}} N_{\mathbf{q}}|M_{\mathbf{q}}|(u_{\mathbf{p}+\mathbf{q}} u_{\mathbf{p}} - v_{\mathbf{p}+\mathbf{q}} v_{\mathbf{p}})^2$$

$$\times n_{\mathbf{p}+\mathbf{q}}(1 - n_{\mathbf{p}}) 2\pi \delta(\hbar \omega_{\mathbf{q}} + \varepsilon_{\mathbf{p}} - \varepsilon_{\mathbf{p}+\mathbf{q}}), \quad (11.238)$$

is a sum of the individual rates, which we treat at the level of the Fermi golden rule.

The difference between the emission and absorption rates,

$$\Gamma_{\mathbf{q}} = \Gamma^{qp}_{\text{abs}} - \Gamma^{qp}_{\text{emiss}}$$

$$= \frac{2}{\hbar} \sum_{\mathbf{p}}' |M_{\mathbf{q}}|^2 (u_{\mathbf{p}+\mathbf{q}} u_{\mathbf{p}} - v_{\mathbf{p}+\mathbf{q}} v_{\mathbf{p}})^2 2\pi \delta(\hbar \omega_{\mathbf{q}} + \varepsilon_{\mathbf{p}} - \varepsilon_{\mathbf{p}+\mathbf{q}})$$

$$\times [N_{\mathbf{q}}(n_{\mathbf{p}+\mathbf{q}} - n_{\mathbf{p}}) - n_{\mathbf{p}}(1 - n_{\mathbf{p}+\mathbf{q}})], \quad (11.239)$$

contains a part which depends on the phonon occupation (through $N_{\mathbf{q}}$) and a term that does not. In combining the absorption and emission terms, we used the time reversal invariance of the coherence factors. The part independent of the phonon occupation defines the effective spontaneous emission rate, whereas the first term defines the total rate at which $N_{\mathbf{q}}$ phonons are absorbed. We define, then,

$$\frac{dN_{\mathbf{q}}}{dt} = -\frac{2}{\hbar} \sum_{\mathbf{p}} |M_{\mathbf{q}}|^2 (u_{\mathbf{p}+\mathbf{q}} u_{\mathbf{p}} - v_{\mathbf{p}+\mathbf{q}} v_{\mathbf{p}})^2 2\pi \delta(\hbar \omega_{\mathbf{q}} + \varepsilon_{\mathbf{p}} - \varepsilon_{\mathbf{p}+\mathbf{q}})$$

$$\times N_{\mathbf{q}}(n_{\mathbf{p}+\mathbf{q}} - n_{\mathbf{p}})$$

$$= -\alpha_{\mathbf{q}} N_{\mathbf{q}}, \quad (11.240)$$

as the net acoustic-attenuation rate.

To evaluate $\alpha_{\mathbf{q}}$, we convert the sum to an integral,

$$\alpha_{\mathbf{q}} = \frac{4\pi}{\hbar}|M_{\mathbf{q}}|^2 g^2(0) \int d\epsilon_{\mathbf{p}} d\epsilon_{\mathbf{p}'} (u_{\mathbf{p}} u_{\mathbf{p}'} - v_{\mathbf{p}} v_{\mathbf{p}'})(n_{\mathbf{p}'} - n_{\mathbf{p}})$$

$$\times \delta(\hbar\omega_{\mathbf{q}} + \varepsilon_{\mathbf{p}} - \varepsilon_{\mathbf{p}'}) \int \frac{d\Omega_{\mathbf{p}}}{4\pi} \frac{d\Omega_{\mathbf{p}'}}{4\pi} \delta(\mathbf{p} - \mathbf{p}' - \mathbf{q}), \quad (11.241)$$

in which momentum conservation is ensured by the angular factor $\delta(\mathbf{p} - \mathbf{p}' - \mathbf{q})$. To simplify Eq. (11.241), we note that the angular factors are identical in the normal and superconducting states. Because the gap is assumed to be isotropic, we can separate out the angular dependence. Let us call this factor A_Ω. We need now an expression for the coherence factors

$$(u_{\mathbf{p}} u_{\mathbf{p}'} - v_{\mathbf{p}} v_{\mathbf{p}'})^2 = \frac{1}{2\varepsilon_{\mathbf{p}}\varepsilon_{\mathbf{p}'}}[\varepsilon_{\mathbf{p}}\varepsilon_{\mathbf{p}'} + \epsilon_{\mathbf{p}}\epsilon_{\mathbf{p}'} - \Delta^2], \quad (11.242)$$

where we have assumed that the gap is a constant $\Delta = \Delta_{\mathbf{p}} = \Delta_{\mathbf{p}'}$. We note that $n_{\mathbf{p}}$ and $\varepsilon_{\mathbf{p}}$ are both even in $\epsilon_{\mathbf{p}}$. Hence, integration over the linear $\epsilon_{\mathbf{p}}\epsilon_{\mathbf{p}'}$ factors vanishes because the limits are even. As a further simplification, we change the variable of integration from $\epsilon_{\mathbf{p}}$ to $\varepsilon_{\mathbf{p}}$ by noting that $d\varepsilon_{\mathbf{p}}/d\epsilon_{\mathbf{p}} = \varepsilon_{\mathbf{p}}/\epsilon_{\mathbf{p}}$. We are left with

$$\alpha_{\mathbf{q}} = \frac{8\pi}{\hbar}|M_{\mathbf{q}}|^2 g^2(0) A_\Omega \int_\Delta^\infty d\varepsilon_{\mathbf{p}} \int_\Delta^\infty d\varepsilon_{\mathbf{p}'} \frac{\varepsilon_{\mathbf{p}}\varepsilon_{\mathbf{p}'}}{\epsilon_{\mathbf{p}}\epsilon_{\mathbf{p}'}} \left(1 - \frac{\Delta^2}{\varepsilon_{\mathbf{p}}\varepsilon_{\mathbf{p}'}}\right)$$

$$\times (n_{\mathbf{p}'} - n_{\mathbf{p}})\delta(\hbar\omega_{\mathbf{q}} + \varepsilon_{\mathbf{p}} - \varepsilon_{\mathbf{p}'}). \quad (11.243)$$

The quantity $N(0)d\epsilon_{\mathbf{p}}/d\varepsilon_{\mathbf{p}} = N(0)\varepsilon_{\mathbf{p}}/\epsilon_{\mathbf{p}}$ is roughly the density of quasi-particle excitations. At the Fermi surface, $\epsilon_{\mathbf{p}} \to 0$ and $d\epsilon_{\mathbf{p}}/d\varepsilon_{\mathbf{p}} \to \infty$. However, over the range of integration, $\varepsilon_{\mathbf{p}}/\epsilon_{\mathbf{p}}$ is finite. In the limit that $\hbar\omega_{\mathbf{q}} \ll \Delta$, $\varepsilon_{\mathbf{p}'} \approx \varepsilon_{\mathbf{p}}$. Consequently,

$$\frac{\varepsilon_{\mathbf{p}}}{\epsilon_{\mathbf{p}}} \frac{\varepsilon_{\mathbf{p}'}}{\epsilon_{\mathbf{p}'}} \left(1 - \frac{\Delta^2}{\varepsilon_{\mathbf{p}}\varepsilon_{\mathbf{p}'}}\right) \to \frac{\varepsilon_{\mathbf{p}}^2}{\epsilon_{\mathbf{p}}^2} \left(\frac{\varepsilon_{\mathbf{p}}^2 - \Delta^2}{\varepsilon_{\mathbf{p}}^2}\right) = 1. \quad (11.244)$$

It is certainly valid to work within this limit, because we are considering phonon absorption as a result of single quasi-particle excitations only. Evaluating the remaining integrand at $\varepsilon_{\mathbf{p}'} = \varepsilon_{\mathbf{p}} + \hbar\omega_{\mathbf{q}}$, we find that

$$\alpha_{\mathbf{q}} = \frac{8\pi}{\hbar}|M_{\mathbf{q}}|^2 g^2(0) A_\Omega \int_\Delta^\infty d\varepsilon_{\mathbf{p}}(n(\varepsilon_{\mathbf{p}} + \hbar\omega_{\mathbf{q}}) - n(\varepsilon_{\mathbf{p}}))$$

$$\approx -\frac{8\pi}{\hbar}|M_{\mathbf{q}}|^2 g^2(0) A_\Omega \hbar\omega_{\mathbf{q}} n(\Delta). \quad (11.245)$$

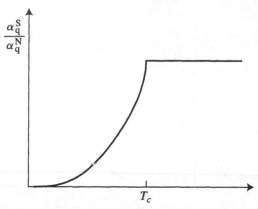

Figure 11.20 Ratio of the ultrasound attenuation in the superconducting state to that in the normal state, as predicted by Eq. (11.247). The fall-off of this ratio in the superconducting state is a signature of Cooper pair formation.

In the normal state, $\Delta = 0$ and, resultantly,

$$\alpha_{\mathbf{q}}^{N} = -\frac{8\pi}{\hbar}|M_{\mathbf{q}}|^2 g^2(0) A_{\Omega} \hbar \omega_{\mathbf{q}} n(0)$$

$$= -\frac{4\pi}{\hbar}|M_{\mathbf{q}}|^2 g^2(0) A_{\Omega} \hbar \omega_{\mathbf{q}}. \tag{11.246}$$

The ratio of $\alpha_{\mathbf{q}}^{s}/\alpha_{\mathbf{q}}^{N}$ reduces to

$$\frac{\alpha_{\mathbf{q}}^{s}}{\alpha_{\mathbf{q}}^{N}} = \frac{2}{e^{\beta\Delta(T)} + 1}. \tag{11.247}$$

At $T = T_c$, $\Delta = 0$, and this ratio yields unity, as illustrated in Fig. (11.20). The calculation of the ultrasonic attenuation rate is in excellent agreement with experimental results [S1964]. Here again, we see that it is the gap that ultimately determines the absorption of sound waves in the superconducting states. This calculation is valid, of course, for longitudinal phonons only. Unlike the longitudinal case in which $M_{\mathbf{q}}$ is the same in the normal as well as in the superconducting state, $M_{\mathbf{q}}$ for transverse phonons changes drastically from the normal to the superconducting state. In this case, the magnetic dependence of $M_{\mathbf{q}}$ leads to a discontinuous drop in the attenuation rate at $T = T_c$, as depicted in Fig. (11.21) [S1964].

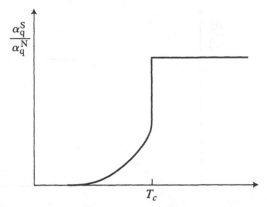

Figure 11.21 Ratio of the superconducting and normal state ultrasound absorption rates for transverse phonons as a function of temperature T.

11.14.3 Spin-Lattice Relaxation

The final calculation we present using the BCS coherence factors is that of the spin-lattice relaxation time, T_1. For historical reasons, this is a key calculation because the prediction that a peak should occur in $1/T_1$ just below T_c predated the experiment by Hebel and Slichter [HS1959]. In fact, experimental confirmation of the enhancement of $1/T_1$ just below T_c was the crowning evidence that led to the wide acceptance of the BCS theory. To define T_1, we consider applying a magnetic field to a collection of nuclear spins. Once the field is turned off, the spins will relax back to achieve the equilibrium magnetization. The time for this process to occur is referred to as T_1. In insulators, it is the coupling between the nuclei and electronic impurities that causes the nuclei to relax. Left alone, nuclei cannot achieve an equilibrium spin temperature. The lattice temperature must be communicated to the nuclei by some agent. In a metal, it is the conduction electrons that provide the coupling. Because the electrons in a superconductor are locked into a coherent state, we expect T_1 to be drastically altered. Consider a nucleus with magnetic moment $\mathbf{S}_N$ located at the origin, $\mathbf{r} = 0$. The net electron spin at $\mathbf{r} = 0$ is given by the local electron spin density, $\mathbf{S} = \hbar\Psi^{\dagger}(0)\sigma\Psi(0)/2$, at the origin where $\Psi(0)$ is the electron wave function at $\mathbf{r} = 0$ and σ are the Pauli spin matrices. The Fermi contact,

$$H_{\text{int}} = \frac{8\pi}{3}\mu_B \mathbf{S}_N \cdot \mathbf{S}, \qquad (11.248)$$

defines the coupling between a nucleus and an electron at $\mathbf{r} = 0$. In Eq. (11.248), μ_B is the Bohr magneton.

To apply this coupling to a calculation of T_1 in a superconductor, we simply need to express $\mathbf{S}$ in terms of the quasi-particle operators. To do this, we introduce the spinor field operators

$$\Psi(\mathbf{r}) = \frac{1}{\sqrt{V}} \sum_{\mathbf{p}} e^{i\mathbf{p}\cdot\mathbf{r}} \begin{pmatrix} a_{\mathbf{p}\uparrow} \\ a_{\mathbf{p}\downarrow} \end{pmatrix}. \tag{11.249}$$

In spinor notation, the local electron spin density at the origin becomes

$$\mathbf{S} = \frac{\hbar}{2V} \sum_{\mathbf{p},\mathbf{p}'} (a^\dagger_{\mathbf{p}'\uparrow} a^\dagger_{\mathbf{p}'\downarrow}) \sigma \begin{pmatrix} a_{\mathbf{p}\uparrow} \\ a_{\mathbf{p}\downarrow} \end{pmatrix}. \tag{11.250}$$

We now express the Fermi contact interaction in terms of $\Psi(\mathbf{r})$. Because the product $\mathbf{S}_1 \cdot \mathbf{S}_2 = S_1^z S_2^z + 2(S_1^+ S_2^- + S_1^- S_2^+)$, we write the Fermi contact,

$$H_{\text{int}} = \frac{4\pi}{3V} S_B' \sum_{\mathbf{p},\mathbf{p}'} S_N^z (a^\dagger_{\mathbf{p}'\uparrow} a_{\mathbf{p}\uparrow} - a^\dagger_{\mathbf{p}'\downarrow} a_{\mathbf{p}\downarrow}) + 2S_N^+ a^\dagger_{\mathbf{p}'\downarrow} a_{\mathbf{p}\uparrow} + 2S_N^- a^\dagger_{\mathbf{p}'\uparrow} a_{\mathbf{p}\downarrow}, \tag{11.251}$$

in the familiar Kondo form, where $\mu_B' = \mu_B / \hbar$.

We are particularly interested in those terms in which the nuclear spin is flipped. Hence, we can focus on either the S_N^- or S_N^+ term. Consider S_N^+. Let H_{int}^+ be that part of the interaction in which the nuclear spin is flipped up. Here again, we introduce the quasi particle representation of

$$a^\dagger_{\mathbf{p}'\downarrow} = u_{-\mathbf{p}'} \gamma^\dagger_{\mathbf{p}'\downarrow} - v_{-\mathbf{p}'} \gamma_{-\mathbf{p}'\uparrow} \tag{11.252}$$

and

$$a_{\mathbf{p}\uparrow} = u_{\mathbf{p}} \gamma_{\mathbf{p}\uparrow} + v_{\mathbf{p}} \gamma^\dagger_{-\mathbf{p}\downarrow}, \tag{11.253}$$

so that H_{int}^+ is transformed to

$$H_{\text{int}}^+ = \frac{4\pi}{3V} \mu_B' S_N^+ \sum_{\mathbf{p},\mathbf{p}'} (u_{-\mathbf{p}'} u_{\mathbf{p}} \gamma^\dagger_{\mathbf{p}'\downarrow} \gamma_{\mathbf{p}\uparrow} - v_{-\mathbf{p}'} v_{\mathbf{p}} \gamma_{-\mathbf{p}'\uparrow} \gamma^\dagger_{-\mathbf{p}\downarrow}$$

$$- u_{\mathbf{p}} v_{-\mathbf{p}'} \gamma_{-\mathbf{p}'\uparrow} \gamma_{\mathbf{p}\uparrow} + u_{-\mathbf{p}'} v_{\mathbf{p}} \gamma^\dagger_{\mathbf{p}'\downarrow} \gamma^\dagger_{-\mathbf{p}\downarrow}). \tag{11.254}$$

The first two terms describe processes in which quasi particles of opposite spin are created and annihilated, whereas the latter terms contain quasi-particle pair production or annihilation. Typically, nuclear Zeeman energies are small relative to the quasi-particle gap energy. Hence, if we focus on the frequency range, $\hbar\omega \approx \mu_B H$, H the external magnetic field, only the former terms are relevant. We have reduced our effective interaction to

$$H_{\text{int}}^+ \simeq \frac{4\pi}{3V}\mu_B' S_N^+ \sum_{\mathbf{p},\mathbf{p}'}(u_{-\mathbf{p}'}u_{\mathbf{p}}\gamma_{\mathbf{p}'\downarrow}^\dagger\gamma_{\mathbf{p}\uparrow} + v_{-\mathbf{p}'}v_{\mathbf{p}}\gamma_{-\mathbf{p}\downarrow}^\dagger\gamma_{-\mathbf{p}'\uparrow})$$

$$(11.255)$$

Matrix elements of this quantity will be nonzero if the initial state is of the form $|\mathbf{p}\uparrow, S_N^z\rangle$ and the final state is $|\mathbf{p}\downarrow, S_N^z + 1\rangle$. Let $\alpha_N = \langle S_N^z + 1|S_N^+|S_N^z\rangle$. The resultant matrix element

$$\Gamma_{\text{SLR}}^{\mathbf{p}\mathbf{p}'} = \frac{8\pi}{3}\mu_B'\alpha_N(u_{-\mathbf{p}'}u_{\mathbf{p}}\sqrt{(1-n_{\mathbf{p}'\downarrow})n_{\mathbf{p}\uparrow}}$$

$$+ v_{-\mathbf{p}'}v_{\mathbf{p}}\sqrt{(1-n_{-\mathbf{p}\downarrow})n_{-\mathbf{p}'\uparrow}})$$

$$= \frac{4\pi}{3}\mu_B'\alpha_N(u_{-\mathbf{p}'}u_{\mathbf{p}} + v_{-\mathbf{p}'}v_{\mathbf{p}})\sqrt{(1-n_{\mathbf{p}'\downarrow})n_{\mathbf{p}\uparrow}} \quad (11.256)$$

leads to the nuclear-spin lattice-relaxation rate

$$T_1^{-1} = \frac{32\pi^3}{9\hbar}S_B^2|\alpha_N|^2\sum_{\mathbf{p},\mathbf{p}'}(u_{-\mathbf{p}'}u_{\mathbf{p}} + v_{-\mathbf{p}'}v_{\mathbf{p}})^2(1-n_{\mathbf{p}'\downarrow})n_{\mathbf{p}\uparrow}\delta(\varepsilon_{\mathbf{p}'} - \varepsilon_{\mathbf{p}}).$$

$$(11.257)$$

At this level of theory, we are ignoring, of course, the Kondo effect. The Kondo regime of a magnetic impurity in a superconductor is a subtle problem, indeed, which we will not consider here.

The procedure from here on out is now standard: 1) convert the sum to an integral by introducing the density of states, 2) evaluate the coherence factors, and 3) cancel all integrals linear in the bare energy, ϵ. From the ultrasonic attenuation calculation, we have that

$$(u_{\mathbf{p}'}u_{\mathbf{p}} + v_{\mathbf{p}}v_{\mathbf{p}'})^2 = \frac{1}{2\varepsilon_{\mathbf{p}}\varepsilon_{\mathbf{p}'}}(\varepsilon_{\mathbf{p}}\varepsilon_{\mathbf{p}'} + \epsilon_{\mathbf{p}}\epsilon_{\mathbf{p}'} + \Delta_{\mathbf{p}}\Delta_{\mathbf{p}'}). \quad (11.258)$$

Here again, we assume that the gap is constant and that the relaxation rate reduces to

$$T_1^{-1} = \frac{64\pi^3}{9\hbar} \mu_B'^2 |\alpha_N|^2 g^2(0) \int_0^\infty d\epsilon_{\mathbf{p}} \int_0^\infty d\epsilon_{\mathbf{p}'} n_{\mathbf{p}} (1 - n_{\mathbf{p}'})$$

$$\times \left(\frac{\varepsilon_{\mathbf{p}} \varepsilon_{\mathbf{p}'} + \Delta^2}{\varepsilon_{\mathbf{p}} \varepsilon_{\mathbf{p}'}} \right) \delta(\varepsilon_{\mathbf{p}'} - \varepsilon_{\mathbf{p}})$$

$$= \frac{64\pi^3}{9\hbar} \mu_B'^2 |\alpha_N|^2 g^2(0) \int_\Delta^\infty d\varepsilon n(\varepsilon)(1 - n(\varepsilon)) \frac{\varepsilon^2 + \Delta^2}{\varepsilon^2 + \Delta^2}. \qquad (11.259)$$

At $T = 0$, the product $n(1 - n) = 0$ because $n = 1$ at $T = 0$.

At $T = 0$, there is a divergence when $\varepsilon = \pm \Delta$. In the vicinity of $\pm \Delta$, we write the integrand as

$$T_1^{-1} \rightarrow \frac{64\pi^3}{9\hbar} \mu_B'^2 |\alpha_N|^2 g^2(0) \int_\Delta^\infty \frac{\varepsilon^2 + \Delta^2}{\varepsilon^2 - \Delta^2} n(\varepsilon)(1 - n(\varepsilon)) d\varepsilon$$

$$\propto \int_\Delta^\infty \frac{\varepsilon + \Delta}{\varepsilon - \Delta} d\varepsilon, \qquad (11.260)$$

which is logarithmically divergent. The coefficient of the ln-divergence is proportional to the gap, Δ. This implies, then, that the ln-divergence starts at $T = T_c$ where the gap begins to arise. In experiments, the gap is anisotropic as a result of its momentum dependence. Consequently, the divergence of T_1 is smeared out slightly over a range comparable to the average value of Δ. Hebel and Slichter [HS1959] were the first to observe this behavior, which is illustrated in Fig. (11.7).

The ln-divergence we have predicted at this low order in perturbation theory might be removed or at least broadened by the summation of all higher-order terms. Hence, it is worthwhile to investigate precisely how robust the ln-divergence is. The full series for the electron-nucleus interaction is

$$\Gamma_{\text{int}}^{\text{full}} = \langle f | H_{\text{int}}^+ | i \rangle + \sum_{\mathbf{p}} \frac{\langle f | H_{\text{int}}^+ | \mathbf{p} \rangle \langle \mathbf{p} | H_{\text{int}}^+ | i \rangle}{E - \varepsilon_{\mathbf{p}} + i\eta} + 0(H^3) + \cdots . \qquad (11.261)$$

In the context of the local-moment problem, we defined the Green function to be

$$G(E + i\eta) = \sum_{\mathbf{p}} \frac{|\mathbf{p}\rangle\langle \mathbf{p}|}{E - \varepsilon_{\mathbf{p}} + i\eta}. \qquad (11.262)$$

Schematically, the perturbation series is of the form

$$\Gamma_{\text{int}}^{\text{full}} = \langle f|H_{\text{int}}^+|i\rangle + \langle f|H_{\text{int}}^+ G\,(E+i\eta)\,H_{\text{int}}^+|i\rangle + \cdots$$

$$= \langle f|\frac{H_{\text{int}}^+}{1 - H_{\text{int}}^+ G\,(E+i\eta)}|i\rangle. \tag{11.263}$$

To determine the divergent part of $\Gamma_{\text{int}}^{\text{full}}$, we convert the sum in the Green function to an integral,

$$G\,(E+i\eta) = N\,(0)\int d\epsilon_{\mathbf{p}}\frac{1}{E - \varepsilon_{\mathbf{p}} + i\eta}$$

$$= N\,(0)\int_{\Delta}^{\infty}\left(\frac{d\epsilon_{\mathbf{p}}}{d\,\varepsilon_{\mathbf{p}}}\right)\frac{d\,\varepsilon_{\mathbf{p}}}{E - \varepsilon_{\mathbf{p}} + i\eta}. \tag{11.264}$$

We have shown previously that $N\,(0)d\epsilon_{\mathbf{p}}/d\,\varepsilon_{\mathbf{p}}$ defines the density of states

$$\rho(\varepsilon_{\mathbf{p}}) = \frac{N\,(0)\varepsilon_{\mathbf{p}}}{\sqrt{\varepsilon_{\mathbf{p}}^2 - \Delta_{\mathbf{p}}^2}} = N\,(0)\frac{\varepsilon_{\mathbf{p}}}{\epsilon_{\mathbf{p}}}, \tag{11.265}$$

which is divergent at $\varepsilon_{\mathbf{p}} = \Delta_{\mathbf{p}}$. Consequently,

$$\lim_{\eta\to 0} G\,(E+i\eta) = \int_{\Delta}^{\infty}\rho(\varepsilon_{\mathbf{p}})\left[P\left(\frac{1}{E - \varepsilon_{\mathbf{p}}}\right) - i\pi\delta(E - \varepsilon_{\mathbf{p}})\right]$$

$$= ReG - i\pi\rho\,(E)\,, \tag{11.266}$$

and the perturbation series becomes

$$\Gamma_{\text{int}}^{\text{full}} = \langle f|\frac{H_{\text{int}}^+}{(1 - H_{\text{int}}^+ ReG\,) + i\pi H_{\text{int}}\rho(E)}|i\rangle. \tag{11.267}$$

In the vicinity of $E = \Delta$, the denominator of Eq. (11.267) is dominated by the $\rho\,(E)$ term. In this limit, $\Gamma_{\text{int}}^{\text{full}} \to 1/\,(i\pi\rho\,(E))$. The square of this quantity exactly cancels the divergent term, $1/(\varepsilon_{\mathbf{p}}^2 - \Delta^2)$, in our previous expression for the relaxation rate.

To see this more clearly, we rewrite the relaxation rate as

$$T_1^{-1} = \frac{64\pi^3}{9\hbar}\mu_B'^2|\alpha_N|^2 g^2(0)\int_{\Delta}^{\infty}d\varepsilon\,n(\varepsilon)\,(1 - n\,(\varepsilon))\left(\frac{\rho\,(\varepsilon)}{\varepsilon}\right)^2$$

$$\times \frac{(\varepsilon^2 + \Delta^2)}{(1 - H_{\text{int}}^+ ReG\,)^2 + \pi^2|H_{\text{int}}^+|^2\rho^2\,(\varepsilon)}, \tag{11.268}$$

which reduces to

$$T_1^{-1} \propto \int_\Delta^\infty d\varepsilon \, (1 - n(\varepsilon)) \, n(\varepsilon) \frac{\varepsilon^2 + \Delta^2}{(1 + \pi^2 |H_{int}^+|^2 g^2(0)) \varepsilon^2 - \Delta^2}. \qquad (11.269)$$

In deriving Eq. (11.269), we ignored the real part of the Green function, as it serves no relevant purpose as far as the convergence is concerned. This expression is completely convergent at $\varepsilon^2 = \Delta^2$. In fact, because $(1 + \pi^2 |H_{int}|^2 g^2(0)) > 1$, the integral does not diverge over the complete integration range. Consequently, summing high-order terms in the perturbation series results in a smearing of the peak in the relaxation rate immediately below T_c.

11.15   JOSEPHSON TUNNELING

Consider two superconductors separated from one another by a thin insulating barrier. Naively, we would expect no appreciable transport of charge between the two superconductors in the absence of an applied voltage, save possibly for single quasi-particle tunneling through the insulating barrier. Josephson [J1962] showed that this naive picture is not correct. In particular, he proved that, in the absence of an applied voltage for a sufficiently thin barrier, Cooper pairs flow coherently between the two superconductors, thereby establishing a supercurrent through the barrier. Further, the transport of Cooper pairs across the barrier does not result in the creation of quasi particles in either superconductor. When an applied voltage is present, the supercurrent oscillates with a well-defined period. The essence of both of these effects, dc and ac Josephson tunneling, rests in the phase coherence that obtains in the superconducting state.

We focus first on the dc Josephson effect. Consider two superconductors separated by a thin insulating barrier. Let H_T represent the Hamiltonian for single particle tunneling across the thin barrier. The specific form of this term is not essential here. The only important feature is that H_T transfers only one electron at a time. We assume at the outset that there is no voltage difference between the two superconductors, and, hence, they are at the same chemical potential. We can derive the Josephson effect by making an analogy with electron transport in a 1d periodic chain in the tight-binding approximation. In this approximation, a single orbital is placed on each lattice site and a hopping term mediates transport among nearest-neighbor sites. In such a system, no energy is required to transport an electron across m lattice sites. Likewise, it requires no energy to translate m Cooper pairs across the barrier in the absence of an external voltage differential between the two superconductors. Let $|\Phi_{2m}\rangle$ represent the many-body state, when $2m$ Cooper pairs are transferred across the barrier. The degeneracy of these states is split by the single-electron tunneling term, H_T. Consequently, we expand the total state of our system as a linear combination

$$|\Psi_\phi\rangle \equiv |\phi\rangle = \sum_m e^{2im\phi}|\Phi_{2m}\rangle \qquad (11.270)$$

over all such pair states. The phase ϕ plays the role of the wave vector k in the 1d periodic tight-binding model. As we have established earlier, the particle number, $2m$, and the phase, ϕ, are conjugate variables.

To compute the energy shift as a result of the tunneling processes, we employ perturbation theory. The first-order term, $\langle\phi|H_T|\phi\rangle$, vanishes identically because ϕ is a sum of all pair states and H_T is a one-body operator. Consequently, the first nonzero term appears in second order. Let

$$\widehat{H}_T^{(2)} = \widehat{H}_T \frac{|I\rangle\langle I|}{E - E_I}\widehat{H}_T \qquad (11.271)$$

represent the tunneling operator at second order with E_I the energy of the intermediate state, $|I\rangle$. The second-order correction to the energy

$$
\begin{aligned}
E_\phi &= \langle\phi|\widehat{H}_T^{(2)}|\phi\rangle \\
&= \sum_{m,m'} e^{2i\phi(m-m')}\langle 2m'|\widehat{H}_T^{(2)}|2m\rangle \\
&= \sum_m e^{2i\phi m}(e^{2i\phi}\langle 2m|\widehat{H}_T^{(2)}|2(m+1)\rangle + e^{-2i\phi}\langle 2m|\widehat{H}_T^{(2)}|2(m-1)\rangle)
\end{aligned}
$$

$$(11.272)$$

is a sum of all matrix elements that differ by a single Cooper pair. We have assumed that $\langle\phi|\phi\rangle = 1$. To simplify this expression, we note that the energy of the intermediate state involves a particle-hole excitation, and, hence, E_I must exceed E by at least 2Δ. Consequently, $E - E_I < 0$. If we regard the tunneling term to be purely real, we simplify the energy shift to

$$E_\phi = -\frac{\hbar J_0}{2}\cos 2\phi, \qquad (11.273)$$

with

$$\hbar J_0 = 4 \sum_m |\langle 2m|\widehat{H}_T^{(2)}|2(m+1)\rangle|. \qquad (11.274)$$

The minus sign in the energy shift arises from the sign of the excitation energy. From the Hamilton equation, Eq. (11.121), it is clear that if the energy shift depends on the phase, then the pair number fluctuates on either side of the barrier. This fluctuation is due entirely to the tunneling processes. We calculate the pair current directly,

$$I = 2e\frac{d\langle 2m \rangle}{dt} = 2e\left\langle \frac{dE_\phi}{d\hbar\phi} \right\rangle = 2eJ_0 \sin\phi, \qquad (11.275)$$

by differentiating the energy shift with respect to the phase. Consequently, in the absence of an applied voltage, a dc supercurrent flows across the barrier. The value of the current ranges from $-2eJ_0$ to $2eJ_0$. A supercurrent of this form was first observed by Anderson and Rowell [AR1963]. If a potential difference V exists across the barrier, then a term of the form $2mV$ must be added to the Hamiltonian. Consequently, from Hamilton's equations, Eq. (11.121), the phase fluctuates in time according to

$$\frac{d\langle \hbar\phi \rangle}{dt} = 2eV. \qquad (11.276)$$

Together, these two equations, Eq. (11.275) and Eq. (11.276), completely determine the behavior of the supercurrent across the barrier. To illustrate, consider the simplest case in which the voltage V is a constant in time. In this case, the phase ϕ varies linearly with time and, as a consequence, the current oscillates as $\sin(2eVt/\hbar)$. Hence, an alternating current flows with a frequency of $2eV/\hbar$.

11.16 SUMMARY

We have shown that the pairing hypothesis of BCS is sufficient to account for all relevant experimental observables of low-temperature superconductors. In fact, the BCS pairing mechanism is the only account available currently to describe the transition to a superconducting state. In contrast to low-T_c materials, superconductivity in the cuprates originates from doping an insulator. Further, the insulator possesses a partially filled band and, hence, falls into the class of Mott insulators in which an absence of transport originates from strong electron repulsions. Consequently, we know *a priori* that we are not justified in starting from Fermi-liquid theory to describe even the normal state properties. Simply stated, the deep phenomenology of the cuprates lies in the physics of doped Mott insulators. Whether a theory as succinct and crystal clear as the BCS account can be formulated for such systems remains to be seen.

PROBLEMS

1. Within the Ginsburg-Landau phenomenological approach, determine the form of the free-energy density when a magnetic field is present. Show that the free-energy difference between the superconducting and normal states is given by

$$F_S - F_N = -\frac{H_c^2(T)}{8\pi}. \tag{11.277}$$

2. Writing the Ginsburg-Landau wave function as $\psi(\mathbf{r}) = \sqrt{n(\mathbf{r})}e^{i\theta(\mathbf{r})}$, show that the current density in terms of the variables θ and $n(\mathbf{r})$ is given by

$$\mathbf{j} = \frac{e\hbar}{m}\left(\nabla\theta - \frac{e\mathbf{A}}{c\hbar}\right)n(\mathbf{r}). \tag{11.278}$$

Now assume that, in the bulk of a material, the current density vanishes. As a consequence, $\hbar\nabla\theta = e\mathbf{A}$. Integrate both sides of this expression around a closed loop in a superconducting ring and show that the resultant magnetic flux enclosed is quantized. What is the correct value of e for a superconductor?

3. Use second-order perturbation theory directly to show that the electron-phonon interaction is negative and given by the second term in Eq. (11.34).

4. Redo the Cooper pair instability calculation for triplet pairing between the electrons.

5. Evaluate $\langle r^2 \rangle$ for a singlet Cooper pair.

6. In the problem of the instability of the superconducting state in the presence of the BCS pairing interaction, determine the form of the growth rate of the pair amplitude as $T \rightarrow T_c$.

7. Evaluate the commutator $[b_k, b_k^\dagger]$, where the b_k's are the Cooper pair annihilation operators. What does the lack of commutativity of the Cooper pair creation and annihilation operators mean?

8. Calculate the average number of particles in a superconductor. Let $|\Psi\rangle$ represent the BCS pair state. Show that the average value

of the number operator, N, is given by

$$\langle \Psi | N | \Psi \rangle = \langle \Psi | \sum_{k,\sigma} a_{k\sigma}^{\dagger} a_{k\sigma} | \Psi \rangle = 2 \sum_{k} |v_k|^2 \qquad (11.279)$$

in the pair state. Also evaluate the fluctuation $\langle (N - \langle N \rangle)^2 \rangle$. You should obtain a simple result involving u_k and v_k only. For what special value of u_k and v_k is the fluctuation maximized? Interpret your result.

9. So far we have ignored any spatial inhomogeneities in the gap. Consider a gap of the form $\Delta_{\mathbf{q}} = \Delta_0 e^{2i\,\mathbf{q}\cdot\mathbf{r}}$, where $q \ll p_F$. Find the new self-consistent condition for Δ_0. At $T = 0$, show that Δ is independent of q for $q < q_c \approx \Delta_0/\hbar v_F$. Near T_c, expand the gap equation to find that

$$\frac{\Delta(T)}{k_B T_c} \approx \frac{8\pi^2}{7\zeta(3)}(1 - T/T_c) - \frac{2}{3}\left(\frac{\hbar^2 p_F}{m k_B T_c}\right)^2 q^2. \qquad (11.280)$$

Then determine the critical value of q that makes the gap vanish.

10. Evaluate the sums explicitly in Eq. (11.210) and show that for $T/T_c \ll 1$, $F_N - F_S \propto 1 - (T/T_c)^2$.

11. An Anderson-type impurity is placed in a superconductor. You are to formulate this problem and develop a criterion for local moment formation. There are a number of assumptions that can be applied. First, when you transform to the quasi-particle basis, ignore all terms that do not conserve spin and particle number. The problem should now be straightforward. You should be able to redo the Anderson problem completely. Discuss clearly when the local moment exists and when it does not.

REFERENCES

[A1957] A. A. Abrikosov, *Sou. Phys. JETP*, **5**, 1174 (1957).

[AR1963] P. W. Anderson, J. M. Rowell, *Phys. Reu. Lett.* **10**, 230 (1963).

[BCS1957] J. Bardeen, L. N. Cooper, J. R. Schrieffer, *Phys. Rev.* **106**, 162 (1957); **108**, 1175 (1957).

[C1956] L. N. Cooper, *Phys. Rev.* **104**, 1189 (1956).

[FP1963] R. A. Ferrell, R. E. Prange, *Phys. Reu. Lett.* **10**, 479 (1963).

[GL1950] G. V. L. Ginsburg, L. D. Landau, *J. Exptl. Theor. Phys.* (USSR), **20**, 1064 (1950).

[GF1987] M. Gurvitch, A. T. Fiory, *Phys. Rev. Lett.* **59**, 1337 (1987).

[HS1959] L. C. Hebel, C. P. Slichter, *Phys. Rev.* **113**, 1504 (1959); L. C. Hebel, *Phys. Rev.* **116**, 79 (1959).

[J1962] B. D. Josephson, *Physics Letters*, **1**, 251 (1962).

[K1962] L. P. Kadanoff and G. Baym, *Quantum Statistical Mechanics* (Benjamin, New York, 1962) pp. 187–190.

[L1956] L. D. Landau, *Sov. Phys. JETP* **3**, 920 (1956); ibid. **8**, 70 (1959).

[L1964] A. I. Larkin, *Sov. Phys. JETP* **19**, 1478 (1964).

[L1965] A. J. Leggett, *Phys. Rev. Ser. A* **140**, 1869 (1965).

[S1964] J. R. Schrieffer, *Theory of Superconductivity* (Benjamin, New York, 1964).

– 12 –

Localization: The Strong, the Weak, and the Defiant

A problem that has always been central to solid state physics is the insulator-metal transition. The question to answer here is, why do some materials conduct and others do not? Metals are characterized by a nonzero dc conductivity $\sigma(0)$ at zero temperature, whereas $\sigma(0) = 0$ in an insulator. There are currently three standard models that describe a transition between these two extremes. Anderson [A1958] was first to point out that scattering from a static, but random, potential can disrupt metallic conduction and lead to an abrupt localization of the electronic eigenstates. Mott [M1949], in contrast, proposed that an insulating state can obtain, even in a material such as NiO, which possesses a half-filled valence band. The insulating state arises from strong electron correlations that induce a gap at the Fermi energy. The closing of the gap, signalling the onset of a metallic state, results typically in the intermediate coupling regime in which the kinetic energy leads to a quasi-particle peak at the Fermi level. Finally, a structural transition in which the lattice periodicity doubles can also thwart metallic transport. While all of these mechanisms are of considerable interest in their own right, our focus in this chapter will be the disorder-driven insulator-metal, or Anderson, transition.

We start by reviewing the essential physics and some of the key controversies surrounding the disorder-induced localization transition. Two controversies we address are the role of perturbation theory and whether the conductivity is continuous in the vicinity of the Anderson transition. To address the latter question, we develop the scaling theory of localization. The two key predictions of this approach are that the localization transition is continuous and that an Anderson transition exists only for $d > 2$. That is, any amount of disorder localizes all the electronic states in d = 1 and d = 2. Finally, the weak-localization analysis makes it profoundly clear that the Anderson transition occurs precisely in the strong-disorder limit in which perturbation theory breaks down. Consequently, no amount of perturbation theory can be harnessed to describe the disorder-induced localization transition. Nonetheless, the essential physical process in weak localization—time-reversed back scattering—appears to be the mechanism behind the Anderson transition. However, even when time-reversal symmetry is intact, there are important exceptions to the standard localization scenario in d ≤ 2. We close this chapter by considering all three known examples: 1) the general class of 1-dimensional models that defeat both strong and weak localization, namely, the random dimer model, 2) insulator-superconductor transitions in thin films, and 3) the newly discovered conducting state in a dilute 2d electron gas.

12.1 PRIMER ON LOCALIZATION

Anderson's original model grew out of a series of electron spin resonance experiments by Feher and collaborators [FFG1955] on donor impurities, such as P and As in Si. They noticed that the electron spin on each ^{31}P nucleus retained its characteristic frequency on a timescale ranging from seconds to minutes. A simple golden rule calculation (of the type performed in the context of the Kondo problem, but in this case it is sufficient to retain only the $O(J^2)$-term) however, predicted that the expected lifetime as a result of interaction with the other impurities would range from 0.1 to $10^{-6}s$ [A1958]. The persistence of localized spin packets in the Feher experiments indicated an absence of spin diffusion among the ^{31}P impurity spins. As the host Si was otherwise extremely pure, Anderson traced the absence of spin diffusion to disorder arising from the random distribution of the dopant impurities. To capture the essence of quantum mechanical transport in the presence of a random potential, Anderson proposed the tight-binding model

$$ H = \sum_n \epsilon_n a_n^\dagger a_n + V \sum_{n,m} a_n^\dagger a_m \qquad (12.1) $$

for conduction in an impurity band in which one orbital and a single site energy ϵ_n are assigned at random to the lattice sites. A constant nearest-neighbor matrix element V mediates transport between nearest-neighbor sites. In Eq. (12.1), $a_n^\dagger(a_n)$ creates (annihilates) an electron on site n. In general, V depends on distance, such as in the spin diffusion problem in which dipolar interactions give rise to a matrix element that decays as $V \propto 1/r^3$. However, including randomness in the matrix elements does not qualitatively change the nature of the transport problem.

The site-disordered tight-binding model in Eq. (12.1) describes a localization-delocalization transition, as can be seen by considering two simple limits. We assume initially that the site energies are chosen from a uniform distribution of width W. When $W = 0$, an ordered system obtains, as all the sites have the same energy. The resultant eigenstates are delocalized Bloch states that remain unscattered over the size of the sample. Transport in this regime is ballistic. However, in the limit that $V = 0$, none of the sites are connected and transport ceases. That is, the resultant eigenstates are localized. Hence, the Anderson model describes a localization-delocalization transition that is governed, at least partially, by the ratio V/W. On physical grounds, it is tempting to argue that in all cases, gradually increasing the ratio V/W will lead to a smooth interpolation between the limit of extreme localization, $V = 0$, and the ballistic regime, $V \gg W$. We know now from the scaling theory of localization, as well as from the early work of Mott and Twose [MT1961] and Borland [B1963], that this is not the case. For d $\leq$ 2, it is now well established (except in some special d $= 1$ cases [DWP1990]) that an infinitesimal amount of disorder precludes the existence of extended states. That is, the smooth interpolation between the ballistic and the localized regimes does not obtain for d $= 1, 2$. In three dimensions, extended states fail to form when W/V exceeds some critical value, $(W/V)_c$. When $W/V < (W/V)_c$, extended and localized states coexist in the energy band but are separated at an energy now known as the *mobility edge*, E_c, such that if the energy E of a particle exceeds E_c, then the particle is extended. Localization obtains in the opposite regime, $E < E_c$.

A natural question that arises in the context of the localization transition is, what happens to the conductivity in the vicinity of the mobility edge? There are two possibilities. Either the conductivity goes to zero continuously as the mobility edge is approached from the metallic side or, as Mott [M1972] proposed, σ decreases to some minimum value and then plummets to zero discontinuously. Both of these possibilities are shown in Fig. (12.1). The Mott minimal conductivity has proven to be one of the most controversial, but wrong, ideas proposed for the localization transition. The intuitive appeal of this idea is immediate,

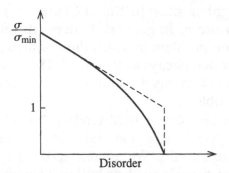

Figure 12.1 Two possibilities for the behavior of the conductivity in the vicinity of the mobility edge of the Anderson transition. The dashed line represents the minimum-metallic conductivity hypothesis of Mott. The solid line, in contrast, depicts the continuous decrease of the conductivity at the Anderson transition. The latter is observed experimentally and predicted by the scaling theory of localization.

however, when one rewrites the Drude dc conductivity,

$$\sigma(0) = \frac{e^2 n_e \tau}{m} \approx \left(\frac{e^2}{\hbar}\right)\frac{\mathbf{a}^{2-\mathrm{d}}}{\pi}\frac{\ell}{\mathbf{a}} = \sigma_0\frac{\ell}{\mathbf{a}}, \qquad (12.2)$$

in terms of the mean-free path, $\ell = v_F \tau$. We have approximated the electron density of a d-dimensional system as $n_e = \mathbf{a}^{-\mathrm{d}}$, where $\mathbf{a}$ is the lattice spacing. Disorder enters through the ratio $\ell/\mathbf{a}$. Clearly, if ℓ has a minimum value, then so will $\sigma(0)$. Within Boltzmann transport theory, ℓ cannot be smaller than the lattice spacing. Consequently, Mott [M1949] postulated that a minimum-metallic conductivity of the form $\sigma_{\min} = b\sigma_0$ must exist. The constant σ_0 is universal, whereas b is generally taken to be between 0.08 and 0.3. The minimum-metallic conductivity hypothesis is that $\sigma(0)$ should decrease as ℓ and, hence, should plummet to zero once $\ell = \mathbf{a}$, as depicted in Fig. (12.1). Experimentally, this prediction has not been borne out, as there are now numerous examples in which conductivities lower than $\sigma_{\min}$ have been measured in the vicinity of the localization transition. Further, the scaling theory of localization predicts a continuous transition. Both of these results completely invalidate the $\sigma_{\min}$ hypothesis. As we will see, the Anderson transition cannot be described within Boltzmann transport theory. Hence, the minimum value that Boltzmann theory imposes on the mean-free path is not correct.

12.2 STRONG LOCALIZATION: THE ANDERSON TRANSITION

The least a theory of Anderson localization should provide is an accurate method for distinguishing between localized and extended states. To do so, we focus on the return probability. Consider placing a particle at the origin of a site-disordered d-dimensional lattice at $t = 0$. As a result of the hopping term, the particle has a finite probability of moving away from the origin. We can ask the question, then, does any probability remain at the origin as $t \to \infty$? If, as $t \to \infty$, the probability at the origin remains nonzero, then the particle has a finite probability of returning to the origin. Consequently, it is localized. On the other hand, if the probability at the origin decays to zero, the particle will be found throughout the lattice and, hence, will be delocalized. The site-return probability is the key quantity in the localization problem [A1958, AAT1973]. In addition, it is the phase relationship between time-reversed closed paths that leads to weak localization.

In the context of the model in Eq. (1), we define $c_n(t)$ to be the probability amplitude that an electron is on site n at time t. The return probability at long times is determined by the square of the site probability amplitudes, $\lim_{t \to \infty} |c_n(t)|^2$. To obtain this quantity, we focus on the Heisenberg equations of motion,

$$i\hbar \dot{c}_n(t) = \epsilon_n c_n + \sum_m V_{nm} c_m, \qquad (12.3)$$

for the probability amplitude $c_n(t)$, where V_{nm} is zero unless sites n and l are nearest neighbors, in which case $V_{nm} = V$. From this equation, it is clear that in the absence of the hopping term, the site probabilities do not evolve from their initial values. Hence, an initially localized distribution will remain localized. What is surprising is that localized solutions exist for the probability amplitudes, even when the hopping term is nonzero. The criterion for localization can be established as follows. We first Fourier transform the equations of motion

$$c_n(E) = \frac{c_n(t = 0)}{E - \epsilon_n} + \sum_{n \neq m} \frac{V_{nm} c_m(E)}{E - \epsilon_n} \qquad (12.4)$$

by defining

$$c_n(E) = \int_{-\infty}^{\infty} e^{iEt/\hbar} c_n(t) dt, \qquad (12.5)$$

with E a general complex variable with units of energy. We consider the localized initial condition, $c_n(t = 0) = \delta_{n0}$. The Fourier-transformed equation is best solved by iteration, leading to

$$c_0(E) = \frac{1}{(E - \epsilon_0)} + \sum_{l \neq 0} \frac{V_{0l} V_{l0}}{(E - \epsilon_0)(E - \epsilon_l)} + \cdots \qquad (12.6)$$

as our equation for the site-probability amplitude at the origin. As a result of the localized initial condition, the linear term in V vanishes. Because of the restriction that $l \neq 0$, the nth term in this series represents a self-avoiding random walk of n steps that starts and ends at the origin. No site except the origin is visited twice. We sum the series in Eq. (12.6) by defining the self-energy, $S_n(E)$, for site n such that

$$c_n(E) = \frac{1}{E - \epsilon_n - S_n(E)}. \qquad (12.7)$$

Random walks of all lengths are summed into the site self-energy. The question of localization at long times ($E \to 0$) now rests on the self-energy.

There is a close connection between the site-probability amplitudes and the diagonal elements of the exact Green function for our original Hamiltonian. As a resolvent, the Green function, $G(E) = (E - H)^{-1}$, contains all information about the exact eigenstates of Eq. (12.1). The matrix elements of the single-particle Green function are

$$(E - \epsilon_n)G_{nm}(E) = \delta_{nm} + \sum_{n \neq l} V_{nl} G_{lm}(E). \qquad (12.8)$$

The form of Eq. (12.8) is illuminating for two primary reasons. First, because the tight-binding states form the atomic basis, the mth eigenstate of the Green function can be written quite generally as a superposition

$$|\phi_m\rangle = \sum_n c_{nm}|n\rangle \qquad (12.9)$$

of these states. The coefficients c_{mn} determine the amplitude that the mth eigenstate overlaps site n. Without the second term in Eq. (12.8), the singularities of the Green function are simple poles with energies $E = \epsilon_m$, and the exact eigenstates reduce to $|\phi_m\rangle = |m\rangle$. This corresponds to the strongly localized regime. When the hopping term is present, the most general statement that can be made about a localized

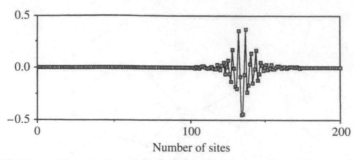

Figure 12.2 The real part of a typical eigenstate in a random binary alloy.

state is that not all the c_{mn}'s are nonzero. In fact, if a sufficiently large number of the site amplitudes vanish, the resulting eigenstate is characterized by an exponentially decaying envelope, $\phi_m \propto e^{-r/\xi_m}$, where ξ_m is the localization length (spatial extent) of the eigenstate. An example of an exponentially localized state in the one-dimensional random binary alloy is shown in Fig. (12.2).

Secondly, the expansion in powers of V for the diagonal elements of the Green function,

$$G_{nn}(E) = \frac{1}{(E - \epsilon_n)} + \sum_{l \neq n} \frac{V_{nl} V_{lm}}{(E - \epsilon_n)(E - \epsilon_l)} + \cdots$$

$$= \frac{1}{E - E_n - S_n(E)}, \tag{12.10}$$

is identical to that for the site-probability amplitudes. Hence, the analytical properties of the Green function are directly related to the fate of the return probability. In general, E can be complex. However, if as E approaches the real axis, $S_n(E)$ is purely real, then the singularities of $G_{nn}(E)$ are once again simple poles, and the eigenstates are localized states. Extended states form only when $G_{nn}(E)$ is complex or, equivalently, when $\mathrm{Im}S_n(E)$ is nonvanishing as E approaches the $\mathrm{Re}E$ axis. That is, the inverse of $\mathrm{Im}S_n(E)$ determines the lifetime of the state at energy E. If the state with energy E has a finite lifetime, the return probability will vanish at long times. For the Anderson model, the self-energies, $S_n(E)$, are a function of the distribution of site energies. Hence, they are themselves random variables. For this reason, the precise quantity that is relevant to the localization problem is the probability distribution of the site self-energies [A1958]. The signature of the absence of extended states is the vanishing of the probability distribution of the site self-energy for all energies along the $\mathrm{Re}E$ axis. The

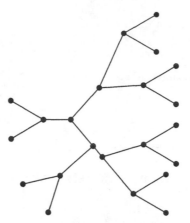

Figure 12.3 A Cayley tree of connectivity $K = 3$.

critical value $(W/V)_c$ determines the amount of disorder required to make the Anderson transition obtain.

The primary hurdle in the Green function analysis of the localization problem is the computation of the site self-energy. The self-energy requires the enumeration of all self-avoiding walks that return to a given site. Hence, on lattices lacking closed loops, such as Cayley trees, the site self-energy can be calculated exactly. A typical Cayley tree is depicted in Fig. (12.3). A further simplification that is employed in the analysis of the probability distribution of the self-energy is to ignore the real part of the self-energy. As we showed in the context of the local moment model in Chapter 6, the contribution from the real part of the self-energy is significant only if the density of states fluctuates wildly. For the localization problem, inclusion of the real part of the self-energy suppresses the critical amount of disorder required for the localization transition. The analysis of the Anderson transition [AAT1973] in which the real part of the self-energy is ignored yields

$$(W/2V)_c = 4K \ln (W/2V)_c \qquad (12.11)$$

for the upper critical value of the disorder beyond which no state is extended. Attempts have certainly been made [EC1970], [SE1973] to extend the self-energy analysis to hypercubic lattices. In such approaches, the self-energy is calculated through self-avoiding walks of M steps. Economou and Cohen [EC1970] argued that a stability analysis of the Mth root of the term of order M in the self-energy can be used to distinguish between localized and extended states. If in the limit that $M \to \infty$, the Mth root of the $O(M)$ term in the self-energy exceeds unity, then the perturbative expansion for $S_n(E)$ diverges and the electronic state at energy E is extended. Although this criterion is not as precise as the vanishing of the probability distribution for the imaginary part of the self-energy, it does appear to provide reliable results for square lattices, as shown by Soukoulis and Economou [SE1973].

12.3 SCALING THEORY

Though the Green function approach provides direct insight into the return probability, much of the physics of the localization transition is suppressed in this view. For example, the dimensional dependence of the localization transition is not easily determined from the Green function approach. That dimensionality might play a strong role in the localization transition can easily be deduced from the fact that a single defect (see Problem 12.1) in an otherwise ordered lattice produces a bound state for d ≤ 2, regardless of the strength of the defect. For d > 2, the defect strength must exceed a critical value for a bound state to form. Although the single defect results *cannot* be applied straightforwardly to the case of an infinite system containing a finite fraction of disordered sites, the strong dimensional dependence does suggest that the localization transition should somehow reflect this behavior. To this end, we focus on the scaling theory of localization [A1979]. The key assumption in this account is that regardless of the dimensionality, a single parameter, the dimensionless conductance defined as

$$g(L) = \frac{2\hbar}{e^2} G(L),$$ (12.12)

is a universal function of the sample size, and it completely characterizes the localization transition. The conductance is $G(L)$, and L is the linear dimension of the sample. For a more complicated Hamiltonian containing random V's along with energy disorder, one-parameter scaling is not as immediately transparent as it is for the simple model in Eq. (12.1) with a single matrix element V and diagonal disorder of width W. Nonetheless, it describes this case, as well.

Two other key ideas anchor the scaling theory of localization: 1) Fermi-liquid theory completely describes the clean system, and 2) the logarithmic derivative of the conductance with respect to system size is a monotonic function of the conductance, g. As we will see, these three postulates lead necessarily to an absence of current-carrying states for d $\leq$ 2 and an absence of a minimum-metallic conductivity.

To develop the scaling account, we consider a hypercubic lattice with edge length L and increase the size of a unit cell of the lattice by a factor p. The new lattice constant is $L_1 = Lp$, where the new unit cell contains p^d old lattice sites. By our initial assumption, the localization properties in the new unit cell are determined entirely by $g(L_1)$. After m transformations, the new linear dimension is $L_m = Lp^m$. In such a sample, $g(L_m)$ characterizes the localization transition. A further consequence of the scaling assumption is that all of the $g(L_i)$'s must be related as a result of their universal length dependence. That is, we maintain that

$$g(pL) = f(p, g(L)). \tag{12.13}$$

Our goal is now to construct the universal function f. If the conductances are related when the sample size is increased, so are the eigenstates. Moreover, the new eigenstates in the scaled system must be constructible by knowing a single parameter, the dimensionless conductance. However, when the sample size is increased, there is a shift in both the energy level spacing, ΔW, as well as in the energy of each eigenstate, ΔE. Consequently, when linear combinations of the eigenstates for the sample of length L_1 are taken to construct those of the sample of length L_2, the crucial quantity of interest is $\Delta E / \Delta W$ [T1982]. If a state is well localized within the initial sample size, changing the edge length from L to L_1 will not induce an energy shift, ΔE. Consequently, ΔE is generally computed by determining how sensitive the eigenstates are to a change in boundary conditions. A common boundary effect used in the computation of ΔE is a change from periodic to antiperiodic boundary conditions. As ΔE is largest for an extended state, and ΔW is determined by the density of states, the ratio $\Delta E / \Delta W$ should be a measure of the dimensionless conductance. This physical argument further supports the conclusion from one-parameter scaling that the dimensionless conductance must be given by the ratio $\Delta E / \Delta W$.

The mean spacing between the energy levels is simply given by the inverse of the number of particles in the system: $\Delta W = (n_e L^d)^{-1}$. Consequently, $g(L) = n_e L^d \Delta E$. What about ΔE? From dimensional considerations, we can write $\Delta E = \hbar / \tau$, where τ represents some physical time associated with transport in our sample. Let us take τ to be the time it takes the particle to diffuse from the center to the edge of the sample [LT1975]. That is, $\tau = (L/2)^2/D_0$, with D_0 the diffusion constant. Consequently, $\Delta E = 2\hbar\sigma/(e^2 n_e L^2)$, where we have used the Einstein relationship, $\sigma = 2e^2 n_e D_0$. Combining these relationships, we obtain

$$g(L) = \frac{2\hbar}{e^2}\sigma L^{d-2} = \frac{2\hbar}{e^2}G(L) \tag{12.14}$$

as the length dependence of our single parameter for the localization problem. In Eq. (12.14), $G(L) = \sigma L^{d-2}$ is the dc conductance. Note that the form given here for the conductance is simply Ohm's law for a metal. The crucial assumption in one-parameter scaling theory, then, is that the dimensionless conductance g relates the eigenstates in a sample of edge length $2L$ to those in an original sample of linear dimension L [A1979].

We define

$$\beta = \frac{d\ln g(L)}{d\ln L} \tag{12.15}$$

as the logarithmic derivative of $g(L)$ [A1979]. As we mentioned in the context of the Kondo problem, $\beta(g)$ is referred to as the *beta function*. The central physics of the localization problem is contained in $\beta(g)$. If in the limit that $L \to \infty$, $\beta(g) > 0$, then $g(L)$ must diverge. The divergence of the conductance is the signature of extended states. By contrast, if $\beta(g) < 0$ as $L \to \infty$, then $g(L)$ must monotonically tend to zero in this limit. Localization obtains in this case. It is evident, then, that the sign of $\beta(g)$ can be used to distinguish between localized and extended states. Invoking the assumption that $\beta(g)$ is a monotonic function of g or $\ln g$ leads necessarily to an absence of metallic behavior for $d > 2$. We proceed as follows. In the weak-disorder limit, the conductance satisfies Ohm's law; that is, $G = \sigma L^{d-2}$. As a result, $\beta(g) = d - 2$ in the large-conductance limit. In the limit of strong disorder, the localization length is much less than the sample size, $\xi \ll L$, and the conductance decays exponentially with L. This implies that $\beta(g) = \ln g$ is the form for the scaling function in the limit of strong disorder, $g \ll 1$. Our assumption of monotonicity guarantees that the strong-disorder form for the conductance interpolates smoothly to the asymptotic value of $d - 2$ at weak disorder. Perturbative expansions in the weak-disorder regime [LN1966] in powers of $1/g$, as well as in the strongly localized regime in powers of g, confirm the assumption of monotonicity and continuous evolution between these two regimes. Consequently, $\beta(g) \le 0$

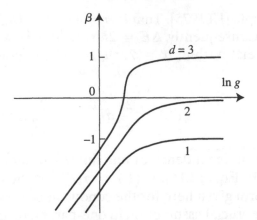

Figure 12.4 Dimensional dependence of the scaling function $\beta(g) = d\ln(g)/d\ln L$ as a function of the conductance, g. For dimensions d = 1, 2, $\beta(g)$ is strictly negative and tends asymptotically to zero for large g in d = 2. This indicates the absence of a metal transition in d = 1, 2. In d = 3, $\beta(g)$ crosses the $\beta(g) = 0$ axis and asymptotically assumes the value of d − 2, therefore signaling the transition to a metallic state.

for d = 1, 2, and all states are localized. A plot of the beta-function illustrating this behavior is shown in Fig. (12.4). As is evident, d = 2 is the marginal dimension in which $\beta(g) = 0$ in the limit of large conductance. In d = 3, $\beta(g)$ crosses the $\ln g$ axis and asymptotically approaches unity, signaling an onset of metallic behavior. Note that the β function is continuous in the vicinity of the d = 3 transition, hence the absence of a minimum metallic conductivity.

We extract the functional form of the conductance in the vicinity of the Anderson transition by considering the slope of the β-function near $\beta(g_c) = 0$. The zero of the beta-function at g_c defines the critical, or fixed point. In the vicinity of the critical point, g_c, we linearize the β-function, such that

$$\beta(g) = \frac{g - g_c}{\nu g_c} = \frac{\delta g}{\nu}, \tag{12.16}$$

with ν a positive number determined by the slope. Near the transition point, the conductance is slowly varying. Consequently, we approximate the logarithmic derivative of the conductance by $(1/g_c)dg/d\ln L$. Combining this result with Eq. (12.16) yields

$$\frac{\delta g}{\nu} = \frac{1}{g_c}\frac{dg}{d\ln L}, \tag{12.17}$$

which we integrate in the interval $[L_0, L]$ to obtain

$$\frac{L}{L_0} = \left(\frac{g - g_c}{g_0 - g_c}\right)^{\nu}, \tag{12.18}$$

where $g_0 = g(L_0)$. The utility of Eq. (12.18) is made evident by first considering the localized regime in which $g_0 < g_c$. Introducing the length scale

$$\xi = L_0\left(\frac{g_c}{g_0 - g_c}\right)^{\nu} = L_0|\epsilon|^{-\nu} \tag{12.19}$$

allows us to conclude that in the localized regime, the conductance

$$g = g_c\left(1 - \left(\frac{L}{\xi}\right)^{\frac{1}{\nu}}\right) \tag{12.20}$$

decreases algebraically with the length of the sample. The length scale ξ diverges at the transition point, g_c. As the localization length must diverge as the fixed point is approached from the insulating side, ξ is identified as the localization length (a key hypothesis of the "gang of four," the colloquial name for the originators of the scaling theory), and ν is its critical exponent. For sufficiently large samples, the conductivity obeys Ohm's law on the metallic side. However, at the critical point, the dc conductivity vanishes as $|\epsilon|^s$ with $s = (d - 2)\nu$. We prove the latter assertion by noting that the scaling hypothesis guarantees that the conductance is a universal function of L/ξ on all length scales. The behavior for $L \ll \xi$ is given by Eq. (12.18). At long length scales where Ohm's law is presumed to apply, the conductance must vary as $(L/\xi)^{d-2}$. Since the conductivity is defined as $\sigma(0) = g/L^{d-2}$, we have that

$$\sigma(0) \propto \xi^{2-d} \propto |\epsilon|^{(d-2)\nu}, \tag{12.21}$$

as advertised, and the conductivity vanishes continuously at the critical point, as opposed to the discontinuous drop predicted by Mott. Consequently, the localization length and conductivity exponents are related: $s = (d - 2)\nu$ [W1976]. From the scaling analysis, $s = 1$ (see Problem 12.5). However, experimentally [T1982], s appears to vary between $1/2$ and 1, depending on the degree of compensation. This evolution is shown in Fig. (12.5). For materials such as $Nb_x Si_{1-x}$ or compensated Ge:Sb, $s = 1$; whereas, for Si:P or uncompensated Ge:Sb, $s = 1/2$. In

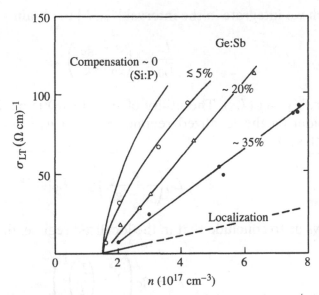

Figure 12.5 Evolution of the conductivity exponent from $s = 1/2$ in the uncompensated semiconductor, Si:P, toward the value of $s = 1$ predicted from scaling theory (dashed line) as the compensation is increased in Ge:Sb: $s = .55$ at 0%, $s = 0.7 \pm 0.2$ at 5%, $s \approx 1$ at 20%, and $s \approx 1$ for 35% compensation. σ_{LT} is the low-temperature conductivity. Compensation increases the disorder and, hence, the increased validity of the disorder model of Anderson for which scaling theory predicts that $s = 1$. The data are taken from G. A. Thomas, et. al., *Phys. Rev.* B **25**, 4288 (1982).

Si:P, each phosphorous atom gives up an electron. The random potential is due to the charged P-ions. Hence, there is a charge for each scattering center. In compensated Ge:Sb, equal amounts of Sb and B are added, which then exchange an electron. Hence, the number of scatterers exceeds the number of charge carriers in the compensated samples. Consequently, disorder plays a larger role in compensated than in uncompensated systems and, hence, the improved agreement with scaling theory. In the uncompensated materials, additional interactions, such as spin-orbit or electron-electron repulsion, are operative. Inclusion of such effects leads to a conductivity exponent of $s = 1/2$ [LR1985].

Though the scaling theory of localization has been highly successful, it does not constitute a mathematical proof that the return probability remains finite in $d = 1, 2$. Currently, there is no rigorous mathematical proof on the fate of the return probability for quantum transport in a d-dimensional random system. Nonetheless, the scaling analysis of the localization problem has received much support from more rigorous theories, numerical simulations, and from experiments (as seen in the preceding paragraph). As mentioned in the previous sections, the Green function calculations of Soukoulis and Economou [SE1973] confirmed an absence of extended states in a d = 2 disordered square lattice. Further support for the scaling theory of localization can be found in the work of Vollhardt and Wölfle [VW1980], who have shown that the dimensional predictions of scaling theory can be reproduced by a self-consistent diagrammatic expansion of the density response function for a system of independent particles moving in a random potential. A further key result of this work is that much of the physics of the localization problem can be obtained by analyzing the maximally crossed diagrams. Such diagrams contain the contribution from coherent backscattering. It is these diagrams that we will focus on in the next section, as they contain the essence of weak-localization physics. Other, more recent, analytical work that has verified the scaling predictions is the random matrix theory approach of Muttalib [M1990]. Based on earlier work by Imry [I1986], Muttalib has shown how the most probable value of the conductance can be calculated from the distribution of eigenvalues for the transfer matrices. One-parameter scaling theory follows immediately [M1990], once it is assumed that the distribution of eigenvalues of the transfer matrices is described by a single parameter. Although this assumption does not appear to hold for strong disorder, it is supported by numerical simulations in the presence of weak disorder [M1990]. Only the weak-disorder limit is of interest here, because it is in this regime that the one-parameter scaling function predicts an absence of a metal transition for d = 1, 2.

12.4   WEAK LOCALIZATION

We have yet to provide a purely physical basis for electron localization in random systems. An analysis of weak localization [A1983] will yield this missing link. The goal of weak localization is to perturbatively evaluate the role of scattering processes that are precursors to Anderson localization. Consider a collection of noninteracting quantum mechanical particles moving in a random uncorrelated distribution of pointlike scatterers. In the weak-disorder limit, the mean-free path, ℓ, of the particle is much larger than the average particle separation, $\mathbf{a} \approx k_F^{-1}$. Hence, in weak localization, $k_F \ell \gg 1$, and we can define the small parameter

$$\gamma = \frac{1}{\pi k_F \ell} \ll 1 \qquad (12.22)$$

as a measure of the strength of the disorder. The goal is to develop a perturbative theory for the conductivity in terms of γ. To this end, we write the conductivity

$$\sigma = \sigma_0 + \delta\sigma, \quad |\delta\sigma| \ll \sigma_0 \qquad (12.23)$$

as a sum of a *zeroth*-order dc conductivity indicative of the metallic regime, σ_0, and a correction $\delta\sigma$ that depends nontrivially on γ. We will treat σ_0 in the Boltzmann transport limit described in Chapter 10. In this limit, all scattering events, even if they involve visitations to the same site, are considered to be uncorrelated. This approximation is clearly inadequate if there is an inherent tendency for repetitive scattering from the same site. Hence, the average scattering rate (or relaxation rate in the linearized Boltzmann equation) with the random impurities, τ^{-1}, is linear in the concentration of defects. As a result, σ_0 has the simple Drude form,

$$\sigma_0 = \frac{e^2 n_e}{m} \tau. \qquad (12.24)$$

Between scattering events, the velocity of the particle is a constant; hence, $\ell = v_F \tau$. Because $v_F = \hbar k_F / m$, the *zeroth*-order conductivity is proportional to $k_F \ell$ or, equivalently, $\sigma_0 \propto \gamma^{-1}$.

To go beyond the Boltzmann limit, we need to consider the phase relationships that enter a scattering process. Quantum mechanically, when a particle makes an excursion from point A to point B, one must include a sum over all paths connecting these two points. Consider first the case in which A and B are spatially separated. The probability for an excursion from A to B,

$$P = \sum_i |A_i|^2 + \sum_{i,j} A_i A_j^*, \qquad (12.25)$$

involves the sum of the square of each amplitude, A_i, as well as the interference terms between the paths, the second term in Eq. (12.25). When the end points are spatially separated, there is no special phase relationship between the paths. Consequently, the second term in Eq. (12.25) averages to zero. Hence, the Boltzmann treatment is adequate.

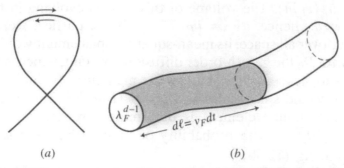

(a)   (b)

Figure 12.6 (a) Time-reversed paths in a closed loop. (b) Volume element for a trajectory of a quantum mechanical particle. The cross-sectional area is λ_F^{d-1}, and the differential length is $d\ell = v_F dt$.

However, when A and B coincide, the result is quite different. Consider the loop shown in Fig. (12.6a). Let A_1 represent the amplitude associated with the clockwise path around this loop and A_2 the amplitude for the time-reversed path. It is loops of this sort that are relevant in repeated scattering from the same impurity. We generate the time-reversed path from the direct path by the transformation $\mathbf{p} \rightarrow -\mathbf{p}$. Because phase coherence is maintained in the traversal of the closed loop, A_1 and A_2 interfere constructively, and as a result, we can set $A = A_1 = A_2$. Consequently, the quantum mechanical probability for traversing the closed loop is $P_{qm} = |A_1 + A_2|^2 = 4|A|^2$ whereas the classical result, $P_{cl} = |A_1|^2 + |A_2|^2 = 2|A|^2$, is a factor of two smaller, implying that backscattering effects are maximized in the quantum mechanical case. Consequently, the return probability (or, equivalently, the likelihood of localization) in a quantum mechanical system always exceeds that in the corresponding classical problem. This state of affairs arises from the wave characteristics of an electron.

The contribution from closed-loop trajectories decreases the conductivity. Hence, $\delta\sigma$ enters with a minus sign. To estimate $\delta\sigma$, we must determine the probability that the diffusing particle traverses a closed loop without the loss of phase coherence. Consequently, the relevant time interval spans the minimum time for a single collision, τ, to the shortest time at which phase coherence is lost, τ_ϕ. Loss of phase coherence can arise from such disparate microscopic phenomena as phonon scattering (an explicitly inelastic process) and spin-flip scattering, which costs zero energy in the absence of a magnetic field. We focus on calculating the probability that an electron traverses the closed tube shown in Fig. (12.6b). In this tube, the electron has a constant wavelength,

$\lambda_F = \hbar/(v_F m)$. The volume of this tube is evolving in time through $d\ell = v_F dt$; hence, $dV = v_F \lambda_F^{d-1} dt$. If the particle were allowed to wander over all space, its mean-square displacement would increase as $D_0 t$, with D_0 the *zero*th-order diffusion constant, which is proportional to σ_0 and, hence, scales as $1/\gamma$. The maximum "volume" element the particle would span scales as $V_{\text{max}} \approx (D_0 t)^{d/2}$. The maximum value of this "volume" element is determined by the dephasing length, $L_\phi = \sqrt{D_0 \tau_\phi}$. The probability of finding the particle in the tube shown in Fig. (12.6b)

$$P_{\text{wl}} = \int_\tau^{\tau_\phi} \frac{dV}{V_{\text{max}}} = v_F \lambda_F^{d-1} \int_\tau^{\tau_\phi} \frac{dt}{(D_0 t)^{d/2}} \tag{12.26}$$

is given by the ratio of dV to V_{max} integrated over the time interval $[\tau, \tau_\phi]$. The probability P_{wl} is proportional to $\delta\sigma/\sigma_0$. Performing the time integral, we observe that the weak-localization correction

$$\frac{\delta\sigma}{\sigma_0} \equiv -\gamma_d \begin{cases} \left(\frac{\tau_\phi}{\tau}\right)^{1/2}, & d = 1 \\[2ex] \hbar \ln\left(\frac{\tau_\phi}{\tau}\right), & d = 2 \\[2ex] \hbar^2 \left(\frac{\tau_\phi}{\tau}\right)^{-1/2}, & d = 3 \end{cases} \tag{12.27}$$

has a strong dimensional dependence, where $\gamma_d = \gamma v_F D_0^{-d/2}/(v_F m)^{d-1}$. The absence of $\hbar$ implies that 1d weak localization is not quantum mechanical in origin, as far as the interference effects are concerned. Of course, factors of $\hbar$ can be present in the inelastic relaxation time. Such quantum effects are different in kind from coherent backscattering. In d = 1, all paths are closed, as there can only be forward and backscattering. Hence, there is no difference between the classical and quantum results in d = 1, as far as coherent backscattering is concerned. Second, to lowest order, the weak-localization correction is linear in γ.

If τ_ϕ is determined by phonon scattering or electron-electron interactions, then from Eqs. (10.98) and (11.67), $1/\tau_\phi \propto T^\eta$ where $\eta > 0$. For acoustic phonons, $\eta = 5$ (see Eq. 10.98), whereas for electron-electron scattering in a Fermi liquid, $\eta = 2$, (see Eq. 11.67 and subsequent discussion). Substitution of this algebraic form for τ_ϕ leads to

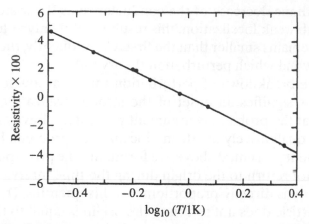

Figure 12.7 Resistivity measurements of Dolan and Osheroff [DO1979] on weakly disordered thin films of palladium.

the temperature dependence

$$\frac{\delta\sigma}{\sigma_0} \equiv -\gamma_d \begin{cases} T^{-\eta/2}, & d = 1 \\ \frac{\eta\hbar}{2}\ln\left(\frac{\hbar_\tau}{k_B T}\right), & d = 2 \\ \hbar^2 T^{\eta/2}, & d = 3 \end{cases} \tag{12.28}$$

of the weak-localization correction. For $d \leq 2$, the temperature dependence is divergent as $T \to 0$. However, the expression we have derived here for $\delta\sigma/\sigma_0$ cannot exceed unity, in accordance with the initial assumption that $|\delta\sigma| \ll \sigma_0$. Hence, there is a lower limit to the temperature at which Eq. (12.28) can be used for $d \leq 2$. The resistivity measurements shown in Fig. (12.7) on weakly disordered palladium films have confirmed the logarithmic temperature dependence of weak localization [DO1979]. In 3d, the weak-localization correction vanishes as $T \to 0$. This is a reflection of the relatively small probability an electron has of returning to the origin in a 3d weakly disordered system.

The logarithmic behavior of the weak-localization correction can also be seen by considering the asymptotic form for the β-function in two dimensions. Figure (12.4) illustrates that in $d = 2$, the β-function is always negative and goes to zero asymptotically as roughly $-1/g$, a result confirmed also from perturbative analyses. Integration of this asymptotic form in the interval $[L_0, L]$ leads to the familiar

$$g_{d=2}(L) = g(L_0) - \ln\frac{L}{L_0} \tag{12.29}$$

logarithmic decrease of the conductance as the system size increases. As with weak localization, this result is valid only as long as the second term remains smaller than the first. Consequently, there is a length cut-off beyond which perturbation theory fails.

The breakdown of perturbation theory at low temperatures or as $L \to \infty$ signifies an onset of the strong-localization regime. It turns out that the probability argument presented above can be modified to tackle qualitatively the strong-localization problem. In the spirit of the argument presented above, we formulate the probability that a particle does not return to the origin during the time interval $[\tau, \tau_\phi]$. The conductivity is directly proportional to this quantity. The probability that the particle does not return to the origin is equal to the product of the probabilities

$$
\begin{aligned}
P_{\text{no return}} &= \prod_{i=1}^{N}\left(1 - \frac{\lambda_F^{d-1} v_F \Delta t_i}{(D_0 t_i)^{d/2}}\right) \\
&= \exp\left(-\sum_i \frac{\lambda_F^{d-1} v_F \Delta t_i}{(D_0 t_i)^{d/2}}\right)
\end{aligned}
\tag{12.30}
$$

that the particle does not return during any differential time element Δt_i. The conversion of the sum to an integral is left as an exercise (see Problem 12.4). The result is analogous to the standard weak-localization result. However, the difference now is that the integral is in the exponent. For a 2d sample, we obtain that the conductivity

$$
\sigma_{d=2} \approx \frac{e^2}{\hbar}\left(\frac{\ell}{L_\phi}\right)^{1/k_F \ell \mathbf{a}}
\tag{12.31}
$$

scales algebraically with the dephasing length, but for a thin 1d wire, the conductivity

$$
\sigma_{d=1} \approx \frac{e^2}{\hbar} L_\phi \exp\left(-\frac{1}{(k_F \mathbf{a})^2}\frac{L_\phi}{2\ell}\right)
\tag{12.32}
$$

decays exponentially with the phase coherence length. Exponential decay of the conductivity in 1d is the standard signature of the localizing effect of disorder. However, for 2d, the analysis presented here is insufficient to describe coherent backscattering effects that lead to exponential localization of the eigenstates.

Our analysis of weak localization has shown that coherent back-scattering plays a significant role in the localization problem. In fact, Vollhardt and Wölfle [VW1980] succeeded in developing a theory of the Anderson localization transition by including self-consistently the weak-localization terms to all orders, thereby reinforcing the fact that time-reversed trajectories are essential to localization. As such, it stands to reason that any external perturbation in a disordered system that disrupts time-reversal symmetry should destroy localization and, as a consequence, enhance the conductivity. Consider, for example, turning on a magnetic field. A magnetic field explicitly breaks the symmetry between $\mathbf{p}$ and $-\mathbf{p}$. The momentum is now replaced by $\mathbf{p} - 2e\,\mathbf{A}$, where $\mathbf{A}$ is the vector potential. The amplitudes along the direct and time-reversed paths are $A_1 \rightarrow Ae^{i\phi}$ and $A_2 \rightarrow Ae^{-i\phi}$, respectively, where

$$\phi = 2\pi \frac{\Phi}{hc/e}. \tag{12.33}$$

The magnetic flux is $\Phi = HR$, where R is the area of the region enclosed by the closed tube in Fig. (12.6b). For diffusive motion, R is proportional to the mean-square displacement that scales as $D_0 t$. As a consequence, the return probability in the presence of a magnetic field

$$P_H = 4|A|^2 \cos^2\left(\frac{eHD_0 t}{\hbar c}\right) \tag{12.34}$$

is an oscillatory function of time. It is left as a homework problem (12.2) to evaluate the corresponding magnetic field correction to the conductivity, $\Delta\sigma(H) = \delta\sigma(H) - \delta\sigma(0)$. This correction is always positive and proportional to $H^2 \tau_{\text{in}}^2$ at large fields and, at weak fields, grows logarithmically as $\ln(H\tau_{\text{in}})$. The critical field determining the crossover from the power law to $\ln H$ is determined by setting $2eHD_0\tau_{\text{in}}/\hbar c = 1$. Hence, the crossover field depends on the temperature. Experimentally, the crossover field is typically on the order of $100G$ or, equivalently, $10mT$. The sensitivity of localization to such small magnetic fields reiterates that time-reversed scattering is the key physical origin of the localization transition.

A similar effect we are also equipped to treat is the breaking of phase coherence by spin-flip scattering. Spin-flip scattering weakens the localization effect, as it provides an effective oscillating magnetic field. To quantify this effect, we rely on the derivation of the spin-flip scattering time in Chapter 7. From Eq. (7.42), we have that

$$1/\tau_s = 1/\tau_s^0 \left(1 - 2J_0 N(0) \ln \frac{T_F}{T} + \cdots \right). \tag{12.35}$$

Substitution of this result into Eq. (12.27) yields the contribution of spin-flip scattering to the conductivity

$$\delta\sigma \approx \gamma\sigma_0 \left(\ln \frac{\tau_s^0}{\tau} - N(0)J_0 \ln \frac{T_F}{T}\right) \tag{12.36}$$

for a d = 2 sample.

Because $J_0 < 0$, the Kondo logarithmic term reduces the magnitude of the weak-localization correction. The reduction of the weak-localization correction by spin-flip scattering will ultimately lead to a disorder-induced suppression of the Kondo effect in thin films as shown by Martin, Wan, and Phillips [MWP1997]. This effect arises because the standard Kondo logarithm comes in with the opposite sign relative to the second term in Eq. (12.35). Consequently, if weak-localization physics dominates, disorder suppresses the Kondo resistivity, as is observed experimentally in thin-film metal alloys of Au(Fe) and Ag(Fe) [BG1995]. Because spin-flip scattering counters weak localization, it is often invoked to explain deviations from the standard localization scenario. Consider for example, the dephasing time, τ_ϕ. In an electronic system, phase coherence is lost by coupling to the environment. As in the case of phonon and electron-electron scattering, all standard dephasing mechanisms [A1983] turn off at low temperatures; consequently, τ_ϕ should diverge at low temperatures. In the late 1980s, several experiments [LG1989] reported that τ_ϕ saturates at low temperatures. The possible role of magnetic impurities cast serious doubts on this result. However, more controlled experiments [MJW1997] have been performed recently, and, as shown in Fig. (12.8), the saturation persists for a wide range of systems and down to temperatures 0(10m K). The apparent saturation of τ_ϕ is one of the true surprises in the physics of metals. In fact, should the saturation of τ_ϕ as $T \to 0$ prove to be an intrinsic effect, as the current experiments seem to indicate, then this will place serious limitations on the degree to which quantum coherence can be controlled in conductors. As of this writing, there is no agreed-upon explanation for the saturation of τ_ϕ or if this effect is intrinsic.

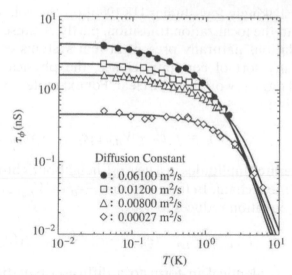

Figure 12.8 Saturation of the electron dephasing time τ_ϕ observed in quasi-one-dimensional gold (Au) wires. The degree of disorder in these samples is denoted by the classical diffusion constant D_0. The saturation time is typically longer and the saturation temperature is lower as the disorder decreases. The data are reprinted from P. Mohanty, E. M. Q. Jariwala, and R. A. Webb, *Phys. Rev. Lett.* **78**, 3366 (1997).

12.5   THE DEFIANT: EXCEPTIONS TO LOCALIZATION

The absence of current-carrying states in d = 1, 2 is one of the truly surprising results in disordered systems. As mentioned in the previous section, when time-reversal symmetry is broken, current-carrying states can survive in low dimensions. However, even when time-reversal symmetry is intact, three important exceptions exist to the localization principle: 1) the random dimer model, 2) insulator-superconductor transitions in thin films, and 3) the newly discovered conducting state in dilute 2d electron gases. In this section, we will review all three. However, our discussion of the latter two will be brief.

12.5.1   Random Dimer Model

In the models discussed thus far, the site energies or matrix elements were strictly statistically independent entities assigned from a single random distribution. In any real physical system, however, the site energies and matrix elements are correlated. Until recently, there have been strikingly few studies [JK1986] on the role of statistical correlations in the localization transition, partly because of the suspicion that correlations naturally present in real systems cannot be determined with any sort of certainty. Hence, the physical significance of proposed models would not be clear. For example, consider the eigenvalue equation

$$Ec_n = \epsilon_n c_n + V_{n,n+1} c_{n+1} + V_{n,n-1} c_{n-1} \qquad (12.37)$$

for the site amplitudes in a nearest-neighbor tight-binding Hamiltonian for a linear chain. In the event that $-\epsilon_n = V_{n,n+1} + V_{n,n-1}$, the eigenvalue equation reduces to

$$Ec_n = V_{n,n+1}(c_{n+1} - c_n) + V_{n,n-1}(c_{n-1} - c_n), \qquad (12.38)$$

which is identical in form to a diffusion equation for the probabilities with random nearest-neighbor hopping rates. This equation is well known to have a diffusion pole at $E = 0$; that is, the state with $E = 0$ is extended [A1981]. Hence, the simple model used here for the correlated disorder has a diffusive mode. Except in the context of structural disorder, the correlation leading to Eq. (12.38) is, at best, artificial at the electronic level. Flores [F1989] has explored additional algebraic correlations between the site energies and matrix elements and found that an infinitesimal fraction of the electronic states possessed zero reflection coefficients and, hence, are extended in one dimension. This result is indeed surprising in light of the scaling arguments presented earlier, which prohibit extended states in one dimension.

To understand what minimal physical condition must be satisfied for 1d systems to possess extended states, we consider the much-studied model introduced by Dunlap, Wu, and Phillips [DWP1990] called the *random dimer model* (RDM). To put the RDM in context, consider the random binary alloy. In the tight-binding model of a random binary alloy, site energies ϵ_a and ϵ_b are assigned at random to the lattice sites with probability p and $1 - p$, respectively. A constant nearest-neighbor matrix element, V, mediates transport among the lattice sites. All states in this model are localized. A typical wave function in a random binary alloy is shown in Fig. (12.2). The RDM can be obtained from the binary alloy by replacing all clusters containing an odd number of ϵ_b's with clusters containing the same number of ϵ_a's. The result is a random lattice in which the b-defects occur randomly but in pairs. The RDM refers to any lattice in which at least one of the

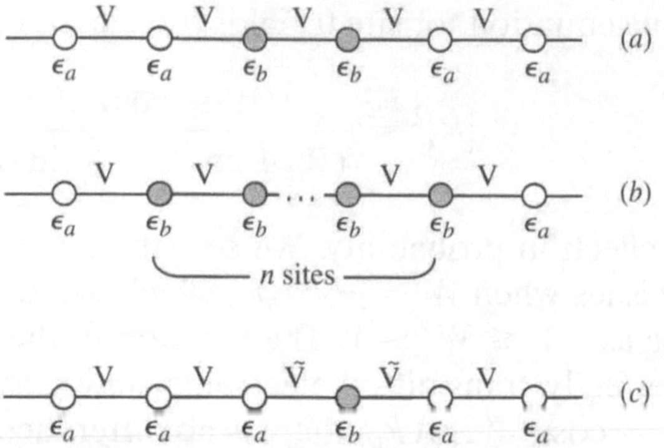

Figure 12.9 (a) Dimer defect in the RDM, (b) n-site defect possessing the dimer symmetry, and (c) defect in the repulsive binary alloy.

site energies is assigned at random to pairs of lattice sites. A typical dimer in the RDM is shown in Fig. (12.9). The surprising feature of the RDM is that $\sqrt{N}$ of the electronic states remain extended over the entire sample, provided that $-1 \leq W \leq 1$, with

$$W = \frac{\epsilon_a - \epsilon_b}{2V}. \tag{12.39}$$

An additional feature of the RDM is that the mean-square displacement of an initially localized particle grows superdiffusively as $t^{3/2}$, provided that $-1 < W < 1$. Diffusion obtains only when the disorder is increased, such that $W = \pm 1$. In all other cases, the particle remains localized at long times. The presence of at least diffusive transport is the key feature that ensures that the RDM does represent a counterexample to the scaling argument prohibiting long-range transport in 1d.

The simplest way to understand the RDM is to calculate the reflection coefficient through a dimer placed in an otherwise ordered lattice with the unit lattice constant. Let us place the dimer on sites 0 and 1 with energy ϵ_b. We assign the energy ϵ_a to all other lattice sites. To compute the reflection (R) and transmission (T) amplitudes, we write the site amplitudes as

$$c_n = \begin{cases} e^{ikn} + Re^{-ikn} & n \leq -1 \\ Te^{ikn} & n \geq 1, \end{cases} \tag{12.40}$$

where k is the dimensionless wave vector. The site amplitude at the origin is determined by substituting Eq. (12.40) into the eigenvalue equation, Eq. (12.37), for sites 1 and -1. The result is that $c_0 = 1 + R = T(\epsilon_- e^{-ik} + V)/V$ with $\epsilon_- = \epsilon_a - \epsilon_b$. Substituting this result into the eigenvalue equation for site 0 yields the closed expression

$$|R|^2 = \frac{(W + \cos k)^2}{(W + \cos k)^2 + \sin^2 k} \tag{12.41}$$

for the reflection probability. We see, then, that the reflection coefficient vanishes when $W = -\cos k$, which will occur for some value of k as long as $-1 \le W \le 1$. The location in the parent-ordered band of the perfectly transmitted electronic state corresponds to the wave vector $k = \cos^{-1} W$. At k_0, there is no difference between the ordered and disordered bands; the densities of states coincide at this point. The vanishing of the reflection coefficient through a single dimer at a particular energy can be understood as a resonance effect. That is, the dimers are acting as resonance cavities. At a particular wave vector, $k_0 = \cos^{-1} W$, the reflection from the second site in the dimer is 180° out of phase from the reflection from the first. At this energy, unit transmission obtains.

Consider the disordered case. Because the transmission coefficient through a single dimer is unity, then the particular distribution of random dimers should not affect the perfectly transmitted states. Indeed, this is true, as can be seen from the following argument. Transport across an arbitrary segment of the RDM can be represented by $\cdots T_a^{n_1} T_b^{2n_2} T_a^{n_3} T_b^{2n_4} \cdots$, where the n_i's are random variables and T_n is the transfer matrix. Notice that the b-transfer matrices all occur an even number of times. This is the dimer constraint. From the eigenvalue equation for the site amplitudes, Eq. (12.37), the transfer matrix is given by

$$\begin{pmatrix} c_{n+1} \\ c_n \end{pmatrix} = \begin{bmatrix} \frac{E - \epsilon_n}{V} & -1 \\ 1 & 0 \end{bmatrix} \begin{pmatrix} c_n \\ c_{n-1} \end{pmatrix} = T_n \begin{pmatrix} c_n \\ c_{n-1} \end{pmatrix}. \tag{12.42}$$

The resonant condition $W = -\cos k$ is equivalent to $E = \epsilon_b$, because E is the energy of the ordered band. At this energy, the transfer matrix across a b-defect reduces to

$$T_b = \begin{bmatrix} 0 & -1 \\ 1 & 0 \end{bmatrix}. \tag{12.43}$$

The square of this matrix (which corresponds to the transfer matrix across a b-dimer) is the negative of the unit matrix. At resonance then, the product of the transfer matrices in the RDM commutes and the

b-dimers can simply be erased at the expense of a sign change. Consequently, the phase shift through a dimer defect is $\Omega = -2k + \pi$. The unscattered state corresponds to a Bloch state in which the dimer sites have been effectively decimated at the cost of a phase shift at the dimer sites. Hence, once an electron passes through a single dimer defect, it will not scatter from any other dimer defects, as all dimers possess the identical resonance condition, regardless of their placement. If dimers are placed at sites $0, 1$ and $2, 3$, the unscattered state is

$$\cdots e^{-2ik}, e^{-ik}, 1, -e^{-ik}, -1, e^{-ik}, 1, e^{2ik}, \cdots. \qquad (12.44)$$

An example of such an unscattered state is plotted in Fig. (12.10).

Of course, no transport would occur if only a single electronic state remained unscattered. To determine the total number of states that extend over the entire sample, we expand R around k_o. To lowest order, we find that in the vicinity of k_o, $|R|^2 \propto (\Delta k)^2$, where $\Delta k = k - k_o$. The time between scattering events, τ, is inversely proportional to the reflection probability. As a result, the mean-free path $\ell = v_F \tau \propto 1/(\Delta k)^2$ in the vicinity of k_o. Upon equating the mean-free path to the length of the system N, we find that the total number (ΔN) of states whose mean-free path is equal to the system size scales as $\Delta N = \sqrt{N}$. Because the mean-free path and the localization length are equal in one dimension, we find that the total number of states whose localization lengths diverge is $\sqrt{N}$. The single-dimer argument presented here has been shown to hold for a random system by Bovier [B1992]. Bovier's argument is particularly elegant and establishes rigorously that the Lyapunov exponent (inverse localization length) vanishes quadratically (as a function of Δk_o) in the vicinity of the resonance, $k_o = \cos^{-1} W$. Consequently, in the RDM, $\sqrt{N}$ of the electronic states remain extended over the total length of the sample.

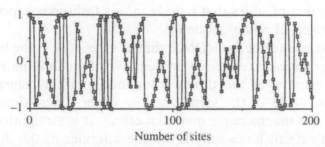

Figure 12.10 The real part of a typical extended state in the random dimer model.

These states ultimately contribute to the transport properties. Because the mean-free path of the extended states in the RDM is at least the system size, such states move through the crystal ballistically with a constant group velocity $[v(k)]$, except when they are located at the bottom or top of the band, where the velocity vanishes. Because all the other electronic states are localized, we determine the diffusion constant by integrating $v(k)\ell(k)$ over the width of states that participate in the transport. The upper limit of the integration is then proportional to the total fraction of unscattered states, or $\ell(k)/\sqrt{N}$ and $\ell(k) \approx N$. When the velocity is a nonzero constant, we obtain that $D \approx \sqrt{N}$. Because the states that contribute to transport traverse the length of the system with a constant velocity, the distance that they cover scales as a linear function of time. Consequently, $N \propto t$ and $D \approx t^{1/2}$. Consequently, the mean-square displacement grows as $t^{3/2}$. At the bottom or top of the band, where the group velocity vanishes, $v(k) \propto k$ and $D \propto O(l)$. This illustrates that the narrow band of extended states in the RDM is sufficient to give rise to delocalization of an initially localized particle. It is certainly a curious feature of the RDM that so few states do so much.

Nonetheless, the RDM is completely compatible with the localization criterion based on the imaginary part of the self-energy. On this account, if as E approaches the real axis, $\mathrm{Im}S(E) \to 0$, then the eigenstates are localized. $\mathrm{Im}S(E)$ will be nonzero only if $G(E)$ has an imaginary part or, equivalently, has a branch point singularity. An infinitesimal number of extended states is not sufficient in the thermodynamic limit to give rise to a branch cut of nonvanishing width. Consequently, $\mathrm{Im}S(E)$ will vanish for all energies in the RDM. The self-energy localization criterion is insensitive to a set of extended states of zero measure. What is surprising is that these states are sufficiently numerous to give rise to long-range transport. Pendry [P1987] has shown that in standard disordered models, isolated states at particular energies remain extended over $\sqrt{N}$ of the lattice sites. However, the number of states that behave in this fashion is exponentially small and, hence, of no consequence.

The essence of the RDM is that at an energy in the band, the defect transfer matrices reduce to the negative of the unit matrix [WGP1992]. Consequently, at the dimer resonance, perfect transmission occurs from one side of the sample to the other. Local dimer correlations create this macroscopic quantum effect. It is straightforward to show that this result holds only if the site energies of the dimer are equal; that is, the dimer is symmetric. A plane of symmetry is a necessary and sufficient condition for a defect to possess a resonant state. The general statement that can be made is as follows: *The standard tendency of disorder to localize electronic states is suppressed at certain energies in the band whenever the defects contain a plane of symmetry.* This is the minimal requirement for extended states to exist in 1d. The RDM is just the simplest model in which the defects possess a plane of symmetry. Any random n-mer will suffice, as shown in Fig (12.9b). In addition, the single defect model shown in Fig. (12.9c) is the off-diagonal dual of the RDM. Similar delocalization occurs in this model, as well [P1993]. Such simple correlated disorder models continue to be an active field of study. In fact, recent experiments in GaAs/AlGaAs heterostructures [B1999] in which RDM correlations were engineered have confirmed the existence of the extended states in the random dimer model.

12.5.2   Insulator-Superconductor Transitions

In the second class of exceptions, localization is thwarted by strong electron interactions. In the first case, we consider the transition from an insulator to a superconductor in thin films. Though single electrons are localized by disorder, electrons forming Cooper pairs can become delocalized in 2d. Experimentally, a direct transition from a superconductor to an insulator in 2d has been observed by two distinct mechanisms. The first is simply to decrease the thickness of the sample [J1989]. This effectively changes the scattering length and, hence, is equivalent to changing the amount of disorder. As a result, Cooper pairs remain intact throughout the transition. However, in the insulator, they are localized, whereas in the superconductor they form a coherent state. To understand how Cooper pairs can be tuned between the insulating and superconducting extremes, consider Fig. (12.11). As illustrated in Fig. (12.11), formation of Cooper pairs is not a sufficient condition for superconductivity. If one envisions dividing a thin film into partitions, as illustrated in Fig. (12.11), insulating behavior obtains if each partition at each snapshot in time has the same number of Cooper pairs. That is, the state is static. However, if the number of pairs fluctuates between partitions, transport of Cooper pairs is possible and superconductivity obtains. As shown in Chapter 11, phase and particle number are conjugate variables in a superconductor. If the phase is sharp, then the particle number is completely indeterminate. Likewise, complete determination of the particle number leads to infinite uncertainty in the phase. However, destruction of phase coherence suppresses

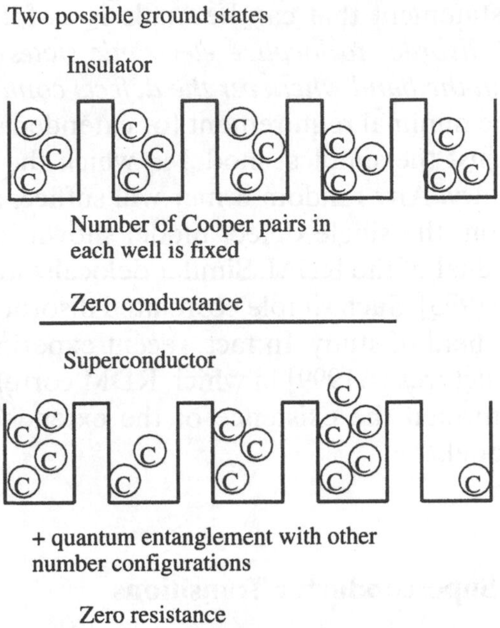

Two possible ground states

Insulator

Number of Cooper pairs in
each well is fixed

Zero conductance

Superconductor

+ quantum entanglement with other
number configurations

Zero resistance

Figure 12.11 Insulating and superconducting ground states of Cooper (C) pairs, illustrating the conjugacy between phase and number fluctuations of the Cooper pairs. In the insulator, Cooper pair number fluctuations cease, leading to infinite uncertainty in the phase. In contrast, in a superconductor, phase coherence obtains, leading thereby to infinite uncertainty in the Cooper pair particle number.

super-conductivity. Hence, a transition from an insulating state of Cooper pairs to a superconducting one obtains when an external parameter is tuned, so that global phase coherence is instated. Film thickness or disorder is one such parameter. At criticality, the resistivity will be independent of temperature. Hence, Cooper pairs on the brink of losing their global phase coherence acquire a finite resistivity at zero temperature, as in a metal. This is a particularly important point because it illustrates that there are three options [J1989, FGG1990] for Cooper pairs when they are confined to a plane: 1) an insulator with infinite resistivity, 2) a superconductor with vanishing resistivity, and 3) a critical metal with finite resistivity at zero temperature. Only the first is possible for single electrons. For Cooper pairs, phase coherence is maintained at sufficiently weak disorder, but above a critical value of the disorder, phase coherence is lost and insulating behavior with a divergent resistivity obtains. Regardless of the mechanism that drives an

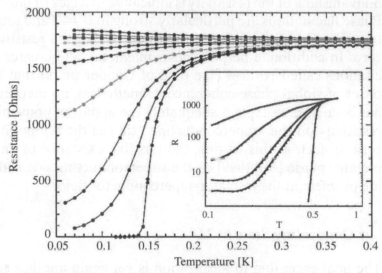

Figure 12.12 Electrical resistance of MoGe thin film plotted vs. temperature at B = 0, 0.5, 1.0, 2.0, 3.0, 4.0, 4.4, 4.5, 5.5, 6 kG. The sample becomes a superconductor at 0.15 K in zero field but for fields larger than about 4.4 kG the sample becomes insulating. At fields lower than this but other than zero, the resistance saturates. The saturation behavior is better shown in the inset for another sample with a higher transition temperature. The inset shows data for B = 0, 1.5, 2, 4, and 7kOe. At higher fields, this sample is an insulator. The data are taken from A. Yazdani and A. Kapitulnik, *Phys. Rev. Lett.* **74**, 3037 (1995).

insulator-superconductor transition, it is the uncertainty relationship between the phase and the particle number that ultimately leads to an evolution between the zero-resistance and divergent-resistance states.

The second means by which a superconducting state can be transformed to an insulator in 2d is by applying a perpendicular magnetic field [YK1995, MK1999, HP1990] on the superconducting side. Shown in Fig. (12.12) is the resistivity as a function of temperature for a magnetic field-tuned insulator-superconductor transition. Above a critical value of the magnetic field, the Cooper pairs are destroyed and insulating behavior obtains by means of single-particle localization. Once again, the resistivity is independent of temperature at the separatrix between these two phases. However, the surprise with this data is the presence of an apparent flattening (see inset, Fig. 12.12) of the resistivity at low temperatures on the "superconducting" side. That is, rather than continuing to drop to zero, as is expected for a superconductor, the resistivity saturates [YK1995, E1996]. Identical behavior has been observed in the thickness-tuned transitions, as well. The nonvanishing of the resistivity is indicative of a lack of phase coherence. Phase fluctuations are particularly strong in d = 2 and are well known to widen the temperature regime over which the resistivity drops to zero. In addition, a perpendicular magnetic field creates resistive excitations called *vortices* (the dual of Cooper pairs) that frustrate the onset of global-phase coherence. Nonetheless, no theoretical account has been able to explain adequately the apparent nonvanishing of the resistivity on the "superconducting" side as the temperature tends to zero. In fact, at this writing, the possible existence of an intervening metallic phase [MK1999] on the superconducting side is the outstanding problem in the insulator-superconductor field.

12.5.3   New Conducting State in 2D

The final exception to localization is yet again another surprise from electrons moving in a plane. In the experiments revealing the possible existence of a new conducting phase, the tuning parameter is the concentration of charge carriers [K1996, PFW1997, S1998, H1998]. Electrons (or holes) were confined to move laterally at the interface between two semiconductors, as in an Si metal-oxide-semiconductor-field-effect transistor (MOSFET) or in a quantum well, as in the case of GaAs. In both systems, the electron density is controlled by adjusting the bias voltage at the gate. For electrons (holes), the more positive (negative) the bias, the higher the electron density. Devices of this sort are identical to those used in the study of the quantum Hall effect (see Chapter 13). As illustrated in Fig. (12.13), when the electron density is slowly increased beyond $\approx 10^{11}/cm^2$, the resistivity changes from increasing (insulating behavior) to decreasing as the temperature decreases, the signature of conducting behavior. At the transition between these two limits, the resistivity is virtually independent of temperature. Though it is still unclear ultimately what value the resistivity will acquire at zero temperature, the marked decrease in the resistivity above a certain density is totally unexpected and, more importantly, not predicted by any theory. Whether we can correctly conclude that a zero-temperature transition exists between two distinct phases of matter is still not settled, however. Nonetheless, the data do possess a feature common to quantum phase transitions such as insulator-superconductor transitions [S2000], namely, scale invariance. In this context, scale invariance simply implies that the data above the flat region in Fig. (12.13) all look alike. This also holds for the data below the flat region in Fig. (12.13). As a consequence, the upper and

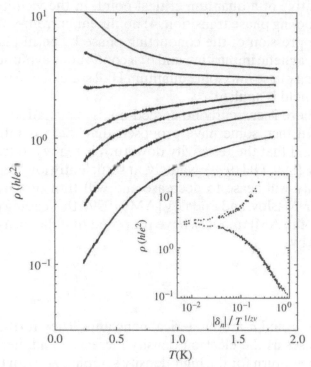

Figure 12.13 Resistivity (ρ) vs. temperature for two-dimensional electrons in silicon in zero magnetic field and at different electron densities (n) (from top to bottom: 0.86, 0.88, 0.90, 0.93, 0.95, 0.99, and 1.10 $\times$ 10^{11} per cm^2). Collapse of the data onto two distinct scaling curves above and below the critical transition density (n_c) is shown in the inset. Here, $\delta = (n - n_c)/n_c$, $z = 0.8$, and $\nu = 1.5$.

lower family of resistivity curves at various densities can all be made to collapse onto just two distinct curves by scaling each curve with the same density-dependent scale factor. The resultant curves have slopes of opposite sign, as shown in the inset of Fig. (12.13). The product of exponents that arises in this case is the correlation length (or localization length) exponent, as well as the dynamical exponent, z. In quantum phase transitions, an energy scale vanishes at the critical point as $\Omega \propto \xi^{-z}$. It is difficult to reconcile such scaling behavior unless the two phases are electrically distinct at zero temperature.

Key experimental features [SK1999] that characterize the transition are as follows: 1) the existence of a critical electron or hole density, n_c, above which the conducting phase appears, 2) a characteristic temperature, typically on the order of half the Fermi temperature, T_F, below which the resistivity on the conducting side decreases roughly as $\rho_1 + \rho_0(T)$ where $\rho_0(T) \approx \rho_0 \exp(-T_0/T)$, 3) critical scaling, indicative of a quantum critical point, in the vicinity of the insulator-conducting phase transition, 4) nonlinear current-voltage (I-V) curves, 5) suppression of the conducting phase by an in-plane magnetic field and magnetic impurities, and 6) a continuous evolution of the new conducting phase into some quantum Hall state once a perpendicular magnetic field is applied.

There is currently no agreed-upon explanation for these phenomena. In fact, some have proposed that there is ultimately no transition and that the resistivity downturn is really a refrigeration problem [AMP1999, DH1999]. Namely, at sufficiently low temperatures, the resistivity will cease to decrease and will turn upward as $T \rightarrow 0$. Altshuler, Maslov, and Pudalov [AMP1999], the leading proponents of this view (the AMP model), have proposed that the turn-up temperature is given by

$$T_{\min} = \frac{pa}{2} \frac{\rho_1^2}{d\rho_0/dT|_{T=T_{\min}}}, \tag{12.45}$$

where p and a are numerical constants. They further claim that $T_{\min}$ increases as the electron density increases, and, hence, it is easiest to see the upturn for the high-density samples. As with the Mott "minimal metallic conductivity," the claim that $T_{\min}$ should increase with electron density is easily falsifiable, as all of the parameters are experimentally available. Shown in Fig. (12.14) is a determination of $T_{\min}$ using the experimental values for ρ_1 and $d\rho_0/dT|_{T=T_{\min}}$ in [P1999]. As the figure clearly shows, $T_{\min}$ predicted by the AMP model *decreases* (roughly as $1/n_e$) rather than increases with density. Hence, the hypothesis that the data are consistent with an incipient conducting phase remains unproven.

Should the data really describe a true conducting phase at $T = 0$, then they raise the natural question, what is so special about the density regime probed? We know definitively that at high and ultra-low densities, a 2d electron gas is localized by disorder. These regimes are distinguished by the magnitude of the Coulomb interaction between the electrons relative to their kinetic energy. As discussed in Chapter 5, because the Coulomb interaction decays as $1/r$ (with r the separation between the electrons) whereas the kinetic energy decays as $1/r^2$, Coulomb interactions dominate at low density. Consequently, in the high-density regime, electrons are essentially noninteracting and insulating behavior obtains, as dictated by the Anderson localization principle. In the ultra-low-density regime where the kinetic energy is almost zero, the Coulomb interaction drives the physics. In this limit, the

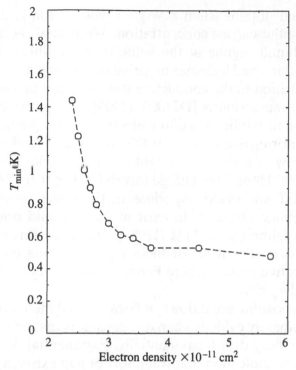

Figure 12.14 Each circle represents a calculation of T_{min} determined by Eq. (12.45), as a function of electron density for each of the eleven Si-MOSFET samples reported in the inset of Fig. 3 in [P1999]. We see explicitly that T_{min} decreases as density increases. This figure is reprinted from P. Phillips, *Phys. Rev.* **64**, 113202/1 (2001).

electrons minimize the Coulomb interaction by being as far away from one another as possible, resulting in a crystalline array or Wigner crystal that is pinned by any amount of disorder and, hence, is an insulator. It is precisely between the Wigner limit and the noninteracting regime that the new conducting phase resides. As illustrated in Fig. (12.15), this density regime represents one of the yet-unconquered frontiers in solid state physics, as nothing definitive is known about the ground

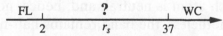

Figure 12.15 Density regimes and what is known definitively about them for a 2d electron gas. At low density, a Wigner crystal (WC) forms, whereas at high density, a Fermi liquid (FL) arises. Both of these phases are localized by disorder. The Wigner crystal melting boundary has been determined by B. Tanatar and D. Ceperley, *Phys. Rev.* B **39**, 5005 (1989). The new conducting phase lies between these two regimes. It is unclear what state of matter forms in this regime.

state that obtains when a Wigner-type crystal is destroyed upon increasing the carrier concentration. We placed the termination of the Fermi liquid regime at the value of r_s at which the compressibility changes sign and becomes negative (see Problem 8.5). Experimentally, the transition to the conducting state is accompanied by a sign change in the compressibility [DJ2000, I2000]. A negative compressibility is a common instability in a dilute electron system, signifying an instability to a uniform-density state. If the compensating charge background is rigid, however, the ions cannot move to mediate a nonuniform electron state. Hence, the charge instability is not realized. Prior to the Si MOSFET and GaAs experiments, there was some anticipation that if a metallic phase could exist in 2d, it would obtain in the perturbative regime ($r_s \ll 1$) [LR1985]. However, this does not appear to be the case. The new conducting phase occurs precisely in the nonperturbative regime where Fermi-liquid theory cannot be assumed to be valid *a priori*.

The possible breakdown of Fermi-liquid theory in the dilute regime is immediately evident when one considers that on the conducting side, the resistivity drops exponentially. Exponential decrease of the resistivity is an indication that some sort of gap exists in the single-particle spectrum. Fermi liquids by definition cannot have a gap in the single-particle spectrum. In fact, no traditional metal has a gap in the single-particle spectrum. The only phase we know of that has a gap in the single-particle spectrum that conducts at zero temperature is a superconductor. In fact, the only excitations proven to survive the localizing effect of disorder in 2d are Cooper pairs. It is partly for these clear physical reasons [P1998] and for other reasons relating to disorder [BK1998] that superconductivity has been proposed to explain the new conducting state in 2d.

Wigner crystals share a class resemblance to their correlated cousin, Mott insulators. The original argument [M1949] for a Mott insulating state was based on a lattice of one-electron atoms. Mott proposed that as the lattice separation increases, the overlap can decrease so drastically that the electronic band reduces to the localized atomic states. In this limit, each atom is neutral, and, hence, no transport of charge is possible. Nonetheless, the band remains half-filled. Consequently, the Mott state is unique in solid state physics in that the chemical potential lies in a gap but partially filled states exist below the Fermi energy. Like the Wigner crystal, the Mott insulating state obtains in the dilute regime where the Coulomb effects dominate.

Dilute electron systems, such as Mott insulators, are susceptible to numerous instabilities, the most notable of which is superconductivity. For example, the entire family of organic conductors [M1997] and the copper oxide high-temperature superconductors [A1992] are Mott insulators in the undoped state. Upon doping, both classes of materials become superconducting. In fact, it appears to be an experimental fact that doped Mott insulators composed from layered materials exhibit superconductivity in the vicinity of a charge or spin-ordered state. Why this state of affairs obtains continues to be one of the lasting puzzles in solid state physics. By analogy, some have proposed [P1998] that a 2d electron gas in the vicinity of the Wigner crystal melting boundary should behave no differently. But this proposal remains unproven at this point, as it is not known whether the charge carriers are $2e$ or e in the new conducting phase. Clearly, further experiments are needed to settle this question regarding the new conducting state in two dimensions and the possible relationship, if any, with the leveling of the resistance in insulator-superconductor transitions at low temperatures.

12.6 SUMMARY

Localization obtains from phase-coherent backscattering. Such processes lead to a nonvanishing of the electron return probability and the subsequent absence of a metallic state for $d \leq 2$. Application of a magnetic field or the introduction of nearest-neighbor correlations of the dimer type can counteract the natural tendency of disorder to localize electronic states. These seemingly unrelated effects share a common origin, however. In the context of a magnetic field, the breaking of the discrete symmetry of time reversal diminishes the contribution of electron backscattering to the conductivity. In the random dimer model, a transmission resonance in the presence of defects that possess a plane of symmetry leads to a vanishing of the reflected or backscattered wave. It is the complete absence of backscattering that gives rise to long-range transport in models of the RDM type. The other exceptions to localization occur in two dimensions and involve the electron interaction. Both problems are currently not settled, either experimentally or theoretically.

PROBLEMS

1. Consider a linear chain with a single defect placed at the mth site. Let the site amplitude be c_n and W the strength of the defect in a

linear chain described by the following evolution equation:

$$Ec_n = \epsilon c_n + V(c_{n+1} + c_{n-1}) + W\delta_{nm}c_m. \qquad (12.46)$$

Use the method of defects to show that a single bound state forms outside the continuous band for $W \neq 0$. To implement this method, use the following procedure: 1) Multiply the eigenvalue equation by e^{ikn} and sum over n. 2) Then multiply the resultant equation by e^{-ikm} and integrate to obtain as a condition for the location of the bound state

$$1 = W\langle G(E)\rangle, \qquad (12.47)$$

where

$$\langle G(E)\rangle = \frac{1}{2\pi}\int_{-\pi}^{\pi}\frac{dk}{E - \epsilon(k)} \qquad (12.48)$$

with $\epsilon(k) = \epsilon + 2V\cos k$. Show that if $W > 0$, the bound state lies above the band, that is, at an energy $E > \epsilon + 2V$, whereas for $W < 0$, the bound state lies below the band. Repeat the same calculation for d $= 2$.

2. In the presence of a magnetic field, the total change of the conductivity, $\delta(H) - \delta(H = 0)$, depends on the difference between P_H and $P_H = 0$. Use Eq. (12.34) to calculate this difference and integrate the result using Eq. (12.26). Define $x = 2HD_0\tau_{in}/c$ as a measure of the field strength. Show that when $x \ll 1$, $\Delta(H) \propto H^2\tau_{in}^2$. In the opposite limit $x \gg 1$, show that $\Delta(H) \propto \ln(H\tau_{in})$.

3. Show that the extended state in the off-diagonal dual of the RDM (see Fig. 12.9c) differs from a perfect Bloch state by only a multiplicative factor $V/\tilde{V}$ at the site of each defect.

4. Convert the sum to an integral in Eq. (12.30) and evaluate the conductivity for a thin film (d $= 2$) and for a wire (d $= 1$). You should obtain Eqs. (12.31) and (12.32), respectively.

5. Assume that in the Ohmic regime, the general form for the β-function is

$$\beta = d - 2 - \frac{A_d}{g}. \qquad (12.49)$$

Integrate this quantity in the interval $[L_0, L]$ and obtain the explicit length dependence for the conductance for d $= 1$ and d $= 3$, respectively. Show that $s = 1$, provided that $\epsilon = d - 2 \ll 1$.

REFERENCES

[AAT1973] R. Abou-Chacra, P. W. Anderson, D. J. Thouless, *J. Phys.* C **6**, 1734 (1973).

[A1979] E. Abrahams, P. W. Abrahams, D. C. Licciardello, T. V. Ramakrishnan, *Phys. Rev. Lett.* **42**, 673 (1979).

[A1981] S. Alexander, J. Bernasconi, W. R. Schneider, R. Orbach, *Rev. Mod. Phys.* **53**, 175 (1981).

[A1982] B. L. Altshuler, A. G. Aronov, D. E. Khmel'nitskii, A. I. Larkin, in *Quantum Theory of Solids*, ed. I. M. Lifshits (MIR Publishers, Moscow) 1982.

[AMP2000] B. L. Altshuler, D. L. Maslov, V. M. Pudalov, *Phys. Stat. Sol.* (b) **218**, 193 (2000).

[A1958] P. W. Anderson, *Phys. Rev.* **109**, 1492 (1958).

[A1992] P. W. Anderson, *Science* **256**, 1526 (1992).

[B1999] V. Bellani, E. Diez, R. Hey, L. Toni, L. Tarricone, G. B. Parravicini, F. Dominguez-Adame, R. Gomez-Alcal, *Phys. Rev. Lett.* **82**, 2159 (1999).

[BK1998] D. Belitz, T. R. Kirkpatrick, *Phys. Rev.* B **58**, 8214 (1998).

[B1982] G. Bergmann, *Phys. Rev.* B **25**, 2937 (1982).

[BG1995] M. A. Blachly, N. Giordano, *Phys. Rev* B **51**, 12537 (1995).

[B1963] R. E. Borland, *Proc. R. Soc. London Ser. A* **274**, 529 (1963).

[B1992] A. Bovier, *J. Phys. A* **25**, 1021 (1992).

[DH1999] S. Das Sarma, E. H. Hwang, *Phys. Rev. Lett.* **83**, 164 (1999).

[DO1979] G. J. Dolan, D. D. Osheroff, *Phys. Rev. Lett.* **43**, 721 (1979).

[DJ2000] S. C. Dultz, H. W. Jiang, *Phys. Rev. Lett.* **84**, 4689 (2000).

[DWP1990] D. H. Dunlap, H.-L. Wu, P. Phillips, *Phys. Rev. Lett.* **65**, 88 (1990).

[EC1970] E. N. Economou, M. H. Cohen, *Phys. Rev. Lett.* **24**, 1445 (1970).

[E1996] D. E. Ephron, A. Yazdani, A. Kapitulnik, M. Beasley, *Phys. Rev. Lett.* **76**, 529 (1996).

[FGG1990] M. P. A. Fisher, G. Grinstein, S. Girvin, *Phys. Rev. Lett.* **64**, 587 (1990).

[FFG1955] G. Feher, R. C. Fletcher, E. A. Gere, *Phys. Rev.* **100**, 1784 (1955).

[F1989] J. C. Flores, *J. Phys. Condens. Matt.* **1**, 8471 (1989).

[GM1998] A. M. Goldman, N. Markovic, *Physics Today* **51**, 39 (1998).

[H1998] Y. Hanien, et al., *Phys. Rev. Lett.* **80**, 1288 (1998).

[HP1990] A. F. Hebard, M. A. Paalanen, *Phys. Rev. Lett.* **65**, 927 (1990).

[I1986] Y. Imry, *Europhys. Lett.* **1**, 249 (1986).

[J1989] H. M. Jaeger, D. B. Haviland, B. G. Orr, A. M. Goldman, *Phys. Rev.* B **40**, 182 (1989).

[I2000] S. Ilani, A. Yacoby, D. Mahalu, H. Shtrikman, *Phys. Rev. Lett.* **84**, 3133 (2000).

[JK1986] R. Johnston, B. Kramer, *Z. Phys.* B **63**, 273 (1986).

[K1996] S. V. Kravchenko, D. Simonian, M. P. Sarachik, W. Mason, J. E. Furneaux, *Phys. Rev. Lett.* **77**, 4938 (1996).

[LN1966] J. S. Langer, T. Neal, *Phys. Rev. Lett.* **16**, 984 (1966).

[LR1985] P. A. Lee, T. V. Ramakrishnan, *Rev. Mod. Phys.* **57**, 287 (1985).

[LT1975] D. C. Licciardello, D. J. Thouless, *J. Phys.* C **8**, 4157 (1975).

[LG1987] J. J. Lin, N. Giordano, *Phys. Rev.* B **35**, 1071 (1987).

[MWP1997] I. Martin, Y. Wan, P. Phillips, *Phys. Rev. Lett.* **78**, 114 (1997).

[MK1999] N. Mason, A. Kapitulnik, *Phys. Rev. Lett.* **82**, 5341 (1999).

[M1997] R. H. McKenzie, *Science* **278**, 820 (1997).

[MJW1997] P. Mohanty, E. M. Q. Jariwala, R. A. Webb, *Phys. Rev. Lett.* **78**, 3366 (1997).

[M1949] N. F. Mott, *Proc. Phys. Soc. London Sect. A* **62**, 416 (1949).

[M1972] N. F. Mott, *Phil. Mag.* **26**, 1015 (1972).

[MT1961] N. F. Mott, W. D. Twose, *Adv. Phys.* **10**, 107 (1961).

[M1990] K. A. Muttalib, *Phys. Rev. Lett.* **65**, 745 (1990).

[P1987] J. B. Pendry, *J. Phys.* C **20**, 733 (1987).

[P1993] P. Phillips, *Ann. Rev. Phys. Chem.* **44**, 115 (1993).

[P1998] P. Phillips, Y. Wan, I. Martin, S. Knysh, D. Dalidovich, *Nature* **395**, 253 (1998).

[PFW1997] D. Popovic, A. B. Fowler, S. Washburn, *Phys. Rev. Lett.* **79**, 1543 (1997).

[P1999] V. M. Pudalov, G. Brunthaler, A. Prinz, G. Bauer, *JETP Lett.* **70**, 48 (1999).

[SK1999] M. P. Sarachik, S. V. Kravchenko, *Proc. Natl. Acad. Sci.* **96**, 5900 (1999).

[S2000] S. Sachdev, *Quantum Phase Transitions* (Cambridge University Press, Cambridge, 1999).

[S1998] M. Y. Simmons, et al., *Phys. Rev. Lett.* **80**, 1292 (1998).

[SE1984] C. M. Soukoulis, E. N. Economou, *Phys. Rev. Lett.* **52**, 565 (1984).

[T1982] G. A. Thomas, Y. Ootuka, S. Katsumoto, S. Kobayashi, W. Sasaki, *Phys. Rev. B* **25**, 4288 (1982).

[T1977] D. J. Thouless, *Phys. Rev. Lett.* **39**, 1167 (1977).

[VW1980] D. Vollhardt, P. Wölfle, *Phys. Rev. Lett.* **45**, 482 (1980).

[W1976] F. Wegner, *Z. Phys. B* **25**, 327 (1976).

[W1997] P. Wochner, X. Xiong, S. C. Moss, *J. Superconductivity*, **10**, 367(1997).

[WGP1992] H.-L. Wu, W. E. Goff, P. Phillips, *Phys. Rev. B* **45**, 1623 (1992).

[YK1995] A. Yazdani, A. Kapitulnik, *Phys. Rev. Lett.* **74**, 3037 (1995).

– 13 –

Quantum Phase Transitions

In the previous chapter, we discussed both the insulator-superconductor and the insulator-metal transitions. As such transitions are disorder- or magnetic field–tuned, thermal fluctuations play no role. Phase transitions of the insulator-superconductor or insulator-metal type are called *quantum phase transitions*. Such phase transitions are not controlled by changing the temperature, as in the melting of ice or the λ-point of liquid helium, but rather by changing some system parameter, such as the number of defects or the concentration of charge carriers. In all such instances, the tuning parameter transforms the system between quantum mechanical states that either look different topologically (as in the transition between localized or extended electronic states) or have distinctly different magnetic properties. As quantum mechanics underlies such phase transitions, all quantum phase transitions obtain at the absolute zero of temperature and thus are governed by a $T = 0$ quantum critical point. Although initially surprising, this state of affairs is expected, because quantum mechanics is explicitly a zero-temperature theory of matter. Of course, this is no surprise to chemists, who have known for quite some time that numerous materials can exhibit vastly distinct properties simply by changing the chemical composition and, most importantly, that such transformations persist down to zero temperature. Common examples include turning insulators, such as the layered cuprates, into superconductors simply by chemical doping, or semiconductors into metals, once again by doping, or ferromagnets, such as $Li(Ho, Tb)_x Y_{1-x} F_4$, into a spin glass [AR1998] by altering x. Certainly, the technological relevance of doped semiconductors, the backbone of the electronics industry, is well established.

Given the ubiquity and importance of quantum phase transitions, it would seem that a theory of such phenomena would be well developed. However, such is not the case. The problem in constructing a theory of quantum phase transitions (QPTs) lies in the fact that they occur at zero temperature. All phase transitions are driven by some type of fluctuation. As thermal fluctuations desist at zero temperature, QPTs must be driven by a distinctly different entity. The only fluctuations that survive at zero temperature are quantum mechanical in origin. Hence, the uncertainty principle lies at the heart of all QPTs. Consequently, present in any microscopic Hamiltonian that admits a QPT must be two noncommuting operators that describe two competing ordering tendencies. In all QPTs, the magnitude of the coupling constant describes the *essential tension* between the states that compete.

A clear example of this *essential tension* is found in the Hamiltonian for the Ising model in a transverse magnetic field [P1979]. Consider a linear chain of spins: each spin is in an eigenstate of S^z, the z-component of the spin. Nearest-neighbor spins interact ferromagnetically with a strength $-J$. Also present is a field that tends to align individual spins along the x-direction. A spin along the $\pm x$−axis is formed from the linear combination $|\text{up}\rangle \pm |\text{down}\rangle$. The Hamiltonian describing this system,

$$H_{\text{spin}} = -J \sum_j (g \hat{\sigma}_j^x + \hat{\sigma}_j^z \hat{\sigma}_{j+1}^z), \qquad (13.1)$$

contains the Pauli matrices for the transverse as well as the z-component of the spins, $\hat{\sigma}_j^x$ and $\hat{\sigma}_j^z$, respectively. These are the two noncommuting operators that compete and ultimately determine the existence of the quantum critical point. Though this Hamiltonian is simple, it is of immense utility, as it can be generalized [S2000] to describe the magnetic system $Li(Ho, Tb)_x Y_{1-x} F_4$. When $g = 0$, neighboring spins lower their energy by pointing parallel to one another. The overall system is thus in an eigenstate of the total S^z operator. The term proportional to g in Eq. (13.1) favors a transverse orientation of the spins. When $g = \infty$, all the spins point along the $+x$ or $-x$ direction and, hence, the system is an eigenstate of S^x. Consequently, both the $g = 0$ and $g = \infty$ limits represent stationary states of the system in which $\langle \hat{\sigma}^z \rangle$ and $\langle \hat{\sigma}^x \rangle$, respectively, have well-defined values. In between these two limits, the system is not in a stationary state, as local excitations exist that tend to destroy the perfect $g = 0$ or $g = \infty$ order. Figure (13.1) illustrates that

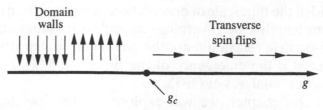

Figure 13.1 Local excitations that occur on either side of the quantum phase transition in the transverse Ising model. As the coupling constant g increases, the magnitude of the transverse field increases with a critical point at $g_c = 1$.

close to $g = 0$, local excitations consist of a region or a domain wall of spins that all point in a direction opposite to the dominant order in the $g = 0$ ground state. Likewise, for $g \gg 1$, the excitations all resemble local transverse spin flips, also shown in Fig. (13.1). Given that the ground states at $g = 0$ and $g = \infty$ are magnetically distinct, as are their quasi-particle excitations, it stands to reason that there should be a critical point at some finite value of g that signals the termination of the $g = 0$ ordered state. In fact, such a critical point does exist at $g = g_c = 1$, as shown by Pfeuty [P1979] in his exact solution of this model in 1d. For $g < 1$, the ground state is qualitatively similar to the $g = 0$ ferromagnet, whereas for $g > 1$, the ground state resembles the $g = \infty$ state with all spins aligned along the transverse direction. The coupling constant g represents the *essential tension* between the two ordering tendencies, with $g = g_c = 1$ defining the quantum critical point. In the $g \ll 1$ limit, $\hat{\sigma}^x$ has no well-defined value, whereas in the opposite extreme, the uncertainty in the value of $\hat{\sigma}^z$ is essentially infinite. Consequently, the product of the uncertainty in the x- and z-components of the spin remains constant as g is varied as dictated by the Heisenberg uncertainty principle. We see, then, that quantum uncertainty drives the phase transition as g is tuned.

At g_c, the ground state is not particularly straightforward. What we know about the ground state at g_c is that it exhibits scale invariance. That is, the correlations between spins are sufficiently long-range that if we increase the length scale over which we observe two spins, nothing changes. As a consequence, the ground-state wave function tells us nothing about how far apart two spins are. This picture is expected to change at finite temperature, however, as finite temperature relaxation of the quasi particles affects the purely quantum mechanical nature of the problem. The natural timescale that enters is $\hbar/k_B T$. Hence, whereas QPTs occur at $T = 0$, remnants of what is quantum mechanical about the phase transition can be seen at finite temperature, provided the timescale of observation satisfies certain constraints. A diverging length scale governing physical correlations and a timescale for quasi-particle relaxation are the key features characterizing all QPTs. In fact, it is the emergence of the timescale $\hbar/k_B T$ that compounds theoretical analyses of QPTs.

In this chapter, we will explore the intertwining of spatial and time dimensions as they determine the behavior of a quantum critical point. Rather than analyze the spin model, we will explore the phase-only model for insulator-superconductor transitions. In this model, the QPT is governed by fluctuations of the quantum mechanical phase of the order parameter. Though we just as easily could have focused on a model in which the pair amplitude fluctuates, the phase-only model is particularly instructive because of its direct applicability to Josephson junction arrays, and further, it serves to illustrate beautifully the central role played by the uncertainty principle. Our analysis will be generally heuristic, except where a calculation must be performed. Emphasis will be placed on the essential tension embodied in the coupling constant that drives the underlying quantum critical point.

13.1 QUANTUM ROTOR MODEL

The simplest way of introducing the phase-only model for an insulator-superconductor transition is to consider an array of superconducting islands placed on a 2d square lattice. All electrons on the grains are locked into Cooper pairs and are characterized by the same phase. Consequently, an essential ingredient of the phase-only model is that the pair amplitude remains frozen for all time. Hence, we associate with the j th grain, a pair amplitude, Δ, a unique phase, θ_j, and a complex wave function or order parameter, $\psi_j = \Delta e^{i\theta_j}$. Further, for simplicity, we will treat the Cooper pairs as bosons. There are two contributions to the energy of such a system: 1) the net energy of each superconducting grain, and 2) the transport energy between two grains. Consider a grain of radius R and with a capacitance C relative to its surroundings. The energy required to remove a Cooper pair with charge $2e$ from such a grain and place it at infinity is $E_C = (2e)^2/2C$, the capacitance-charging energy. There are two important features of E_C. First, E_C scales as the square of the charge and, hence, must be proportional to the N^2, where N is the total number of Cooper pairs on the grain. Second, E_C is inversely proportional to the capacitance. Hence, the capacitance term dominates as the grain size decreases. Transport of Cooper pairs between grains i and j requires a term of the form, $b_i^\dagger b_j$, where b_j is the annihilation operator for a Cooper pair on grain j. Consequently, our Hamiltonian is of the form

$$H = E_C \sum_i (\hat{n}_i - n_0)^2 - t \sum_{\langle ij \rangle} (b_i^\dagger b_j + b_j^\dagger b_i), \quad (13.2)$$

where $\hat{n}_i = b_i^\dagger b_i$ is the number operator for the ith grain, n_0 sets the average density on each grain, and $-t$ is the hopping matrix element between nearest-neighbor grains. The summation, $\langle ij \rangle$, is restricted to nearest-neighbor sites only. Known as the *Bose-Hubbard model*, Eq. (13.2) is closely related to its fermion analog discussed in Chapter 9. That Eq. (13.2) possesses two distinct limits can be seen by first setting the kinetic term to zero. In this limit, all Cooper pairs are frozen on their respective grains. In the opposite regime, $E_C = 0$, free transport of charge is permitted.

To see that the free transport and frozen charge regimes correspond to superconducting and "insulating" states, we make a transformation to phase-angle variables. To this end, we set n_0 to be an integer. In this case, the translation $n_j \rightarrow n_j + n_0$ is permitted. Second, we note that in a superconducting condensate, a conjugacy relationship exists between the phase and the particle number. This conjugacy allows us to equate the particle number for a pair on each grain with the phase momentum for a pair: $\hat{n}_j = \partial/i\partial\theta_j$. Third, we set $b_j^\dagger = \sqrt{n_0} e^{i\theta_j}$. Combining these transformations results in the phase-only model

$$H = E_C \sum_j \left(\frac{1}{i} \frac{\partial}{\partial \theta_j} \right)^2 - J \sum_{\langle ij \rangle} \cos (\theta_i - \theta_j) = H_0 + H_1, \quad (13.3)$$

also known as the *quantum rotor model*. We set $tn_0 = J$ because the second term in Eq. (13.3) is of the Josephson-tunneling form. The physics of this model is now immediately transparent. From the first term in Eq. (13.3), we see that the charging effects lead to fluctuations in the phase, whereas from the second term, phase coherence maximizes transport of Cooper pairs between grains. Consequently, we define the coupling constant, $g = E_C/J$, which embodies the *essential tension* between superconductivity and the loss of phase coherence. In the limit in which $g = 0$, there is no cost to charge a grain, and Cooper pairs transport freely via Josephson tunneling. Such a system has no

phase fluctuations and, hence, is a superconductor. In the supercon-
ductor, $\langle \psi_j \rangle \neq 0$. In the opposite regime, $g = \infty$, charging effects
dominate and the phase is randomized, thereby destroying super-
conductivity. In this limit, the system is in a ground state of kinetic energy and there is an almost equal probability of observing
all possible phase angles. As charges are not permitted to transport freely in this regime, it is natural to expect an insulator in
this limit. However, as we will see, this is not exactly so. We will
show [DP2000] that while this phase is gapped, it still admits a
nonzero dc conductivity as $T \rightarrow 0$. What is clear at present, however,
is that the phase-disordered regime is certainly not a superconductor.
Consequently, the phase-only model displays a superconductor-to-
nonsuperconductor transition somewhere between the limits $g = 0$
and $g = \infty$. The physical states demarcated by the quantum critical point, g_c, are depicted in Fig. (12.11). The superconductor-to-
nonsuperconductor transition occurs when Cooper pairs no longer
move freely between the grains. In this limit, the certainty that results in the particle number within each grain is counterbalanced by
the complete loss of phase coherence. In a superconductor, phase certainty gives rise to infinite uncertainty in the particle number. Consequently, the product of the number uncertainty times the uncertainty
in phase is the same on either side of the transition, as dictated by
the Heisenberg uncertainty principle. Once again, we see that it is ultimately quantum uncertainty that drives the phase transition in the
quantum rotor model. This model is relevant to Josephson junction
arrays in which superconductivity is destroyed once E_C/J exceeds a
critical value, as argued first by Anderson [A1964].

13.2 SCALING

Near a quantum critical point, we anticipate that we can express all
physical quantities in terms of the deviation from criticality, $\delta = (g - g_c)$. In the context of the scaling theory of localization, we established that the localization length diverges at the transition as
$\xi \approx |\delta|^{-\nu}$, where ν is the correlation length exponent. Divergence
of the correlation length at criticality signifies that spatial fluctuations
in two distant regions in the sample are related. However, because
quantum critical points are driven by quantum rather than thermal
fluctuations, an additional correlation "length" must enter the problem. Quantum fluctuations occur in time rather than in space. Consequently, the new correlation "length" represents the timescale over

which the system fluctuates coherently. The correlation time is related to the correlation length

$$\xi_\tau \approx \xi^z \tag{13.4}$$

through the dynamical exponent, z. Because the exponent z is not necessarily unity, the dynamical exponent is a measure of the asymmetry between spatial and temporal fluctuations.

To explore further the significance of the dynamical exponent, we consider the partition function

$$Z = Tr[\exp(-\beta H_{model})] \tag{13.5}$$

for the model Hamiltonian,

$$H_{model} = \frac{p^2}{2m} + V(x) = T + V \tag{13.6}$$

describing a single particle moving in an arbitrary 1d potential, $V(x)$. The trace is performed over all single-particle momentum eigenstates. We evaluate the partition function for this Hamiltonian

$$Z = \lim_{M \to \infty} Tr[\exp(-\beta(T+V)/M)]^M$$

$$\approx \lim_{M \to \infty} Tr[\exp(-\Delta\tau T)\exp(-\Delta\tau V)]^M \tag{13.7}$$

by breaking up the imaginary time interval, $0 \le \tau \le \beta$, into discrete time segments,

$$\Delta\tau = \beta/M, \tag{13.8}$$

and inserting $M - 1$ complete sets of momentum eigenstates. The correct quantum mechanical problem is recovered in the limit in which the time slice $\Delta\tau \to 0$. The subsequent integrations over the intermediate momentum states are standard and can be found on pages 42 and 43 of Feynman and Hibbs [FH1965]. In the limit that $\Delta\tau \to 0$, the final partition function is expressed as an integral over all paths, $x(t)$, with the associated action

$$S_{model} = \int_0^\beta d\tau \left(\frac{m}{2}\dot{x}^2(\tau) - V(x(\tau)) \right). \tag{13.9}$$

The continuous integration in the action over imaginary time results explicitly from taking the $\Delta\tau \to 0$ limit. The analogous treatment can be performed for the quantum rotor model once we make the identification that

$$p_\theta = \frac{1}{i}\frac{\partial}{\partial\theta}. \tag{13.10}$$

Consequently, the quantum rotor Hamiltonian is recast as

$$H = E_C \sum_i p_{\theta_i}^2 - J \sum_{\langle ij \rangle} V(\theta_i - \theta_j) \tag{13.11}$$

with its associated action

$$S = \int_0^\beta d\tau \left(\frac{1}{4E_C} \sum_i (\partial_\tau \theta_i)^2 - J \sum_{\langle ij \rangle} \cos(\theta_i - \theta_j) \right), \tag{13.12}$$

obtained by discretizing the partition function and performing the trace over the momentum quantum rotor eigenstates.

The time integration in the action of the quantum rotor model has two consequences. First, the effective dimensionality of the problem has increased. In addition to the spatial dimensions, we must contend with system evolution along the imaginary time axis. Consequently, the effective dimensionality of the 2d quantum rotor model is $(2 + 1)$. Hence, a quantum mechanical problem in 2d spatial dimensions has the complication of a 3d classical problem. A second consequence is that time in quantum mechanics is imaginary and associated with inverse temperature. Hence, divergence of the correlation time at criticality is associated with a vanishing energy scale. This can be seen immediately from a scaling analysis of the kinetic energy. The kinetic energy is proportional to τ^{-2} and, hence, in the vicinity of the critical point scales as $\xi_\tau^{-2} \approx \xi^{-2z}$, thereby vanishing at criticality. This is an important result. In Newtonian mechanics, the kinetic energy is generally associated with a second gradient with respect to position, $\nabla^2\theta$, thereby scaling as ξ^{-2}. The asymmetry between space and time is immediately apparent as a result of the ξ^{-2z} scaling of the kinetic energy in the imaginary-time Euclidean formalism. However, there is no asymmetry [H1976] between space and time if the number of time derivatives equals the number of spatial derivates in the Euclidean and Newtonian representations, respectively, of the kinetic energy. In such cases $z = 1$, as is the case with the isotropic quantum rotor model (also known as the *(2 + 1)-dimensional XY model)*. Note, the result that $z = 1$ in the quantum rotor model holds only when n_0 is an integer (see Problem 13.1). Only the $z = 1$ case will be treated here.

Experiments are always performed at finite temperature, however. At finite temperature, the correlation time, ξ_τ, is finite. The finite temperature behavior is still controlled by the quantum critical point, provided that the timescale for observation, τ, satisfies the constraint $\tau < \hbar/k_B T$, as advertised in the introduction. Experimental data obtained with this constraint in mind should exhibit universal behavior in terms of ξ_τ and ξ. To see how this works, consider first the case of zero temperature where the standard Widom [W1965] correlation length scaling applies. Consider the observable O that depends explicitly on the variable, X. We assume $X \approx \xi^{d_X}$ and, consequently, $\tilde{X} \approx \xi^{-d_X} X$ is scale invariant. Near the critical point,

$$O(X,\delta) = \xi^{d_O} \tilde{O}(\tilde{X}), \tag{13.13}$$

where d_O is the scaling dimension of the observable, O. This equation has profound experimental consequences. It implies that the family of curves $O(\delta, X)$ all collapse onto a single curve $\tilde{O}$ when $\xi^{-d_O} O(X,\delta)$ is plotted as a function of $\tilde{X}$, as opposed to X. Further, because the quantity $\tilde{X}$ is scale invariant, so is any function of $\tilde{X}$. Hence, $\tilde{O}$ does not depend on the distance from the critical point and, consequently, neither does $\xi^{-d_O} O(X,\delta)$. This guarantees that critical exponents can be extracted accurately. At finite temperature, this procedure must be altered because the system is finite in the time dimension. All variables must be rescaled using Eq. (13.4) and noting that $\xi_\tau < 1/T$. For example, as a result of the finite size, the new scale-invariant quantity is $\tilde{X}_T \approx \xi_\tau^{-d_X/z} X \approx T^{d_X/z} X$. Consequently, the finite temperature form for any observable is

$$O(\delta, T, X) = T^{-d_O/z} \tilde{O}(\xi_\tau T, \tilde{X}_T). \tag{13.14}$$

In the case when X is the electric field, E, the scaling dimension is $d_E = -(z + 1)$. This result follows because the electric field is proportional to an energy divided by a length. Near the critical point, all energies scale as ξ_τ^{-1} and lengths are proportional to ξ. Consequently, $E \approx \xi^{-(z+1)}$ and we anticipate data collapse for the conductivity with an exponent of $-\nu(z + 1)$ for electric field scaling in the vicinity of a quantum critical point, as illustrated in Fig. (13.2) for the nominal insulator-superconductor transition in MoGe thin films. In these films,

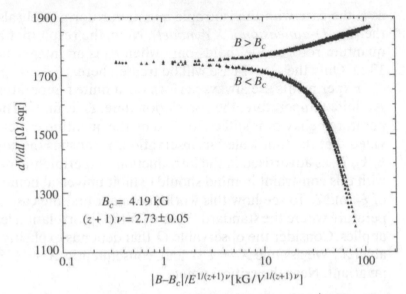

Figure 13.2 Data collapse of the resistivity per square (dV/dI) for the insulator-superconductor transition in MoGe thin films as a function of an applied electric field. The tuning parameter is the magnetic field, B, with a critical value of $B_c = 4.19\,\mathrm{kG}$. From the temperature scaling, $z\nu = 1.36$, whereas $(z+1)\nu = 2.73$ collapses the electric field dependence. The data are taken from A. Yazdani and A. Kapitulnik, *Phys. Rev. Lett.* **74**, 3037 (1995).

the tuning parameter between the insulating and superconducting states is the magnetic field. From the data collapse for both temperature and electric field, $z = 1$ and $\nu = 1.36 \pm 0.05$.

13.3   MEAN-FIELD SOLUTION

The Hamiltonian for the phase-only model contains a nonlinear term in the phase angle difference. As a result of this nonlinearity, this model is intractable in 2d. However, we can obtain a mean-field solution that has the correct qualitative features by minimizing the free energy appearing in the partition function. Rather than write the partition function in terms of the action, we adopt a form that is amenable to perturbation theory. The partition function,

$$Z = Tr \exp\left[-\beta(H_0 + H_1)\right] = Tr \exp\left[\rho\right], \qquad (13.15)$$

is a trace of the density matrix for the total system. Let us define $\rho_0 = \exp\left(-\beta H_0\right)$ and its associated partition function, $Z_0 = Tr\rho_0$, with H_0, the kinetic energy term of the quantum rotor Hamiltonian. Following

Feynman [F1972], we vary $e^{\beta H_0}\rho$ to obtain

$$\frac{\partial}{\partial\beta}(e^{\beta H_0}\rho) = -e^{H_0\beta}H_1\rho. \tag{13.16}$$

The formal solution to this equation is

$$\rho = \rho_0 T \exp\left[-\int_0^\beta d\tau \hat{H}_1(\tau)\right], \tag{13.17}$$

where T represents the time-ordered product. Consequently, the partition function is transformed to

$$Z = Z_0\left\langle T \exp\left(-\int_0^\beta d\tau \hat{H}_1(\tau)\right)\right\rangle, \tag{13.18}$$

with $\langle\cdots\rangle \equiv \langle\rho_0\cdots\rangle/Z_0$ and $\hat{H}_1(\tau)$ the interaction representation of the operator H_1. The partition function is now in a form in which a perturbation series in powers of the interaction can be obtained. Our derivation here will closely mirror the original treatment by Doniach [D1981]. To this end, we define a two-component vector, $\mathbf{S}_i = (\cos\theta_i, \sin\theta_i)$ and its Fourier components,

$$\mathbf{S_k} = \frac{1}{\sqrt{N}}\sum_i \mathbf{S}_i e^{i\mathbf{k}\cdot\mathbf{R}_i}. \tag{13.19}$$

Our working form for the partition function is, then,

$$Z = Z_0\left\langle T \exp\left(-\int_0^\beta d\tau \sum_\mathbf{k} J_\mathbf{k}\hat{\mathbf{S}}_\mathbf{k}(\tau)\cdot\hat{\mathbf{S}}_\mathbf{k}^*(\tau)\right)\right\rangle, \tag{13.20}$$

where

$$J_\mathbf{k} = \frac{J}{N}\sum_{\langle ij\rangle} e^{i\mathbf{k}\cdot\mathbf{R}_{ij}} = J_0(\cos k_x + \cos k_y), \tag{13.21}$$

$J_0 = J\alpha$, and α is the number of nearest neighbors.

Still, an intermediate step is needed. In terms of the discrete Matsubara frequencies, $\omega_l = 2\pi lT$, $l = 0, \pm 1, \pm 2, \cdots$, we recast the spin vectors as

$$\hat{\mathbf{S}}_\mathbf{k}(\tau) = \frac{1}{\sqrt{\beta}}\sum_l e^{i\omega_l\tau}\mathbf{S}_{\mathbf{k}\omega_l} \quad 0 \le \tau \le \beta, \tag{13.22}$$

which are normalized over the imaginary time interval, $0 \leq \tau \leq \beta$. This allows us to simplify the exponential appearing in the partition function,

$$\exp\left[-\int_0^\beta d\tau \sum_{\mathbf{k}} J_{\mathbf{k}} \hat{\mathbf{S}}_{\mathbf{k}}(\tau) \cdot \hat{\mathbf{S}}_{\mathbf{k}}^*(\tau)\right] = \prod_{\mathbf{k},l} \exp\left[-J_{\mathbf{k}}|\mathbf{S}_{\mathbf{k},l}|^2\right], \quad (13.23)$$

to a product. We decouple the quadratic spin interaction in Eq. (13.23) by using the complex form of the Hubbard-Stratonovich transformation [H1959, S1958],

$$e^{-\lambda|a|^2} = \int \int \frac{d(\Re\psi)(\text{Im}\psi)}{\pi} e^{-[|\psi|^2 + \sqrt{\lambda}(a\psi^* + a^*\psi)]}, \quad (13.24)$$

a common trick in many-particle physics used to decouple interaction terms appearing in the partition function. Resulting from this transformation is an added integration over an auxiliary field that can generally be performed by steepest-descent methods. Upon using this transformation in Eq. (13.20), we arrive at a simplified partition function

$$Z = \frac{Z_0}{D} \int \prod \mathscr{D}\psi_{\mathbf{k}}(\tau) e^{\mathscr{F}[\psi]}, \quad (13.25)$$

where $\mathscr{F}$ is the free energy

$$F[\psi_{\mathbf{k}}(\tau)] = \sum_{\mathbf{k}} \int_0^\beta d\tau \frac{\psi_{\mathbf{k}}^*(\tau) \cdot \psi_{\mathbf{k}}(\tau)}{J_{\mathbf{k}}}$$

$$- \ln\left[\left\langle T \exp\left[-2\int_0^\beta d\tau \sum_{\mathbf{k}} \psi_{\mathbf{k}}(\tau) \cdot \hat{\mathbf{S}}_{-\mathbf{k}}(\tau)\right]\right\rangle\right] \quad (13.26)$$

and $D = \pi \prod_{\mathbf{k}} J_{\mathbf{k}}$. In obtaining this form for the free energy, we rescaled the auxilliary field $\psi_{\mathbf{k}} \rightarrow \psi_{\mathbf{k}}/\sqrt{J_{\mathbf{k}}}$.

In the mean-field approximation, we assume that ψ is constant. Further, we are free to choose any orientation of ψ. For convenience, we orient ψ along the x-axis and, consequently, $\psi_{\mathbf{k}} = \sqrt{N}\delta_{\mathbf{k},0}\psi_x$. Substitution of this form for the auxilliary field into Eq.(13.26) results in the mean-field expression

$$F[\psi_{\mathbf{k}}]_{\text{MF}} = \frac{\beta N \psi_x^2}{J_0} - \ln\left[\left\langle T \exp\left[-2\int_0^\beta \psi_x \sum_i S_i^x(\tau) d\tau\right]\right\rangle\right] \quad (13.27)$$

for the free energy. Assuming that ψ is a small parameter, we are free to expand the exponential:

$$\left\langle T \exp\left[-2\int_0^\beta \psi_x \sum_i S_i^x(\tau)d\tau\right]\right\rangle = 1 - 2\psi_x \left\langle \int_0^\beta \sum_i S_i^x(\tau)d\tau \right\rangle$$

$$+ 2\psi_x^2 \left\langle T \int_0^\beta \int_0^\beta d\tau d\tau' \sum_{ij} S_i^x(\tau)S_j^x(\tau') \right\rangle + \cdots.$$

$$(13.28)$$

Physically, the terms in this expansion represent phase fluctuations of the order parameter. Should such terms lead to a sign change in the free energy, a phase transition occurs. To evaluate the averages, we expand in the free-rotor eigenstates

$$\langle m|\theta \rangle = \frac{1}{2\pi}e^{im\theta}. \tag{13.29}$$

In this basis,

$$H_0\langle m_i|\theta_i \rangle = -m_i^2 E_C\langle m_i|\theta_i \rangle, \tag{13.30}$$

leading immediately to a compact form

$$Z_0 = Tre^{-\beta H_0} = \sum_{\{m_i\}} e^{-\beta E_C \sum_i m_i^2} \tag{13.31}$$

for the *zeroth*-order partition function. An additional consequence of this choice of basis is that averages linear in S_i^x vanish identically:

$$\langle m_i|\cos\theta_i|m_i \rangle = \int_0^{2\pi} \frac{d\theta_i}{2\pi} e^{-im_i\theta_i} \cos\theta_i \, e^{im_i\theta_i} = 0. \tag{13.32}$$

Consequently, to lowest order in ψ, only the diagonal elements ($i = j$) in the second-order term in Eq. (13.28) yield a nonzero result. Upon substituting Eq. (13.28) into Eq. (13.27) and differentiating with respect to ψ, we obtain

$$\frac{\beta}{J_0} = 2\left\langle T \int_0^\beta \int_0^\beta d\tau d\tau' S_i^x(\tau)S_i^x(\tau') \right\rangle$$

$$= 2\int_0^\beta d\tau d\tau' C_0(\tau - \tau') \tag{13.33}$$

as the saddle-point equation to lowest order, with

$$C_0(\tau - \tau') = \frac{1}{Z_0} \sum_{\{m_i\}} e^{-\beta E_C \sum_i m_i^2} \langle\{m_i\}|T \hat{S}_i^x(\tau)\hat{S}_i^x(\tau')|\{m_i\}\rangle \quad (13.34)$$

the phase fluctuation correlation function. This is the key quantity that enters the mean-field theory of the phase transition in the quantum rotor model.

The only matrix elements of S_i^x that are nonzero are of the form

$$\langle m_i | \cos \theta_i | m_i \pm 1 \rangle = \frac{1}{2}, \quad (13.35)$$

and the correlation function is given by

$$\langle\{m_i\}|T \hat{S}_i^x(\tau)\hat{S}_i^x(\tau')|\{m_i\}\rangle = \frac{1}{2} e^{E_C|\tau-\tau'|} \cosh(2m_i E_C(\tau - \tau')). \quad (13.36)$$

There are two important limits of this expression. At low temperatures such that $E_C \beta \gg 1$, all the rotors can be treated as being in their ground states; that is, $m_i = 0$. In this case, the correlation function reduces to an exponential

$$C_0(\tau - \tau') = \frac{1}{2} e^{-E_C|\tau-\tau'|} \quad \beta E_C \gg 1, \quad (13.37)$$

and the mean-field stability condition

$$1 = \frac{2J_0}{E_C}\left[1 - \frac{1 - e^{\beta E_C}}{\beta E_C}\right] \quad \beta E_C \gg 1 \quad (13.38)$$

follows, once the time integrations in Eq. (13.33) are performed. To leading order in $1/\beta$, we find that the stability condition for the coupling constant, $g = E_C/J_0$, is given by

$$g = 2\left(1 - \frac{T}{E_C}\right) + O(T^2/J). \quad (13.39)$$

Hence, the critical value of the coupling constant at $T = 0$ is $g_c = 2$. For $g > g_c$, quantum fluctuations dominate, and the system is in the phase-disordered regime.

At high temperature, the full expression for the correlation function, Eq. (13.36), must be retained in performing the time integration in Eq. (13.33):

$$\frac{1}{2} \int_0^\beta d\tau \int_0^\beta d\tau' e^{-E_C|\tau-\tau'|} \cosh 2m_i E_C |\tau - \tau'|$$

$$= \frac{1}{2} \sum_\sigma \left[\frac{\beta}{E_C(1 + 2\sigma m_i)} + \frac{1 - e^{-\beta E_C(1+2\sigma m_i)}}{E_C^2(1 + 2\sigma m_i)^2} \right], \quad (13.40)$$

with $\sigma = \pm 1$. The main contribution from the resultant sum over m_i arises from $m_i < 1/\beta E_C$. However, because $\beta E_C \ll 1$, we expand Eq. (13.40), keeping only the terms independent of m_i. Consequently, the sum over m_i in Eq. (13.34) is canceled by the factor of Z_0 in the denominator, resulting immediately in the stability condition

$$k_B T = J_0 - \frac{E_C}{3} + O((E_C/J)^2) \quad (13.41)$$

at high temperatures. The critical temperature is, then, $T_c = J_0/k_B$. The mean-field phase boundary connecting high- and low-temperature regimes as a function of the coupling constant g is shown in Fig. (13.3). Though this is a mean-field phase diagram, it captures the correct

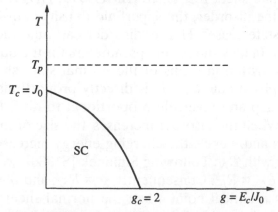

Figure 13.3 Mean-field phase diagram from the quantum rotor model. The solid line represents a continuous line of second-order critical points. The mean-field values for transition temperature and the quantum critical point are $T_c/J_0 = 1$ and $g_c = 2$, respectively. Beyond the solid curve, phase coherence is lost, signaling an absence of the superconducting state. T_p indicates the temperature scale for the onset of the pairing amplitude.

qualitative features of the phase fluctuation–induced destruction of superconductivity in 2d systems. The more exact description of the finite temperature line in Fig. (13.3) is due to Berezinskii [B1971] and Kosterlitz and Thouless [KT1973]. The essence of the Berezinskii-Kosterlitz-Thouless (BKT) transition is that the ordered phase is characterized by bound vortex-antivortex pairs. Unbinding of vortices at criticality leads to dissipation and the onset of a resistive state. The phase-only model can be understood in terms of the vortex-antivortex unbinding transition as a result of the inherent duality [F1990] between charges and vortices within this model. This duality dictates that on the superconducting side of the transition, vortices are bound, whereas Cooper pairs are condensed in a state with a rigid phase. On the disordered side, it is the vortices that condense and the Cooper pairs that remain localized. In the context of a charged superfluid in 2d, the condition for vortex unbinding,

$$k_B T_{\text{BKT}} = \frac{\pi \hbar^2 n_s(T)}{2m^*}, \tag{13.42}$$

depends only on the superfluid density, $n_s(T)$, and the effective mass, m^*, of the charge carriers. Note that the mean-field condition, $T_c = J_0$, is quite similar to T_{BKT}, since the Josephson coupling, J_0, is proportional to the superfluid density. Because disorder depletes the superfluid density, the T_{BKT} line should be inversely proportional to the normal state sheet resistance [BMO1979, HN1979]. Above a critical value of the disorder, the superfluid density vanishes and the phase-ordered state ceases. Hence, disorder can tune the vortex unbinding transition. In fact, the system parameters for the quantum rotor model can be rewritten in terms of the normal state sheet resistance, R_N. The Josephson coupling, J, is directly proportional to the zero temperature gap and inversely proportional to disorder, scaling roughly as R_N^{-1}. When the disorder increases, the size of the Josephson grains decreases and, hence, the charging energy increases. As a result, E_C increases with R_N. Following Simanek [S1980], we make the approximation, $E_C \propto R_N$. Consequently, $g \propto R_N^2$, and the phase-only model possesses a critical point when the normal sheet resistance exceeds a critical value, consistent with experimental observations [HP1990, J1989]. While it is most natural to envision disorder as the mechanism by which R_N increases, other external perturbations, such as an applied magnetic field, suffice, as well. The magnetic field–tuned destruction of the superconducting state in MoGe thin films is illustrated in Fig. (13.2).

Because pairs are assumed at the outset, T_{BKT} or the mean-field T_c shown in Fig. (13.3) is certainly below the temperature at which the pair amplitude first becomes nonzero. The temperature scale for pair formation is indicated by T_p in Fig. (13.3). Above T_c and g_c, the system is phase disordered. This is a crucial point. In $2d$, the energy scales for pair formation and phase coherence are decoupled. This state of affairs arises from the severity of phase fluctuations in $2d$. Also, in the vicinity of g_c, the temperature drops continuously to zero, indicating that the phase transition is second order. Once the temperature is raised at g_c, classical fluctuations obtain. Directly above g_c, the dynamics are controlled by the temperature. This regime is referred to generally as *quantum critical*, as the system is torn between the ordered and disordered phases. Understanding this competition requires a computation of the correlation length as a function of g. To facilitate this, we resort to an effective Landau description of the quantum rotor model.

13.4 LANDAU-GINZBURG THEORY

The final step in our presentation of the quantum rotor model is to recast the corresponding free energy functional in terms of a Landau theory. To this end, we consider the form for the free energy in Eq. (13.26). To start, we write the auxiliary field as a sum over the discrete Matsubara frequencies,

$$\psi_{\mathbf{k}}(\tau) = \frac{1}{\sqrt{\beta}} \sum_l e^{i\omega_l \tau} \psi_{\mathbf{k},\omega_l}, \qquad (13.43)$$

where the Fourier-transformed field is given by

$$\psi_{\mathbf{k},\omega_l} = \frac{1}{\sqrt{\beta}} \int_0^{\beta} \psi_{\mathbf{k}}(\tau) e^{-i\omega_l \tau}. \qquad (13.44)$$

Next, we expand the logarithm in Eq. (13.27) keeping in mind that the odd powers of ψ vanish as a result of Eq. (13.32). Consequently, the free energy

$$\mathcal{F}[\psi] = \sum_{\mathbf{k},l,l'} \psi_{\mathbf{k},l}^* \psi_{\mathbf{k},l'} \left[J_{\mathbf{k}}^{-1} \delta_{l,l'} - 2C_0(\omega_l, \omega_l') \right] + O(\psi^4) + \cdots \qquad (13.45)$$

contains only even powers in ψ. The frequency-dependent spin-correlation function

$$C_0(\omega_l, \omega_{l'}) = \frac{1}{\beta} \int_0^\beta \int_0^\beta d\tau d\tau' C_0(\tau, \tau') e^{i\omega_l \tau} e^{i\omega_{l'}\tau}$$

$$= \delta_{ll'} \frac{E_C}{E_C^2 + \omega_l^2} \approx \delta_{ll'} \frac{1}{E_C}\left(1 - \frac{\omega_l^2}{E_C^2}\right) \quad (13.46)$$

has a simple form at low temperatures, where $C_0(\tau, \tau') \propto \exp E_C|\tau - \tau'|$. The analogous correlation function can also be evaluated for the fourth-order term. However, its exact coefficient, which we represent heuristically as U, is not particularly important in this context. As our focus is the long-wavelength physics, we expand $J_{\mathbf{k}}$, retaining only the quadratic terms, $J_{\mathbf{k}} \approx J_0(1 - (ka)^2/2)$, with a the lattice constant. Combining these results with those of Eqs. (13.45) and (13.46), we see immediately that the free energy

$$\mathcal{F}[\psi] = \sum_{\mathbf{k},l} \psi^*_{\mathbf{k},l} \psi_{\mathbf{k},l}\left[\left(\frac{1}{J_0} - \frac{2}{E_C}\right) + \frac{(ka)^2}{2J_0} + 2\frac{\omega_l^2}{E_C^3}\right]$$

$$+ \frac{U}{2\beta} \sum_{\substack{\omega_1,\ldots,\omega_4 \\ \mathbf{k}_1,\ldots,\mathbf{k}_4}} \delta_{\omega_1 + \cdots \omega_4, 0} \delta_{\mathbf{k}_1 + \cdots \mathbf{k}_4, 0} \psi^*_{\omega_1,\mathbf{k}_1} \psi^*_{\omega_2,\mathbf{k}_2} \psi_{\omega_3,\mathbf{k}_3} \psi_{\omega_4,\mathbf{k}_4}$$

$$(13.47)$$

is of the Landau-Ginzburg form. Noting that the mean-field solution for the zero-temperature quantum critical point is $J_0^{-1} = 2/E_C$ or, equivalently, $(1/J_0 - 2/E_C) = (g - g_c)/E_C$, implying that the constant coefficient of the quadratic term is proportional to the inverse correlation length. This term plays the role of $a(T)$ in Eq. (11.14) in Chapter 11. In addition, the quadratic frequency dependence arises from the two derivatives with respect to θ in Eq. (13.30). In the time domain, the quadratic frequency dependence translates into a second derivative with respect to time, as shown in Eq. (13.12). Consequently, in real space, the free-energy density

$$\mathcal{F}[\psi] = \int d^2r \int d\tau \left[|\nabla\psi(\mathbf{r},\tau)|^2 + |\partial_\tau\psi(\mathbf{r},\tau)|^2 + \delta|\psi(\mathbf{r},\tau)|^2\right.$$

$$\left. + \frac{U}{2}|\psi(\mathbf{r},\tau)|^4\right] \quad (13.48)$$

takes the familiar Landau-Ginzburg form upon appropriate rescalings of the coupling constants, the Matsubara frequencies, and the ψ fields. We have defined $\delta = g - g_c$ and set $\hbar = c = 1$. Absent from the earlier formulation of Landau-Ginzburg theory presented in Chapter 11, Eq. (11.14), but present in Eq. (13.48), is the explicit integration over imaginary time and the quantum fluctuation term proportional to ∂_τ^2. Symmetry between space and time is guaranteed, as there are an equal number of time and spatial derivates in the free energy. Consequently, we identify

$$\epsilon_k = \pm \sqrt{k^2 + \delta} \tag{13.49}$$

as the dispersion relationship for the single-particle spectrum. In the superconducting phase, $\delta = 0$, and we recover a linear dispersion relationship on momentum, as is expected for bosons. However, the dispersion relationship is gapped, $\delta > 0$, when quantum fluctuations dominate. The critical point demarcates the transition between gapped $(g > g_c)$ and free $(g < g_c)$ quasi-particle excitations. On both sides of the phase transition, Cooper pairs are intact. Consequently, the phase transition inherent in the Landau-Ginzburg theory is not of the pair-breaking type but rather is driven by phase fluctuations of the order parameter, $\psi(\omega, \mathbf{k})$.

The role of the fourth-order term is to introduce quasi-particle scattering into the correlation length. The simplest way to quantify this effect is to use Hartree-Fock theory. In this context, we can implement this procedure by performing an expansion in the number components, N, of the field, ψ_μ, μ the index for each component. For the rotor problem, ψ_μ is a two-component field. The Hartree-Fock procedure amounts to taking the $N \to \infty$ limit. Should the corrections in powers of $1/N$ prove to be small, then the large-N expansion offers a simple way of treating quartic terms in the free energy once we replace U by U/N. As we will see, the $1/N$ expansion cannot reproduce the finite temperature transition. However, it is instructive in delineating the distinct regimes that encompass the quantum critical point.

To obtain the correlation length, we exponentiate the free energy and use a slightly modified form of the Hubbard-Stratanovich [H1959, S1958] transformation

$$\int \mathcal{D}\lambda \exp\left\{ \frac{-N}{2U} \sum_{\mathbf{k}\omega} |\lambda(\omega, \mathbf{k})|^2 + \frac{1}{\sqrt{\beta}} \sum_{\mathbf{k}_1, \mathbf{k}_2, \omega_1, \omega_2} \lambda(\omega_1, \mathbf{k}_1) \psi_\mu(\omega_1, \mathbf{k}_1) \right.$$

$$\left. \times \psi_\mu(-\omega_1 - \omega_2, -\mathbf{k}_1 - \mathbf{k}_2) + \text{hc} \right\}$$

$$= \exp\left\{ \frac{UT}{2N} \sum_{\substack{\omega_1,\ldots,\omega_4 \\ \mathbf{k}_1,\ldots,\mathbf{k}_4}}^{\prime} \psi_\mu^*(\omega_1, \mathbf{k}_1) \psi_\mu^*(\omega_2, \mathbf{k}_2) \psi_\nu(\omega_3, \mathbf{k}_3) \psi_\nu(\omega_4, \mathbf{k}_4) \right\} \tag{13.50}$$

to decouple the quartic term, where the prime represents the constraints, $\delta_{\omega_1+\cdots+\omega_4,0}$ and $\delta_{k_1+\cdots+k_4,0}$. In the saddle-point approximation, the auxiliary field is a constant: $\lambda(\omega,k) = \sqrt{\beta}\lambda\delta_{\omega,0}\delta_{k,0}$. Substitution of this expression into the partition function and minimization with respect to λ results in the saddle-point equation

$$\lambda = \frac{U}{\beta} \sum_{\omega_n} \int \frac{d^2k}{(2\pi)^2} \frac{1}{\delta + k^2 + \omega_n^2 + \lambda} \tag{13.51}$$

for λ in the large-N limit. Let $m^2 = \delta + \lambda$ be the square of the inverse correlation length. The critical point is now determined by $m = 0$. Once the integral over momentum is performed in Eq. (13.51), Poisson summation techniques can be used to evaluate (see Problem 4) the sum over the Matsubara frequencies. The resultant self-consistent condition

$$m^2 = \delta + \frac{UT}{2\pi} \ln\left(\frac{\sinh \sqrt{\Lambda^2 + m^2}/2T}{\sinh m/2T} \right) \tag{13.52}$$

admits the solution $m = 0$, provided that $T = 0$ and the renormalized tuning parameter, $\Delta = \delta + U\Lambda/4\pi$, vanishes. Here, Λ is the momentum cutoff. Hence, mean-field theory fails to recover the correct finite-temperature transition (of the BKT type) for the vanishing of the inverse correlation length. Nonetheless, as a function of temperature, three distinct regimes emerge

$$m = \begin{cases} T \exp\left\{ -\dfrac{2\pi|\Delta|}{UT} \right\} & \Delta < 0 & \text{ordered phase} \\[2mm] 2\ln \dfrac{\sqrt{5}+1}{2} T & |\Delta| \ll T & \text{quantum critical,} \\[2mm] \dfrac{4\pi\Delta}{U} & \Delta > 0 & \text{quantum disordered} \end{cases} \tag{13.53}$$

which characterize all QPTs. The phase diagram illustrating each of these regimes is shown in Fig. (13.4). The ordered phase occurs to the left ($\Delta < 0$) of the critical point, $\Delta = 0$, and is characterized by a vanishing of the inverse correlation length below some temperature. The solid line in Fig. (13.4) represents the T_{BKT} line of second-order critical points. For $N = 2$, the ordered state is a superconductor. The

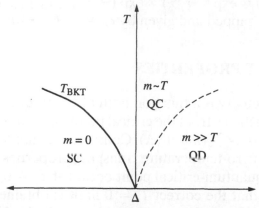

Figure 13.4 Finite-temperature phase diagram for the quantum rotor model as a function of temperature and the distance from the quantum-critical point, $\Delta = g - g_c = 0$. The solid line, T_{BKT}, represents the line of second-order critical points determined by Berezinskii-Thouless-Kosterlitz. Below the solid line, the inverse correlation length, m, vanishes and phase coherence obtains. The dashed line represents a crossover from the quantum-critical (QC) region in which the dynamics are determined by temperature, $m \propto T$, to the phase-disordered (QD) regime, where quantum fluctuations dominate, $m \gg T$. The behavior of the inverse correlation length in each of these regimes is given by Eq. (13.53).

ordered state for $N = 1$ corresponds to a ferromagnet and, hence, is in the universality class of the 2d Ising model. For $N \geq 3$, the finite temperature transition line is suppressed to $T = 0$, thereby indicating the absence of an ordered state at finite temperature. The failure of the $1/N$ expansion to yield a finite-temperature transition is immediately evident from Eq. (13.53), as m vanishes only at $T = 0$ for $\Delta < 0$. Directly above $\Delta = 0$, the system crosses over to an intermediate regime in which thermal rather than quantum fluctuations determine the dynamics. In this regime, the inverse correlation length scales linearly with temperature with a coefficient that is universal. This regime is referred to as *quantum critical* because the system is trying to decide which of the two ground states is preferable. A fundamental characteristic of all systems in such an indeterminate state is the linear scaling of all energy scales or, equivalently, relaxation rates with temperature. Hence, in the quantum-critical regime, quasi-particle excitations are not well defined, as discussed in the Introduction. As Δ is increased, the system crosses over from the quantum-critical to the quantum-disordered regime. In this limit, temperature is subdominant, and quantum fluctuations of the order parameter determine the physics. Consequently,

$\langle \psi^*(\mathbf{r})\psi(\mathbf{r})\rangle \propto \exp{(-r/\xi)} \propto \exp{(-mr)}$. The energy excitations in this regime are gapped and given by $\epsilon_k = \sqrt{k^2 + m^2}$.

13.5   TRANSPORT PROPERTIES

The conductivity is a function both of the frequency and the temperature. The distance from the critical point also determines the conductivity. Hence, $\sigma = \sum(\hbar\omega, T, \Delta)$. Consider the simple case of $\Delta = 0$. How should the zero-temperature transport properties be determined? Because the quantum-critical point occurs at $T = 0$, it seems reasonable to assume that the correct $T = 0$ limit is obtained from $\sum(\hbar\omega, T = 0, 0)$ and the dc conductivity by taking the limit of zero frequency. This, of course, assumes that the frequency and temperature dependence are decoupled. If, in fact, they are decoupled, then the two limits, $\omega \to 0, T = 0$ and $\omega = 0, T \to 0$, commute.

In a pioneering paper, Damle and Sachdev [DS1997] showed that in the vicinity of a quantum-critical point, these two limits do not commute after all. Their argument is based on the observation that in the vicinity of a quantum-critical point, the conductivity is a universal function of $\hbar\omega/k_B T$. This signifies that there are two distinctly different $T \to 0$ limits of $\sum(\hbar\omega/k_B T)$. The limit $\sum(\hbar\omega \to 0, T = 0) = \sum(\infty)$, whereas $\sum(\omega = 0, T \to 0) = \sum(0)$. The essential point of Damle and Sachdev is that $\sum$ is a monotonic function, and, hence, $\sum(0) \neq \sum(\infty)$. Experimentally, all measurements of the dc transport properties are performed in the limit, $\hbar\omega \ll k_B T$. Hence, it is the limit $\sum(0)$ that is relevant to experiments. All of the early theoretical work on insulator-superconductor transitions [WZ1990, C1991, KZ1993, W1994] was based on the limiting form $\sum(\infty)$. This limit corresponds to the coherent regime or high-frequency limit in which relaxation from the externally applied electric field determines the transport properties. The opposite regime is different in kind, however. In the limiting form $\sum(0)$, collisions between thermally excited quasi particles dominate all relaxation processes. To describe this regime, one must formulate a quantum kinetic equation for the quasi-particle scattering directly analogous to the Boltzmann approach formulated for phonon-dominated transport in a metal. The key surprise here is that the transport properties of a $T = 0$ quantum-critical point cannot be understood without an analysis of the finite temperature relaxation processes.

To illustrate how the noncommutativity of the frequency and temperature tending to zero limits drastically affects the conductivity, consider the quantum-disordered regime. In this regime, $m \gg T$, and thermally excited quasi particles exist above the gap. However, from naive considerations, the presence of a gap indicates that the dc conductivity should vanish. But this is not so. The constraint $m \gg T$ in the quantum-disordered regime signifies that the relevant momenta satisfy the constraint $k < \sqrt{mT} < m$. Consequently, the dispersion relation can be written as

$$\epsilon_k \approx m + \frac{k^2}{2m} \approx m \quad m \gg T, \tag{13.54}$$

and the quasi-particle statistics become Boltzmannian in the quantum-disordered regime. Hence, the population of quasi particles is exponentially small. However, this exponential smallness of the population of quasi particles also gives rise to an exponentially small probability of scattering. The exact calculation [DP2000] reveals that in the quantum-disordered regime, the scattering rate

$$\frac{1}{\tau} = \pi T e^{-m/T} \tag{13.55}$$

is independent of momentum and governed by the same exponential that attenuates the population of quasi particles. From the Einstein relation, the conductivity is a product of the density of quasi particles and the scattering time. The product of exponentials cancels [DP2000], giving rise to a conductivity

$$\sigma_{qd} = \frac{2}{\pi} \frac{4e^2}{h} \tag{13.56}$$

that is finite for $m \ll 1$. Farther away from the critical point, the coefficient is modified by the form of the interactions. However, the dc conductivity remains finite, nonetheless. In addition, all logarithmic contributions to the scattering rate [DP2000] vanish at low temperatures and, hence, do not affect the dc conductivity at $T = 0$. The key to this argument is that once the quasi particles obey Boltzmann statistics, scattering events are rare. Hence, quasi particles roam considerable distances without scattering. Consequently, the generic phase diagram for the phase-only model, Fig. (13.5), contains an ordered phase that is a superconductor and a phase-disordered regime that is "metallic." The "insulator" is a metal with a finite dc conductivity. This is one of the key surprises that has come out of the noncommutativity of the frequency and temperature tending to zero limits. We refer to this phase as a *Bose metal* because the fundamental excitations are bosons.

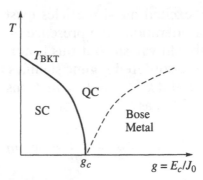

Figure 13.5 Phase diagram for the destruction of phase coherence in an array of Josephson junctions. For this system, g is given by the ratio of the charging energy, E_c, to the Josephson coupling, J_0, T_{BKT} is the Berezinskii-Kosterlitz-Thouless temperature below which phase coherence obtains, and g_c defines the critical value of the phase disorder needed to destroy the superconducting phase. QC refers to the quantum-critical regime in which the inverse correlation length is linear in the temperature. Once phase coherence is destroyed, a metallic state ensues. The metallic state arises entirely from the lack of commutativity between the frequency and temperature tending to zero limits of the conductivity.

At finite frequency, $\Sigma(\omega, T)$ has a Lorentzian-type peak at $\omega = 0$, with a width of order $1/\tau$. Consequently, the constraint on the experimental observation of the Bose metal is that $\omega \ll 1/\tau$. For a temperature of 0.1 K, the relaxation time is $10^{10} \exp(-m/T)\sec^{-1}$, where $m \gg T$. Typical experimental frequencies under which the dc conductivity is measured are on the order of 2–27 Hz. Hence, the Bose metal can be observed, provided that $T > 0.05m$. Typically, $T > 0.1m$. Consequently, there does not appear to be any experimental constraint regarding the frequency that prohibits the observation of the Bose metal phase. However, the Bose metal described here is fragile, as all other scattering mechanisms, for example, disorder, will destroy the perfect cancellation leading to Eq. (13.56). Consequently, as of this writing, one of the open problems with the insulator-superconductor transition remains the origin of the intervening metallic phase at low temperatures.

13.6   EXPERIMENTS

The applicability of the phase-only model to insulator-superconductor transitions in thin films hinges on the existence of Cooper pairs on both sides of the transition. Consequently, tunneling experiments that directly measure the single-particle spectrum are ideally suited for testing the key prediction of the phase-only model. Such experiments on granular films [B1994] reveal that the single-particle energy gap remains finite and unattenuated from the bulk superconducting value on the insulating side of the transition. Consequently, the phase-only model offers an accurate description of the experimentally observed IST in granular thin films.

What about homogeneously disordered thin films? Here the situation is quite different. Shown in Fig. (13.6) is the critical magnetic field needed to destroy the superconducting state in PbBi/Ge as a function of the single-particle energy gap. As is evident, H_c scales as a linear function of the gap. Such a correspondence is expected within BCS theory (see Eq. 11.157), provided that H_c corresponds to H_{c2}, the Cooper pair-breaking field. In fact, Valles and co-workers [HCV1995] have determined that at H_c, the density of states at the Fermi energy is 80 percent of its value in the normal state, as would be the case if $H_c \approx H_{c2}$. Consequently, a large fraction of the sample contains no Cooper pairs. This observation is also consistent with the series of experiments by Dynes and co-workers [V1992] in which the energy gap was observed to vanish at the nominal critical field, H_c. For MoGe, Yazdani and Kapitulnik [YK1995] also found that $H_c \approx H_{c2}$. Collectively, these observations imply that electron-like quasi particles resulting from Cooper pair-breaking populate the insulating side of the transition in homogeneously disordered thin films. Consequently, physics beyond the phase-only model is necessary to describe these systems.

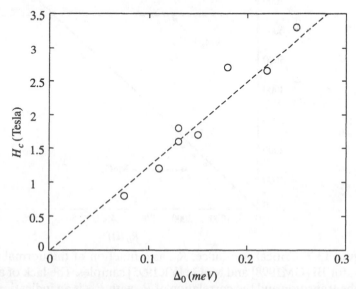

Figure 13.6 Critical magnetic field H_c versus the energy gap, Δ_0, for PbBi/Ge films. Redrawn from S.-Y. Hsu's thesis [H1995].

That this state of affairs obtains could also have been inferred from
the experimental fact that in a vast array of homogeneously disordered
thin films, the resistivity at criticality, R_c, is equal to the resistivity of
the normal state, R_N. A typical plot of R_c versus R_N for Bi [GM1998]
and MoGe [YK1995] is shown in Fig. (13.7). The normal state resistance, R_N, was extracted [YK1995, GM1998] from transport measurements at $T > T_p$ (see Fig. 13.3). Hence, only electron-like excitations
abound in this temperature range. The critical resistance, R_c, is extracted much below T_p and represents the resistivity at the single cross-
ing point of $R(T)$ versus tuning parameter and as $T \rightarrow 0$. In both the
Bi and MoGe films, a magnetic field induces the IST. As is evident from
the data shown for both Bi and MoGe, $R_c \approx R_N$. Unless a transition to
the normal ground state occurs at R_c, there is no reason for the correspondence $R_c \approx R_N$. Because R_N is associated with high-temperature
physics in which no Cooper pairs exist, the fact that $R_c \approx R_N$ signifies that the insulating state must have electron-like quasi particles in
these homogeneously disordered thin films. Further, in all cases, R_c
slightly exceeds R_N and ranges between 900Ω and 8000Ω. These values certainly differ significantly from the boson quantum of resistance,
$h/4e^2 = 6500\Omega$, predicted from the phase-only models, further indicating that physics beyond the phase-only model is at work.

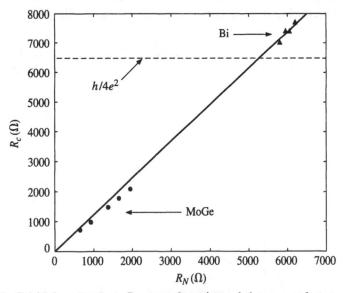

Figure 13.7 Critical resistance, R_c, as a function of the normal state resistance,
R_N, for Bi [GM1998] and MoGe [YK1995] samples. The lack of a universal value
at the transition and the correlation of R_c with R_N is an indication that the transition to the insulating state has electron-like quasi particles, indicative of the high-
temperature normal state.

To conclude, phase-only physics seems to be adequate to describe granular films. However, for most homogeneously disordered films, the experiments point to an insulating state populated with electron-like quasi particles. Consequently, pair-breaking seems to be the dominant mechanism that drives the IST. Consequently, the 3d XY model is no longer adequate to describe the IST in homogeneously disordered films, as gapless electronic excitations exist in the insulator. In this case, the critical resistance need not be defined by the quantum of resistance for $2e$ bosons, $R_Q = h/4e^?$, as is observed experimentally Progress in including normal electrons into the IST could be made along the lines pursued recently by Feigelman and Larkin [FL1998].

13.7 SUMMARY

Quantum phase transitions are governed by a $T = 0$ quantum critical point. Further, as temperature plays a crucial role, the corresponding field theory for QPTs is $(d + 1)$ dimensional, the extra dimension arising from the integration over imaginary time. As a consequence, QPTs are in the universality class of the corresponding classical problem in one higher spatial dimension. The quantum rotor model is the simplest model that captures the *essential tension* between the competing ground states near the superconductor-to-nonsuperconductor quantum-critical point. This model can be recast as a Landau-Ginzburg theory. A key surprise with $T = 0$ quantum-critical points is that the transport properties must be determined by collisions between thermally excited quasi particles. In the context of the quantum rotor model, inclusion of quasi-particle scattering results in a metallic phase (albeit a fragile one as it is destroyed by disorder) in the quantum-disordered regime.

Directly above the quantum-critical point, the dynamics are determined by temperature. Hence, it is tempting to associate resistivities that are linear in temperature with the possible existence of a quantum-critical point, as in the case of the cuprates. In fact, superconductivity in the copper oxides represents the ultimate challenge in understanding QPTs. The generic phase diagram for the cuprates is shown in Fig. (13.8). Because the tuning parameter is the density, the termination of antiferromagnetism and the onset of d-wave superconductivity signals that a zero-temperature quantum-critical point must be present. However, unlike the quantum rotor model or the Ising model in a transverse field, no physically transparent Hamiltonian exists that lays plain the origin of the *essential tension* between the

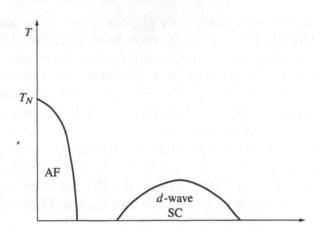

Figure 13.8 Heuristic-phase diagram for the doping dependence of the cuprates. Undoped, the cuprates are antiferromagnetic Mott insulators. Antiferromagnetic order ceases above the Neel temperature, T_N, and beyond a moderate oxygen-doping level. Superconductivity with $d_{x^2-y^2}$ symmetry exists under the dome for a range of doping typically between $0.07 < x < 0.25$.

competing ordered states that evolve from the Mott insulator. Indeed, it is *Mottness* that makes high T_c so intractable.

PROBLEMS

1. In the quantum rotor model

$$H = E_C \sum_i (\hat{n}_i - n_0)^2 - t \sum_{\langle ij \rangle} (b_i^\dagger b_j + b_j^\dagger b_i), \qquad (13.57)$$

show that when n_0 is not an integer, the corresponding action contains the additional term

$$\delta S = i n_i \sum_j \int_0^\beta d\tau \partial_\tau \theta_j (\tau). \qquad (13.58)$$

This version of the quantum rotor model is known as the *incommensurate Bose-Hubbard model*. What consequences does this have on the value of the dynamical exponent?

2. Using Eq. (13.35), prove the result in Eq. (13.36).

3. Use the low temperature form of $C_0(\tau, \tau')$, Eq. (13.37), to establish the final equality in Eq. (13.46).

4. Perform the integration in Eq. (13.51) and then use the Poisson summation formula to arrive at Eq. (13.53).

REFERENCES

[AR1998] G. Aeppli, T. F. Rosenbaum, in *Dynamical Properties of Unconventional Magnetic Systems*, ed. A. T. Skjeltorp and D. Sherrington. (Kluwer Academic Publishers, Netherlands, 1998), p. 107.

[A1964] P. W. Anderson, in *Lectures on the Many Body Problem*, ed. E. R. Caianiello (Academic, New York, 1964), Vol. 2, p. 127.

[B1994] R. P. Barber, Jr., L. M. Merchant, A. La Porta, R. C. Dynes, *Phys. Rev.* B **49**, 3409 (1994).

[B1971] V. L. Berezinskii, *Zh. Eksp. Teor. Fiz.* **61**, 1144 (1971) [*Sov. Phys.-JETP* **34**, 610 (1971)] .

[BMO1979] M. R. Beasley, J. E. Mooij, T. P. Orlando, *Phys. Rev. Lett.* **42**, 1165 (1979).

[C1991] M. C. Cha, M. P. A. Fisher, S. M. Girvin, M. Wallin, A. P. Young, *Phys. Rev. B* **44**, 6883 (1991).

[DS1997] K. Damle, S. Sachdev, *Phys. Rev. B* **56**, 8714 (1997).

[DP2000] D. Dalidovich, P. Phillips, *Phys. Rev. B* **64**, 525 (2001).

[D1981] S. Doniach, *Phys. Rev. B* **24**, 55063 (1981).

[FL1998] M. V. Feigelman, A. I. Larkin, *Chem. Phys.* **235**, 107 (1998).

[FH1965] R. P. Feynman, A. R. Hibbs, *Quantum Mechanics and Path Integrals* (McGraw-Hill, New York, 1965), pp. 42–43.

[F1972] R. P. Feynman, *Statistical Mechanics* (W. A. Benjamin, Reading, Mass., 1972), p. 66.

[F1990] M. P. A. Fisher, *Phys. Rev. Lett.* **65**, 923 (1990).

[GM1998] For a review, see A. M. Goldman, N. Markovic, *Physics Today* **51**, (11) 39 (1998).

[HN1979] B. I. Halperin, D. R. Nelson, *J. Low Temp. Phys.* **36**, 5999 (1979).

[HP1990] A. F. Hebard, M. A. Paalanen, *Phys. Rev. Lett.* **65**, 927 (1990).

[H1976] J. A. Hertz, *Phys. Rev.* B **14**, 1165 (1976).

[H1995] S.-Y. Hsu, Ph.D. Thesis, Brown University, 1995.

[HCV1995] S.-Y. Hsu, J. A. Chervenak, J. M. Valles, Jr., *Phys. Rev. Lett.* **75**, 132 (1995).

[H1959] J. Hubbard, *Phys. Rev. Lett.* **3**, 77 (1959).

[J1989] H. M. Jaeger, D. B. Haviland, B. G. Orr, A. M. Goldman, *Phys. Rev. B* **40**, 182 (1989).

[KZ1993] A. P. Kampf, G. T. Zimanyi, *Phys. B* **47**, 279 (1990).

[KT1973] J. M. Kosterlitz, D. J. Thouless, *J. Phys. C* **6**, 1181 (1973).

[P1979] P. Pfeuty, *Phys. Lett. A* **72**, 245 (1979).

[S2000] S. Sachdev, *Quantum Phase Transitions* (Cambridge University Press, Cambridge, England, 1999).

[S1980] E. Simanek, *Phys. Rev. Lett.* **45**, 1442 (1980).

[S1958] R. L. Stratanovich, *Sov. Phys. Dokl.* **2**, 416 (1958).

[V1992] J. M. Valles, Jr., R. C. Dynes, J. P. Garno, *Phys. Rev. Lett.* **69**, 3567 (1992).

[W1994] M. Wallin, E. S. Sorensen, S. M. Girvin, A. P. Young, *Phys. Rev. B* **49**, 12115 (1994).

[WZ1990] X. G. Wen, A. Zee, *Intl. J. Mod. Phys. B* **4**, 437 (1990).

[W1965] B. Widom, *J. Chem. Phys.* **43**, 3892 (1965).

[YK1995] A. Yazdani, A. Kapitulnik, *Phys. Rev. Lett.* **74**, 3037 (1995).

– 14 –

Quantum Hall Effect

When an electron gas is confined to move at the interface between two semiconductors and a magnetic field is applied perpendicular to the plane, a new state of matter [TSG1982] arises at sufficiently low temperatures. This state of matter is unique in condensed-matter physics, in that it has a gap to all excitations and exhibits fractional statistics. It is generally referred to as an *incompressible quantum liquid* or as a *Laughlin liquid* [L1983], in reference to the architect of this state. Though the Laughlin state is mediated by the mutual repulsions among the electrons, it is the presence of the large perpendicular magnetic field that leads to the incompressible nature of this new manybody state. The precursor to this state is the integer quantum Hall state. In this state, disorder and the magnetic field conspire to limit the relevant charge transport to a narrow strip around the rim of the sample. The novel feature of this rim or edge current is that it is quantized in integer multiples of e^2/h [KDP1980]. The equivalent current in the Laughlin state is still quantized but in fractional multiples of e^2/h. We present in this chapter the phenomenology and the mathematical description needed to understand the essential physics of both of these effects.

14.1 WHAT IS QUANTUM ABOUT THE HALL EFFECT?

To understand what is quantum about the Hall effect, we must first define the concept of the Hall voltage [H1879]. To this end, we consider a conducting metal slab in which a current, I, is directed along the x-direction. The current produces a voltage drop and an electric field, E_x, and current density, j_x, along the x-axis, as illustrated in Fig. (14.1a). We will assume that the current density and the electric field are linearly related (Ohm's law), $j_x = \sigma_o E_x$, and the constant of proportionality is given by the Drude conductivity derived in Chapter 10. In the presence

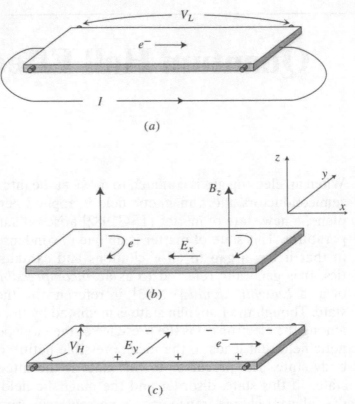

Figure 14.1 (*a*) Standard geometry for the Hall effect. A current, *I*, is passed through a thin sample in the *x*-direction, leading to a voltage drop and an electric field, E_x, along the *x*-axis. The sample is pierced in the positive *z*-direction by a uniform magnetic field B_z. (*b*) The resultant sideways motion in the *y*-direction of the electrons in response to the magnetic field. Electrons accumulate on one face of the material, creating a transverse electric field E_y. (*c*) A positive ion excess is established until the Hall field, E_y, cancels the Lorentz force.

of a magnetic field oriented upward along the *z*-axis, the electrons will be deflected sideways in the negative *y*-direction (Fig. 14.1b) by the Lorentz force. If the sample is infinite in the *y*-direction, the current will no longer be directed solely along the *x*-axis. Rather, the electrons will move at some angle relative to the *x*-axis. This angle defines the Hall angle. However, if the sample is constrained in the *y*-direction, electrons deflected by the magnetic field will ultimately run into the edges of the sample. As they accumulate there, they will produce an electric field, E_y, which will point in the positive *y*-direction. Further accumulation of electrons at the edges of the sample ceases when the electric field E_y is sufficiently large to cancel the Lorentz force. Cancellation of the Lorentz force in equilibrium (Fig. 14.1c) signifies that electrons will transport purely along the x-axis. The additional electric field, E_y, which facilitates the cancellation, can be thought of simply as an induction field that arises from the magnetic flux of the charges at the boundaries. It is the field E_y that is referred to as the *Hall field*. In a uniform system at equilibrium, the Hall field is perpendicular to the current. For a strictly *2d* sample of width W, the Hall voltage V_H is related to the Hall field through $E_y = V_H/W$. The total current is given by $I = W j_x$.

To understand the functional dependence of E_y as a function of the magnetic field, we develop a semiclassical or high-temperature theory. In this regime, we can determine the magnitude of the Hall field by applying Newton's second law

$$F = -e\left(\mathbf{E} + \frac{1}{c}\mathbf{v} \times \mathbf{B}\right) \tag{14.1}$$

to our constrained metal slab, where $\mathbf{v}$ and $\mathbf{B}$ are the electron velocity and the magnetic field, respectively. Cancellation of the forces in the y-direction requires that

$$E_y = \frac{v_x B_z}{c}. \tag{14.2}$$

As in Chapter 10, we invoke the relaxation time approximation for the electron velocity, $v_x = -eE_x\tau/m$, where τ is the collision time. The other fundamental timescale in this problem is set by the cyclotron frequency,

$$\omega_c \equiv \frac{eB_z}{mc}. \tag{14.3}$$

With the relaxation time approximation, we can rewrite the Hall field

$$E_y = -\omega_c \tau E_x \tag{14.4}$$

in terms of the cyclotron frequency. Because the current density,

$$j_x = \frac{e^2 \tau n_e E_x}{m}, \tag{14.5}$$

is proportional to τE_x, the transverse resistivity defined as

$$\rho_H = \frac{E_y}{I} = \frac{E_y}{W j_x} \tag{14.6}$$

is independent of the relaxation time. In Hall experiments, however, the measured quantities are the transverse voltage, V_H, and the longitudinal voltage, $V_L = E_x L$, where L is the longitudinal distance across which the voltage changes by V_L. Hence, rather than working with resistivities, it is more convenient in the Hall effect to consider the resistance. The Hall resistance

$$R_H = \frac{V_H}{I} = \frac{W E_y}{W j_x} = -\frac{B_z}{e c n_e} \tag{14.7}$$

is linearly related to the applied magnetic field, whereas the longitudinal resistance,

$$R_L = \frac{V_L}{I} = \frac{L E_x}{W j_x} = \frac{L}{W} \frac{1}{\sigma_o}, \tag{14.8}$$

is independent of the magnetic field. Experimentally, these relationships agree well with Hall's measurements in 1879 [H1879]. Note that R_H, as we have calculated it here, is independent of the details of the electron-scattering processes. In general, there is a weak dependence on such processes.

In 1980, von Klitzing and collaborators [KDP1980] noticed striking deviations from the resistances given by Eqs. (14.7) and (14.8) at sufficiently high magnetic fields and ultra-low temperatures. He performed his remarkable experiments on a 2d electron gas confined at the interface between SiO_2 and Si in an Si MOSFET. They found that the Hall voltage, as the magnetic field increased, exhibited distinct flat or plateau regions that were highly reproducible from sample to sample, as shown in Fig. (14.2). From the value of the Hall voltage at the plateau regions, they deduced that the Hall conductance (the inverse of the resistance) must be quantized in units of e^2/h. The quantization of the conductance in the form

$$\sigma_H = -\frac{n e^2}{h}, \tag{14.9}$$

where n is an integer, was found to hold for one part in 10^7. Equally surprising was the behavior of the longitudinal resistance. He found that the longitudinal resistance vanished exactly at the Hall plateaus, indicating the onset of dissipationless transport. The presence of both quantization and perfect conduction indicates that something quite fundamental is at the heart of these experiments. At much higher magnetic fields and lower temperatures, Tsui, Störmer, and

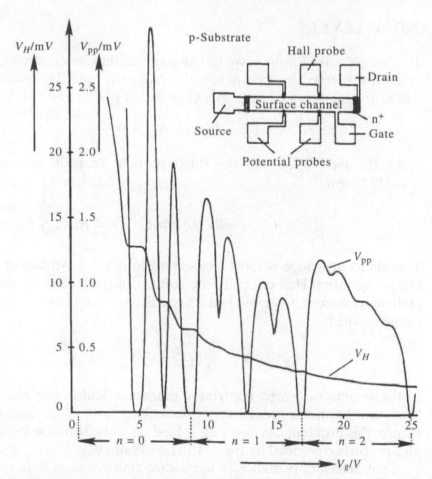

Figure 14.2 Hall voltage, V_H, and the potential drop between the potential probes, V_{pp}, as a function of gate voltage, V_g, at $T = 1.5$ K. The magnetic field was held fixed at 18T and the source drain current at $1\mu A$. Shown in the inset is a top view of the device with the length of $L = 400\mu$m, a width of $W = 50\mu$m, and a distance between the potential probes of $L_{pp} = 130\mu$m. The Hall plateaus occur at integer values of the filling in each Landau level indicated with the index, n. At the plateaus in the Hall voltage, the longitudinal voltage goes to zero, indicating the presence of dissipationless transport. Reprinted from K. Von Klitzing, et al. Phys. Rev. Lett. **85**, 49411980).

Gossard [TSG1982] found that the plateau regions in the Hall voltage were more plentiful than had been thought possible. They found that the plateau regions can occur when the conductance is a fractional multiple of e^2/h, indicating the presence of fractionally charged excitations. We will show that the integer Hall effect can be understood simply as a quantization of the edge current, whereas the fractional effect arises from a fundamentally new correlated many-body state, the Laughlin state [L1983].

14.2 LANDAU LEVELS

To start, we solve for the wave functions describing an electron moving in a plane pierced by a perpendicular magnetic field. Following Landau's original treatment, we orient the vector potential

$$A_y = Bx \qquad A_x = 0 \tag{14.10}$$

along the y-direction and the single-particle Schrödinger equation takes the form

$$-\frac{\hbar^2}{2m}(\partial_x^2 + (\partial_y - ieBx/\hbar c)^2)\psi(x,y) = E\psi(x,y). \tag{14.11}$$

This choice of gauge is most convenient to describe transport in the integer quantum Hall effect. In the context of the fractional quantum Hall effect, however, we will find it expedient to work in the symmetric gauge in which

$$\mathbf{A} = \frac{B}{2}(y\hat{x} - x\hat{y}). \tag{14.12}$$

In the symmetric gauge, applying a magnetic field in the z-direction leads to a harmonic-oscillator problem along both the x- and y-axes. Hence, this problem can easily be solved once the solution to the simpler problem described by Eq. (14.10) is obtained.

Translational invariance in the y-direction suggests that we write the wave function as

$$\psi_{n,k}(x,y) = e^{iky}f_n(x). \tag{14.13}$$

Substitution of $\psi_{n,k}(x,y)$ into Eq. (14.11) reveals that $f_n(x)$ is a solution to a harmonic oscillator equation

$$\frac{\hbar\omega_c}{2}(-\ell^2\partial_x^2 + (x/\ell - \ell k)^2)f_n(x) = \epsilon_n f_n(x), \tag{14.14}$$

where, unlike the cyclotron frequency, the length scale

$$\ell \equiv \sqrt{\frac{\hbar c}{eB}} = \frac{250\text{Å}}{\sqrt{B}} \tag{14.15}$$

is independent of the effective mass and is changed entirely by varying the magnetic field. Known as the *magnetic length*, ℓ is roughly 250Å for a field of $B = 1T$. From the harmonic oscillator ground state, we generate a Gaussian family of wave functions

$$\psi_{n,k}(x,y) = e^{iky} H_n(x/\ell - \ell k)e^{-\frac{(x-x_k)^2}{2\ell^2}}, \qquad (14.16)$$

which are extended in the y-direction but localized in x and centered at $x_k = \ell^2 k$. In Eq. (14.16), H_n is a Hermite polynomial. Each state indexed by n is known as a *Landau level*. The energy of each Landau level is

$$\epsilon_n = \hbar\omega_c \left(n + \frac{1}{2}\right) \qquad (14.17)$$

and, hence, is independent of k. As a result, several iso-energetic states compose each Landau level. For a field of $B = 1T$, the zero-point energy is on the order of 10^{-4}eV, or 1.34 K. We will see the effects of quantization in the discrete Landau levels if the temperature is lower than that determined by the zero-point energy. Our estimate of 1.34 K is a bit in error, as we have not used the semiconductor-effective mass. For GaAs, $m^* = 0.06m$; hence, the zero-point energy increases by a factor of 16, as does the temperature at which quantization effects in Landau levels is experimentally observable.

As a result of the degeneracy, each Landau level can hold many electrons. The degeneracy is determined by the distinct number of k values that generate a state within the same Landau level. We note that the states comprising each Landau level are centered at $x_k = \ell^2 k$, where k can take on a range of values consistent with the confinement of the system in the y-direction. Let L and W be the spatial extents of the sample in the x- and y-directions, respectively. If we write the wave vector k as

$$k_m = \frac{2\pi m}{W}, \qquad (14.18)$$

with m an integer, the maximum number of states allowable in each Landau level is obtained by solving the condition $L = \ell^2 k_{N_{max}}$ or, equivalently,

$$N_{max} = \frac{LW}{2\pi\ell^2} = \frac{eBLW}{hc}. \qquad (14.19)$$

The right-hand side of this expression has a simple physical interpretation. The total magnetic flux in each Landau level is a product of the magnetic field times the area of the sample, BLW. This quantity must be equal to the number of electrons in each level times the flux quantum, hc/e. We see, then, that $N_{\max}$ is also the number of electrons in each Landau level. Consequently, we associate with each Landau level

$$n_B = \frac{1}{2\pi\ell^2} = \frac{eB}{hc} \tag{14.20}$$

as the number of states per unit area. Physically, $1/n_B$ is the irreducible area each state occupies in a Landau level. For $B = 1T$, the irreducible area corresponds to a square with sides of about $0.6 \times 10^{-9}m$, roughly ten times the Bohr radius. Note the area $1/n_B$ is invariant from one Landau level to the next. As a result, the total number of filled Landau levels is given by

$$\nu = \frac{n_e}{n_B} = 1.96 \frac{n_e}{B}, \tag{14.21}$$

where n_e is the number of electrons per unit area. In the integer quantum Hall effect, ν is an integer.

It should now be clear that if the number of electrons in the system is an integral multiple of n_B, then the conductance is quantized. Under such conditions, the electron density $n_e = nn_B$, where n is an integer. As the reciprocal of the Hall resistance, the Hall conductance is given by $\sigma_H = -ecn_e/B = ecnn_B/B = -ne^2/h$.

We can formulate a more penetrating argument for the quantization by appealing to the vanishing of the Lorentz force. If the system is translationally invariant, then the vanishing of the Lorentz force signifies that we can switch to a reference frame that moves at a velocity $\mathbf{v}$ relative to the laboratory frame, such that $\mathbf{v} \times \mathbf{B} = -c\mathbf{E}$. In this reference frame, the velocity is given by $v_i = cE_j / B_k \, \epsilon_{ijk}$, where ϵ_{ijk} is the totally antisymmetric unit tensor defined by

$$\epsilon_{123} = \epsilon_{231} = \epsilon_{312} = 1 \quad \text{even permutation}$$
$$\epsilon_{321} = \epsilon_{213} = \epsilon_{132} = -1 \quad \text{odd permutation.} \tag{14.22}$$

The total current along the i-axis is given by Qev_i, where Q is the total charge in the system. If n Landau levels are occupied with $N_{\max}$ electrons in each, then $Q = nN_{\max}$. Hence, the current density is given by

$$j_i = \frac{eQv_i}{LW} = \frac{ecQE_j}{BLW}\epsilon_{ij} = \sigma_{ij}\, E_j, \tag{14.23}$$

where σ_{ij}, the coefficient of E_j, is the transverse current. This current is antisymmetric with respect to permutation of the indices x and y. Recall that the total magnetic flux $BLW = N_{max}hc/e$. As a consequence,

$$\sigma_{xy} = \frac{ecQ}{BLW} = \frac{ecnN_{max}}{N_{max}hc/e} = \frac{ne^2}{h}. \tag{14.24}$$

Because the transverse current is antisymmetric, $\sigma_{xy} = -\sigma_{yx}$. In the moving reference frame, the diagonal conductance, $\sigma_{xx} = 0$ as a result of the vanishing of the longitudinal electric field. There is a fundamental physical reason for the vanishing of σ_{xx}, however. Because the Fermi level lies in the gap between the highest-occupied and the lowest-unoccupied Landau levels, σ_{xx} vanishes. Alternatively, the allowable phase space for scattering states vanishes when the Fermi level lies in a gap; hence, $\rho_{xx} = 0$, as well. We see then that the conductance in the quantum Hall regime is a purely off-diagonal tensor with elements

$$\sigma = \begin{bmatrix} 0 & \frac{ne^2}{h} \\ -\frac{ne^2}{h} & 0 \end{bmatrix}. \tag{14.25}$$

Whether σ_{xy} or σ_{yx} are identified as the proper Hall conductance simply depends on the axis system used to orient the electric and magnetic fields.

14.3   THE ROLE OF DISORDER

Although the preceding argument is simple, it does not apply to dirty systems in which translational invariance is broken. Further, it cannot explain the origin of fractional values of the conductance. In fact, it is easy to see that without disorder, we cannot account for the plateau nature of the quantum Hall effect. In a translationally invariant system, all the electronic states are extended. If an integral number of Landau levels are occupied, then the Fermi level lies in the gap between the highest-occupied and lowest-unoccupied Landau levels. As the magnetic field decreases, the Fermi level remains constant until the next Landau level is filled, at which point it jumps discontinuously. This would suggest that the Hall conductance should decrease monotonically as a function of magnetic field, as in the classical case. From whence, then, do the plateaus come?

It turns out that disorder saves us. As we showed in Chapter 12, disorder changes both the spatial extent and the energy of electronic states. Hence, the degenerate band of states comprising each Landau

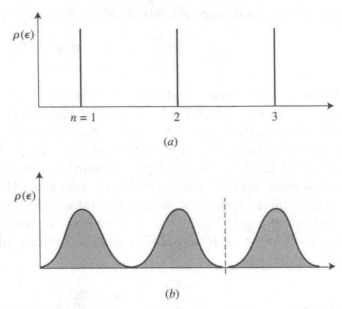

Figure 14.3 (*a*) Density of states of Landau levels in a magnetic field. (*b*) Broadening of the Landau levels as a result of disorder. The dashed line shows the position of the Fermi level.

level can be thought of as being broadened into a band of states that we describe approximately as having a Lorentzian lineshape centered at the unperturbed energy of each Landau level. This is illustrated in Fig. (14.3). Intuitively, the further an electronic state moves away from the unperturbed energy of each Landau level, the more affected it is by the disorder and, hence, the more it has a tendency to be localized. This can be seen by treating the disorder perturbatively. Hence, we arrive at the simple picture that the states close to the center of the Landau level are less localized than those at the edge of the Lorentzian distribution. We showed in Chapter 12, however, that current-carrying states do not survive, for even an infinitesimal amount of disorder in $d = 2$. If this state of affairs persists in the presence of a magnetic field, we arrive at the conclusion that the conductance should vanish in quantum Hall systems, as well. However, a magnetic field is present. As we showed in Chapter 12, magnetic fields break time-reversal symmetry and, hence, disrupt the phase coherence needed to localize electronic states. In $d = 2$, field theoretic [P1984] as well as numerical studies [AA1981; P1981; T1983] show that the scaling theory of localization does in fact break down and current-carrying states obtain. As expected, they remain clustered at the unperturbed energy of each Landau level. All other states are localized. A sharp mobility edge demarcates the separation in energy between the extended and localized states.

From the simple picture that extended states form only at the center of each Landau level and all the other states are localized, we can explain the origin of the quantum Hall plateaus. Because the current is carried only by the states at the center of each Landau level, the current should jump discontinuously as the Fermi level is tuned through the center of each Landau level. Further, the current should remain constant if the occupation of the extended states remains unchanged. That is, although increasing the magnetic field causes the chemical potential to move away from the magical place where the extended states are located, the conductance does not change because the chemical potential now resides in a region where the states are localized. The plateaus correspond to the range of magnetic fields for which the population in the extended states is fixed. The presence of precisely flat steps in the Hall voltage attests to the extreme localization of all states in a Landau level except for the narrow region of extended states located at the center. It is for this reason that the quantum Hall effect is fundamentally rooted in disorder. Paradoxically, disorder does not affect the value of the Hall conductance. Specifically Ando and Aoki [AA1981] and Prange [P1987] showed to lowest order in the drift velocity, $v_x = cE_x/B_z$, that although an isolated δ-function impurity binds an electron state, the extended states carry just enough extra current to compensate for the loss.

14.4   CURRENTS AT THE EDGE

Thus far, we have argued that disorder localizes all electronic states except for those in a narrow window around the unperturbed energy of each Landau level. Once the chemical potential moves into this region, increasing the field further has no effect on the conductance because all other states are localized. Quantum Hall plateaus originate, then, from the separation in energy between extended and localized states. The only possible deviations from perfectly flat plateaus might originate from thermally activated transport from a localized state to an extended state at the center of a Landau level. At low temperatures, such processes contribute negligibly to the transport.

We have yet to explain, however, why the quantization of the conductance in integral multiples of e^2/h is so precise. As noted by Laughlin [L1980], the precise quantization of the conductance suggests that the quantum Hall effect must be due to a fundamental principle

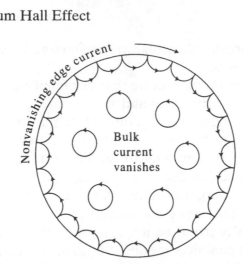

Figure 14.4 Cyclotron orbits in a quantum Hall system. The circular orbits are caused by the Lorentz force. In the bulk of the sample, clockwise and counterclockwise pieces of cyclotron orbits overlap and cancel, leading to a vanishing of the current in the bulk. At the edges, the orbits are truncated and give rise to an edge current.

devoid of any material parameters, such as geometry. To this end, Laughlin formulated a gauge principle to explain the quantization of the Hall conductance.

To understand the essence of this argument, we first make a general observation regarding the current in 2d systems in a magnetic field. As stated earlier, electrons in a magnetic field move in circular orbits as a result of the Lorentz force. As illustrated in Fig. (14.4), in the bulk of a sample, clockwise and counterclockwise pieces of neighboring cyclotron orbits overlap, leading to a vanishing of the current in the bulk. The situation is quite different at the edges of the sample, however. At the edge, the orbits are truncated in response to the confining potential created by the boundary. Once an electron is reflected by the boundary, it still attempts to move in a circular orbit. This induces a skipping-type motion of an electron at the boundary of the sample, as shown in Fig. (14.4). Such motion generates an edge current that flows in the clockwise direction for a magnetic field oriented along the positive z-direction. The chirality of the edge current is determined, then, by the direction of the magnetic field. Though the above argument is valid strictly when the magnetic length much exceeds the wavelength of the electron, $\ell \gg \hbar/p_F$, the chirality of the edge current can be established quite generally from the presence of an induction field at the boundary [W1990]. That the current arises from states at the edge is a truly novel feature of quantum Hall systems. The chirality

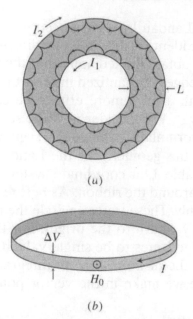

(a)

(b)

Figure 14.5 (a) A quantum Hall disk pierced by a magnetic field pointing out of the page. The arrows indicate the direction of the edge currents. (b) A quantum Hall ribbon in which the magnetic field is everywhere perpendicular to the surface. This geometry is mathematically equivalent to a rectangle with periodic boundary conditions in one direction. The circumference of the ribbon is L and its width is W. The Hall voltage is $\Delta V = E_0 W$.

of the current at the edge is also at the heart of why the edge states remain extended. As we learned in Chapter 12, backscattering is essential for localization to obtain. There can be no backscattering for a chiral edge state. Hence, they resist localization by a random potential.

Since the current is carried entirely by the edges, the geometry of our system cannot matter. We consider, then, a quantum Hall disk with a hole punched into the center, as depicted in Fig. (14.5). This geometry is equivalent to the one used by Halperin [H1982] in his reformulation of the original Laughlin argument. The disk is pierced with a uniform magnetic field in the positive z-direction. Truncation of the cyclotron orbits at the outer rim of the disk leads to an edge current that flows in the clockwise direction. However, at the inner radius, confinement leads to a current in the opposite direction. Clearly, then, if the outer and inner edges of our annulus are at the same chemical potential, no net current will flow in the system. Let's assume that the Fermi levels of the inner and outer edges differ by an amount eE_0. This difference might be due to asymmetries in the confinement potentials at the inner and outer edges as well as to any electro-chemical potentials that might be present. If n Landau levels are occupied, then the total potential drop is neE_0. The identical argument leading to Eq. (14.23) can now be invoked, and we obtain immediately that the net current between the inner and outer edges is quantized in units of e^2/h.

However, with a little more effort, we can extract the same result a different way. We consider now the second geometry, shown in Fig. (14.5b). In formulating the gauge argument, we will find it easier to work with this geometry, as the Landau gauge used previously is directly applicable. Our coordinate system is chosen so that the y-coordinate runs around the ribbon. As before, the current is carried by the edge states only. These states encircle the ribbon, preserving phase coherence as they return to the origin. For the wave functions characterizing the edge states to be single-valued, the flux enclosed upon one trek around the disk must be an integral multiple of 2π. Hence, whatever change we make in the vector potential should satisfy the condition that

$$A = \frac{nhc}{eL}, \tag{14.26}$$

where L is the circumference of the ribbon. For small changes in the vector potential, the current in our system is gauge invariant. Consider now the energy

$$\epsilon_\alpha = \langle \Psi_\alpha | H | \Psi_\alpha \rangle \tag{14.27}$$

of a particular single-particle state, Ψ_α, where H is given by the left-hand side of Eq. (14.11). For short-hand notation, we have defined $\alpha = (n, k)$. We are interested in the derivative of the ϵ_α with respect to A. To simplify this derivative, we use the Hellman-Feynman [H1937] theorem

$$\frac{\partial E(\lambda)}{\partial \lambda} = \left\langle \Psi_\alpha \left| \frac{\partial H(\lambda)}{\partial \lambda} \right| \Psi_\alpha \right\rangle, \tag{14.28}$$

where λ is simply some variable in the Hamiltonian. Varying the single-particle energy with respect to A,

$$\frac{\partial \epsilon_\alpha}{\partial A} = \frac{-e}{mc} \left\langle \Psi_\alpha \left| \mathbf{p} - \frac{eA}{c} \right| \Psi_\alpha \right\rangle$$

$$= L\frac{I_\alpha}{c}, \tag{14.29}$$

defines the current density carried by the state α as it traverses the disk. To evaluate the derivative, we note that in the presence of an

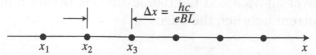

Figure 14.6 The location of the centers for the electron states that compose each Landau level. All states are assumed to be degenerate, with their centers given by $\ell^2 k_m = 2\pi\hbar cm/eBL$. The difference between two centers is hc/eBL. Threading the sample with one flux quantum transforms the mth state into the $m - 1$st. This results in the transfer of charge from one edge of the sample to the other.

electric field, the single-particle energies scale linearly with $eE_0 x_k$, where x_k locates the center of the states constituting each Landau level. If we modify the gauge term in the Hamiltonian such that $\mathbf{A} = Bx\hat{y} \to Bx\hat{y} + \Delta A\hat{y}$, then the location of each center is shifted by $x_k \to x_k - \Delta A/B$. Consequently, the single-particle energies are translated to $\epsilon_{n,k} \to \epsilon_{n,k} - eE_0 \Delta A/B$, and

$$\frac{\partial \epsilon_\alpha}{\partial A} = -\frac{eE_0}{B} \tag{14.30}$$

is independent of the state index. As illustrated in Fig. (14.6), Laughlin's gauge principle [L1980] follows from the fact that the difference between the location of the centers of two neighboring states in the same Landau level,

$$\Delta x_k = x_k^{m+1} - x_k^m = \ell^2(k_{m+1} - k_m) = \frac{2\pi\ell^2}{L}$$
$$= \frac{hc}{eBL} = \frac{\Delta A}{B}, \tag{14.31}$$

is directly related to the change in the vector potential. Consequently, if A is changed by a single flux quantum, the location of the mth center,

$$x_k^m \to x_k^m - \frac{\Delta A}{B} = x_k^m - \frac{hc}{eBL} = x_k^{m-1}, \tag{14.32}$$

is now coincident with the location of the m–1st center. Hence, when one flux quantum is threaded through the ribbon, the states in a Landau level all shift over by one, leading to the net transfer of a single electron (per Landau level) from one edge of the sample to the other. Quantization of the gauge leads to quantization in the charge transfer! This is the Laughlin gauge principle. It illustrates beautifully the topological [T1982] nature of charge transport in the quantum Hall effect. To calculate the current, we substitute Eq. (14.30) into Eq. (14.29) and sum over all occupied Landau levels. We obtain immediately that the net current between the edges,

$$I = \frac{c}{L} \sum_{n,k} I_{n,k}$$

$$= -\frac{ecn_e E_0}{LB} = -\frac{ecnn_B V_H}{B} = -\frac{ne^2 V_H}{h}, \qquad (14.33)$$

is an integer multiple of e^2/h with $V_H = E_0 W$, the Hall voltage. The negative sign in the current corresponds to counterclockwise motion on the ribbon. The Laughlin argument lays plain that the quantization of the Hall current or Hall conductance arises primarily from the restriction that the extended states must be single-valued as they traverse the edge of the sample. For a system in a magnetic field, the single-valuedness of the eigenstates manifests itself as a condition on allowable gauge transformations. It is this condition coupled with the integer filling of Landau levels that leads to the quantization. It is also paramount that the gauge transformation be carried adiabatically, so that the system remains in its ground state as the flux penetrating the system is changed.

As the plateau transitions are driven by changing the magnetic field or the filling, they constitute a genuine QPT of the kind studied in the previous chapter. In fact, because disorder is central to the story of the integer quantum Hall effect, the underlying plateau transitions represent one of the clearest examples of quantum-critical phenomena in a disordered system. Progress in understanding the underlying field-theoretical description of the plateau transitions is based largely on the Chalker-Coddington tunneling network model [CC1988, MT1999, Z1999]. Numerical simulations of this model [LWK1993, LC1994] have yielded a correlation-length exponent of $\nu = 2.3$, which is consistent with experimental observations. However, this exponent is yet to be predicted by a rigorous theoretical account.

14.5   LAUGHLIN LIQUID

In the original experiments of Tsui, Störmer, and Gossard [TSG1982], the Hall conductance exhibited a sharp plateau at the value of $e^2/3h$. In later experiments, fractional quantization of the Hall conductance was also observed for values of $4/3, 5/3, 7/3, 1/5, 2/5, 3/5, 7/5, 8/5$, and so on. The common ingredient in these sequences is the presence of the odd denominator. The original challenge in explaining these experiments, however, lay not so much in accounting for the fractional value of the Hall conductance but rather in explaining the nature of the electronic state that exhibited the fractional quantization. For example, a noninteracting model in which the lowest Landau level is fractionally occupied at a filling of ν will exhibit a conductance $\nu e^2/h$. However, when a Landau level is fractionally occupied, electrons can scatter into the empty states and, hence, longitudinal transport will not be dissipationless; as a consequence, $\sigma_{xx} \neq 0$. In addition, the persistence of a plateau at fractional filling indicates that somehow an appropriately partially occupied Landau level is stable to even the lowest-lying excitations. That is, an energy gap separates the ground state from all excited states. In the absence of electron interactions, the energy cost to add an additional electron to a partially filled Landau level is essentially zero. This would suggest that, as the field is changed, sharp plateaus in the conductance should desist in the noninteracting model for a partially filled Landau level.

Consequently, the fractional quantum Hall state cannot be understood without including the role of electron interactions. However, at the outset, it is not clear what this state should look like. Thus far, we have introduced two electronic states that arise fundamentally from electron-electron interactions: the Wigner crystal and the superconducting state. Although the vanishing of the longitudinal resistance suggests that the fractional quantum Hall state bears some resemblance to a superconducting state, it is not clear how such a state would survive a large perpendicular magnetic field. What about the Wigner crystal? As mentioned in Chapter 5, because a magnetic field freezes the electron zero-point motion, Wigner crystallization is stabilized. In fact, in the limit in which the interparticle separation is large relative to the cyclotron radius (magnetic length), electron correlations dominate, and the conditions for Wigner crystallization become favorable. However, a Wigner crystal does not exhibit dissipationless transport, as a threshold voltage must be applied before transport obtains. We suspect, then, that the resolution of the fractional quantum Hall state lies elsewhere.

Indeed, it does. After Laughlin [L1983, L1987], we consider an interacting electron gas in the presence of a perpendicular magnetic field

$$ H = \sum_i \left[\frac{1}{2m} \left(\mathbf{p}_i - \frac{e}{c} \mathbf{A}_i \right)^2 + V(\mathbf{r}_i) \right] + \frac{1}{2} \sum_{i \neq j} \frac{e^2}{|\mathbf{r}_i - \mathbf{r}_j|}, \quad (14.34) $$

where $V(\mathbf{r})$ is the compensating neutralizing potential from the ions, and where the sum over i and j are over the electrons. Fractional quantization of the Hall conductance is observed at a field of $15T$. At this field, the cyclotron energy is roughly three times the Coulomb energy, e^2/ℓ. Hence, it should be a fairly good approximation to use the noninteracting eigenstates of the lowest Landau level as a starting basis for constructing the true many-body wave function. In general, the noninteracting eigenstates are products of a polynomial in the electron coordinate times a Gaussian. For an interacting system, we expect that the true many-body state will involve differences of the electron coordinates. Let us define the complex electron coordinate, $z = x + iy$. We consider the ansatz,

$$\Psi_N(\mathbf{r}_1, \cdots, \mathbf{r}_N) = \prod_{1 \le j < k \le N} f(z_j - z_k) \exp\left(-\sum_{j=1}^{N} \frac{|z_j|^2}{4\ell^2} \right), \quad (14.35)$$

where $f(z_j - z_k)$ is a polynomial in the electron coordinate. After studying the two-electron problem described by Eq. (14.34), Laughlin found that $f(z_1 - z_2) = (z_1 - z_2)^{2p+1}$, where p is an integer. Because the exponent $2p + 1$ is odd, the wave function is antisymmetric with respect to interchange of two electrons. Analogous results were also obtained for the equivalent three-electron problem [L1983]. Hence, Laughlin proposed that

$$\Psi_m(\mathbf{r}_1, \cdots, \mathbf{r}_N) = \prod_{1 \le j < k \le N} (z_j - z_k)^m \exp\left(-\sum_{j=1}^{N} \frac{|z_j|^2}{4\ell^2} \right) \quad (14.36)$$

must accurately describe the ground state of a fractional quantum Hall system. Though this wave function was originally argued to be a variational state (where m must be determined), the overlap of this wave function with the exact eigenstate for small clusters of electrons is typically greater than 99% for interaction potentials of the form $u(r) = 1/r$, $-\ln r$, and $\exp(-r^2/2)$. This would suggest that the variational character of Ψ_m is minimal, and m must be determined from a fundamental principle. It turns out that Ψ_m is an eigenstate of the total angular momentum operator with eigenvalue

$$M = \frac{N(N-1)m}{2}. \quad (14.37)$$

That M is the total angular momentum of the operator L_z follows from expanding the product over $z_j - z_k$ in Eq.(14.35) and realizing that there are, at most, $N(N-1)m/2$ factors of z_i in each term. If the differential form of the L_z operator is applied to such a product, the eigenvalue M results. As a consequence, we can think of the Laughlin state as being a superposition of states within the lowest Landau level with the same angular momentum. This removes completely the variational character of the Laughlin state.

To understand precisely what physics the Laughlin state describes, we write the square of this state,

$$|\Psi_m|^2 = \exp[-\beta\Phi], \tag{14.38}$$

in terms of a Boltzmann weight, where Φ is the interaction energy

$$\Phi(z_1,\cdots,z_N) = -2\sum_{1\le j<k\le N} \ln|z_j - z_k| + \frac{1}{2m\ell^2}\sum_{j=1}^{N}|z_j|^2 \tag{14.39}$$

and $\beta = m$ plays the role of the inverse temperature. Eq. (14.39) is exactly the interaction energy for a one-component plasma (such as an electron gas) consisting of N identical charges with charge $\sqrt{2}$. The second term in Eq. (14.39) represents the interaction energy with the neutralizing background, $U_b(z_i) = |z_i|^2/(2m\ell^2)$. To see this clearly, we note that the potential for the compensating background should satisfy a Poisson equation, $\nabla^2 U_b(z_i) = 4\pi n_e$, where n_e is given by $\nu n_B = \nu/(2\pi\ell^2)$ (see Eq. 14.21). Performing the differentiation in the Poisson equation reveals that the compensating charge density per unit area,

$$n_e = \frac{1}{2\pi\ell^2 m} \equiv \frac{\nu}{2\pi\ell^2}, \tag{14.40}$$

is exactly the uniform electron charge density in the lowest Landau level if we identify the Landau filling factor ν with $1/m$. The Laughlin state accurately describes the ground state of an N-electron system for density and magnetic field strengths such that the filling in the lowest Landau level is given by $\nu = 1/m$. As a system with a uniform electron density, the Laughlin state is distinct from an electron-crystal state, such as a Wigner crystal. The importance of the Laughlin state in the development of the fractional quantum Hall effect cannot be overestimated.

We are now poised to explore the excitations of the Laughlin liquid. For filling fractions ν different from $1/m$, excitations emerge. Consider changing the filling by piercing the sample at z_0 with an infinitely thin magnetic solenoid. Although ν is now slightly less than $1/m$, the electrons will attempt to stay in the state Ψ_m. However, they cannot do this without diminishing the charge density at the insertion point of the magnetic solenoid. We can simulate such a depletion by excluding the electrons from z_0. Consequently, we anticipate that

$$
\Psi_m^+(z_0; z_1, \cdots, z_N) = \prod_{j=1}^{N} (z_j - z_0)\Psi_m(z_1, \cdots, z_N)
$$

$$
= A_{z_0}^+ \Psi_m(z_1, \cdots, z_N) \tag{14.41}
$$

might describe the wave function for the new many-body state with a "quasi hole" at z_0. Indeed it does, as shown by Laughlin [L1983]. A more quantitative argument for the quasi particle wave function in Eq. (14.41) stems from noting that if the solenoid carries flux ϕ, each single-particle state is changed accordingly,

$$
z^n e^{-\frac{|z|^2}{4\ell^2}} \rightarrow z^{n+\alpha} e^{-\frac{|z|^2}{4\ell^2}}, \tag{14.42}
$$

to accommodate the additional flux. Here $\alpha = \phi/\phi_0$, with $\phi_0 = hc/e$ the flux quantum. If the solenoid carries one quantum of flux, then the prefactor of the Gaussian is z^{n+1}. That each single particle state is now multiplied by an extra factor of z supports the ansatz for the quasi-particle wave function, Eq. (14.41). To utilize the plasma analogy, we square the quasi-particle wave function to find that, aside from a background normalization factor of $|z_0|^2$, the energy of the many-body state with a "quasi hole,"

$$
\Phi_{qp}(z_0; z_1, \cdots, z_N) = \Phi(z_1, \cdots, z_N) - \frac{2}{m}\sum_{j=1}^{N} \ln|z_j - z_0|, \tag{14.43}
$$

is that of a one-component plasma interacting with a charge fixed at z_0. The magnitude of the charge is $1/m$ or, in electron units, e/m. Likewise, if the solenoid were to extract a single flux quantum, a "quasi electron" would be created with charge $-e/m$. Numerically, the energy to create or destroy quasi particles in a three-electron fractional quantum Hall state is roughly 4 K at a field of $15T$ [L1983]. Improved estimates of the gap in the fractional quantum Hall state were obtained by Girvin, McDonald, and Platzman [GMP1985]. In fact, their work was pivotal in establishing that all collective excitations from the Laughlin state have a finite energy gap. Hence, the Laughlin state does satisfy the criterion of having a gap to all excitations. The excitation energy can be thought of as the Coulomb energy required to place a particle of charge $\pm e/m$ in the quantum liquid. The distance over which the charge acts is proportional to the magnetic length $\propto \ell$. Hence, the excitation energy should scale as $e^2/\ell \propto \sqrt{B}$ and thus vanish in the absence of an applied magnetic field. It is the presence of the energy gap in the excitation spectrum that makes the Laughlin state incompressible and ultimately leads to dissipationless transport.

Nonetheless, some experiments have revealed the existence of gapless excitations [ASH1983, MDF1985] in 2d quantum Hall systems. These excitations are believed to live on the edge of quantum Hall systems, as they are explicitly excluded from the bulk by Laughlin's gauge argument. For integer quantum Hall states, the edge excitations are well described by Fermi-liquid theory, as electron interactions are relatively unimportant in the integer effect. However, in the fractional effect, the situation is entirely different, as we have seen. Quantum mechanical states confined to move at the edge of a fractional quantum Hall system are essentially 1d, strongly correlated electron systems. We have shown in Chapter 10 that electron interactions in a 1d system give rise to Luttinger rather than Fermi-liquid behavior. As a result of the chirality of the edge current, edge states in the fractional quantum Hall effect are chiral Luttinger liquids [W1990]. They exhibit all the properties indicative of Luttinger liquids discussed in Chapter 9, including an excitation spectrum that vanishes algebraically in the vicinity of the Fermi energy. Figure (9.6) confirms the algebraic dependence of the excitation spectrum in the edge of the $\nu = 1/3$ quantum Hall state, thereby putting the chiral Luttinger liquid model for the edge states in the fractional quantum Hall effect on firm experimental footing.

Consider for the moment the problem of the statistics associated with interchanging [A1985, F1992] two quasi holes or two quasi electrons in a fractional quantum Hall state. The wave function describing such pair excitations, which we will locate at $z = z'$ and $z = z''$, is analogous to Eq. (14.41), except the product $A_{z'}^{\pm} A_{z''}^{\pm}$ now multiplies the Laughlin state Ψ_m. The problem we address is, under interchange of two quasi particles such that

$$A_{z'}^{\pm} A_{z''}^{\pm} \Psi_m(z_1, \cdots, z_N) = e^{i\phi} A_{z''}^{\pm} A_{z'}^{\pm} \Psi_m(z_1, \cdots, z_N), \quad (14.44)$$

what phase, ϕ, does the wave function incur? For interchange of electrons, $\phi = \pi$ and for bosons, $\phi = 2\pi$ or, equivalently, 0. We will show now that the phase change for the interchange of two quasi particles in the Laughlin state is $1/m$ and, hence, fractional.

Quite generally, the wave function of a particle traversing a closed loop in the presence of a vector potential will acquire the phase

$$i\gamma = i\frac{q}{\hbar c}\oint d\ell \cdot \mathbf{A}. \tag{14.45}$$

We will take the vector potential to be the generator of the magnetic field B felt by the particle. Consequently, the line integral that determines the phase is simply equal to the field times the surface area enclosed by the path. Let R be the radius of the loop. As a result, the phase change is $\gamma = q\pi R^2 B/\hbar c$. Recalling that the magnetic field is related to the electron density through $\rho = \nu e B/\hbar c$, and the quasi particle charge is $e\nu$, we find that the total phase encountered is $\gamma = 2\pi N$, where N is the number of electrons in the system. This is a key result, as the phase is proportional to the number of electrons enclosed in the loop. Now let us redo the argument, assuming that amidst the sea of electrons lies a quasi hole of charge e/m. In this case, we must take into consideration the depletion in the density as a result of the quasi hole. For a quasi hole, the depletion is given by the filling ν or $1/m$. Hence, the phase is now given by $2\pi(N - 1/m)$. Let us assume that $|z'| < |z''|$. As a consequence, when the quasi hole at z'' is moved in a loop that encircles the quasi hole at z', the phase change is $2\pi(N - 1/m)$. The $2\pi N$ factor does not change the sign of the wave function. As a consequence, the extra phase that is gained is $2\pi/m$. However, to interchange two quasi particles, we need only traverse a half circle and then translate the particles. As the translations do not result in any phase change, the net phase change under the interchange of two quasi particles is π/m. For $m = 1$, we recover the result for electrons. However, for any fractional filling, the phase change is a fraction. Hence, quasi particles in the Laughlin liquid exhibit both fractional charge and fractional statistics. Particles obeying fractional statistics [W1982] are termed *anyons* because they can acquire *any* phase change upon interchange.

In closing this section, we introduce a seemingly innocent rewriting [J1989],

$$\Psi_m(\mathbf{r}_1, \cdots, \mathbf{r}_N) = \prod_{1 \le j < k \le N} (z_j - z_k)^{m-1} \Psi_1(\mathbf{r}_1, \cdots, \mathbf{r}_N), \tag{14.46}$$

of the Laughlin state. Here $\Psi_1(\mathbf{r}_1, \cdots, \mathbf{r}_N)$ is simply the Laughlin state with $m = 1$ and, hence, corresponds to the wave function for a completely filled Landau level. The prefactor in this expression resembles the flux attachment prefactor in the "quasi hole" wave function. The Laughlin state can be thought of as a completely filled Landau level in which each electron has physically attached to it $m - 1$ flux quanta. Though in no real sense can electrons bind flux in the manner that pages are bound in a book, the flux attachment scheme offers a pictorial way of thinking about the fractional quantum Hall effect. Further, it has the advantage that the fractional quantum Hall effect can be viewed as a variant of the integer quantum Hall effect with fictitious particles bearing electric charge e and $m - 1$ units of magnetic flux. As a computational tool, the flux attachment scheme or composite particle picture is of extreme utility in generating the hierarchy of fractional states in the lowest Landau level and, as a consequence, forms the basis for current field theoretical approaches to the quantum Hall problem [F1991, LF1991, HLR1993, KLZ1992]. In fact, in the context of the even-denominator states, the composite fermion approach has played a central role. The even-denominator state at $\nu = 1/2$ is particularly odd [W1993] in that it does not appear to be a true fractional state with a vanishing longitudinal resistance. The flux attachment scheme at $\nu = 1/2$ leads to a composite fermion liquid in which the average magnetic field felt by the composite fermions vanishes. In such a system, a Fermi surface should emerge as a result of the apparent vanishing of the field. Recent work [R1998, SM1997] on this problem suggests that the $\nu = 1/2$ state is a compressible state composed of fermionic excitations that possess a dipole moment.

An additional problem of current interest is the observation of anisotropic transport when the magnetic field is sufficiently low, such that the filling fraction exceeds $\nu > 7/2$. In the vicinity of the half-integer states with $\nu > 7/2$, the measured [D2000, L1999] values of the longitudinal resistivity depend on the direction of the current, that is, $\rho_{xx} \neq \rho_{yy}$. Anisotropic transport is inconsistent with a uniform electron state. In principle, the ground state of a $2d$ electron gas is expected to be nonuniform when the magnetic field is sufficiently low, so that no fractional quantum Hall states are accessible, but not so low that disorder destroys the gap established by the cyclotron frequency, ω_c. In this intermediate range of magnetic fields, the ground state is governed by the competition between the short-range exchange and long-range direct Coulomb interactions. Because these interactions are of opposite sign, the electron gas can break into domains, [FPA1979, KFS1996, MC1996, RHY1999, J1999, SMP2000]. Current theoretical work is directed at formulating effective low-energy theories [FK1999, MF2000, BF2001, B2001, L2001] and the transport properties in the presence of a charge modulation in the low-field limit.

14.6 SUMMARY

We presented here the essential physics behind the quantum Hall effect. The integer effect arises from the boundary condition that the current-carrying states or edge states be single valued when an integer number of Landau levels are filled. In the fractional Hall state, electrons partially occupying the lowest Landau level condense into a strongly correlated state (the Laughlin state) whose energy is lower than the corresponding Wigner crystal. This state is incompressible in that it has a gap to all excitations and exhibits fractionally charged excitations with charge $\pm e/m$. The excitation energy for such charged excitations scales as $1/\ell$ and, hence, vanishes in zero field. It is for this reason that while the Laughlin state is mediated by strong electron correlations, the incompressibility stems from the large perpendicular magnetic field. Quantum Hall problems of current interest include the apparently Fermi liquid–like state at $\nu = 1/2$ as well as the onset of an anisotropic state at sufficiently low magnetic fields, so that the filling exceeds $\nu = 7/2$.

REFERENCES

[ASH1983] S. J. Allen, Jr., H. L. Störmer, J. C. M. Hwang, *Phys. Rev. B* **28**, 4875 (1983).

[AA1981] H. Aoki, T. Ando, *Solid St. Comm.* **38**, 1079 (1981).

[BF2001] D. G. Barci, E. Fradkin, cond-mat/0106171.

[B2001] D. G. Barci, E. Fradkin, S. A. Kivelson, V. Oganeysan, cond-mat/0105448.

[A1985] D. P. Arovas, in *Geometric Phases in Physics*, ed. A. Shapere, F. Wilczek (World Scientific, Singapore, 1985), p. 284.

[CC1988] J. T. Chalker, P. D. Coddington, *J. Phys. C* **21**, 2665 (1988).

[D2000] R. R. Du, D. C. Tsui, H. L. Störmer, L. N. Pfeiffer, K. W. Baldwin, K. W. West, *Solid State Commun.* **109**, 389 (1999).

[KFS1996] M. M. Fogler, A. A. Koulakov, B. I. Shklovskii, *Phys. Rev. B* **54**, 1853 (1996).

[F1992] S. Forte, *Rev. Mod. Phys.* **64**, 193 (1992).

[F1991] E. Fradkin, *Field Theories of Condensed Matter Systems*, (Addison-Wesley, Redwood City, USA, 1991).

[FK1999] E. Fradkin, S. Kivelson, *Phys. Rev. B.* **59**, 8065 (1999).

[FPA1979] H. Fukuyama, P. Platzman, P. W. Anderson, *Phys. Rev. B* **19**, 5211 (1979).

[GMP1985] S. M. Girvin, A. H. MacDonald, P. M. Platzman, *Phys. Rev. Lett.* **54**, 581 (1985).

[H1879] E. H. Hall, *Am. J. Math.* **2**, 287 (1879).

[H1982] B. I. Halperin, *Phys. Rev. B* **25**, 2185 (1982).

[HLR1993] B. I. Halperin, P. A. Lee, N. Read, *Phys. Rev. B* **47**, 7312 (1993).

[H1937] H. Hellmann, *Einführung in die Quantenchemie* (Franz Denticke, Leipzig, Germany, 1937) p. 285; see also, R. P. Feynman, *Phys. Rev.* **56**, 340 (1939).

[J1989] J. Jain, *Phys. Rev. B* **40**, 8079 (1989).

[J1999] T. Jungwirth, A. H. MacDonald, L. Smrcka, S. M. Girvin, *Phys. Rev.* **60**, 15574 (1999).

[KLZ1992] S. Kivelson, D.-H. Lee, S.-C. Zhang, *Phys. Rev. B* **46**, 2223 (1992).

[KDP1980] K. von Klitzing, G. Dorda, M. Pepper, *Phys. Rev. Lett.* **45**, 494 (1980).

[L1980] R. B. Laughlin, *Phys. Rev. B* **23**, 5632 (1980).

[L1983] R. B. Laughlin, *Phys. Rev. B* **27**, 3383 (1983); ibid, Phys. Rev. Lett. **50**, 873 (1983).

[L1987] R. B. Laughlin, in *The Quantum Hall Effect* ed. E. Prange, S. M. Girvin (Springer, Berlin 1987), p. 233.

[L2001] A. Lopatnikova, S. H. Simon, B. I. Halperin, X.-G. Wen, cond-mat/0105079.

[LWK1993] D.-H. Lee, Z. Wang, S. A. Kivelson, *Phys. Rev. Lett.* **70**, 4130 (1993).

[LC1994] D. K. K. Lee, J. T. Chalker, *Phys. Rev. Lett.* **72**, 1510 (1994).

[L1999] M. P. Lilly, K. B. Cooper, J. P. Eisenstein, L. N. Pfeiffer, K. W. West, *Phys. Rev. Lett.* **82**, 394 (1999).

[LF1991] A. Lopez, E. Fradkin, *Phys. Rev. B* **44**, 5246 (1991).

[MF2000] A. H. MacDonald, M. P. A. Fisher, *Phys Rev B* **61**, 5724 (2000).

[MT1999] J. B. Marston, S.-W. Tsai, *Phys. Rev. Lett.* **82**, 4906 (1999).

[MDF1985] D. B. Mast, A. J. Dahm, A. L. Fetter, *Phys. Rev. Lett.* **54**, 1706 (1985).

[MC1996] R. Moessner, J. T. Chalker, *Phys. Rev. B* **54**, 5006 (1996).

[P1981] R. E. Prange, *Phys. B* **23**, 4802 (1981).

[P1987] R. E. Prange in *The Quantum Hall Effect*, ed. R. E. Prange, S. M. Girvin (Springer, Berlin, 1987), p. 1.

[P1984] A. Pruisken, *Nucl. Phys. B* **235** (FS11), 277 (1984).

[R1998] N. Read, *Phys. Rev. B* **58**, 16262 (1998).

[RHY1999] E. H. Rezayi, F. D. M. Haldane, K. Yang, *Phys. Rev. Lett.* **83**, 1219 (1999).

[SM1997] R. Shankar, G. Murthy, *Phys. Rev. Lett.* **79**, 4437 (1997).

[SMP2000] T. D. Staneseu, I. Martin, P. Philips, *Phys. Rev. Lett.* **84**, 1288 (2000).

[T1982] D. Thouless, M. Kohmoto, M. denNijs, M. Nightingale, *Phys. Rev. Lett.* **49**, 405 (1982).

[T1983] S. A. Trugman, *Phys. Rev. B* **27**, 7539 (1983).

[TSG1982] D. C. Tsui, H. L. Störmer, A. C. Gossard, *Phys. Rev. Lett.* **48**, 1599 (1982).

[W1990] X. G. Wen, *Phys. Rev. B*, **41**, 12838 (1990).

[W1982] F. Wilczek, *Phys. Rev. Lett.* **49**, 957 (1982).

[W1993] R. L. Willett, R. R. Ruel, M. A. Paalanen, K. W. West, L. N. Pfeiffer, *Phys. Rev. B* **47**, 7344 (1993).

[Z1999] M. R. Zirnbauer, hep-th/9905054.

Index